The Natural Science of the Human Species

Also by Konrad Lorenz

King Solomon's Ring

Man Meets Dog

On Aggression

Studies in Animal and Human Behavior

Civilized Man's Eight Deadly Sins

Behind the Mirror: A Search for a Natural History of Human Knowledge

The Year of the Greylag Goose

Here Am I—Where Are You?: The Behavior of the Greylag Goose

The Natural Science of the Human Species

An Introduction to Comparative Behavioral Research
The "Russian Manuscript"
(1944–1948)

Konrad Lorenz

Edited from the posthumous works by
Agnes von Cranach

Translated by Robert D. Martin

With 12 drawings by the author

The MIT Press
Cambridge, Massachusetts
London, England

First MIT Press paperback edition, 1997

This work originally appeared in 1992 under the title *Die Naturwissenschaft vom Menschen: Eine Einführung in die vergleichende Verhaltensforschung. Das "Russische Manuskript" (1944–1948)*. The translation from the original German edition was subsidized by Inter Nationes, Bonn.

This book was set in Baskerville by Graphic Composition, Inc., and was printed and bound in the United States of America.

Library of Congress Cataloging-in-Publication Data
Lorenz, Konrad, 1903–1989
[Naturwissenschaft vom Menschen. English]
The natural science of the human species : an introduction to comparative behavioral research : the "Russian Manuscript" (1944–1948) / edited from the posthumous works by Agnes von Cranach; translated by Robert D. Martin ; with 12 drawings by the author.
p. cm.
Includes bibliographical references and index
ISBN 0-262-12190-5 (HB), 0-262-62120-7 (PB)
1. Psychology, Comparative. 2. Psychology, Comparative—Research—Methodology—History. 3. Psychology and philosophy. I. Cranach, Agnes von. II. Title.
BF671.L8513 1995
156—dc20 94-48788
CIP

Contents

Editor's Foreword *ix*
Translator's Foreword *xvii*
What Is Our Goal? *xxv*
Part One: Introduction to Comparative Behavioral Research

I
Philosophical Prolegomena

1
Natural Science and Idealistic Philosophy *5*

2
Induction *27*

3
The Consistent Hierarchical System of the Natural Sciences *39*

4
On the Possibility of a Synthesis Between the Natural Sciences and the Humanities *53*

II
Biological Prolegomena

5
General Attempts to Define Life *83*

6
The Unique Historical Origin of Organisms and the Phylogenetic Approach *99*

7
The Organism as an Entity and Analysis on a Broad Front 137

8
Finality 151

9
The Mind-Body Problem 157

III
Historical Origins and Methods of Comparative Behavioral Research

10
Preconditions 179

11
Vitalism 185

12
Mechanism 195

13
The Implications of the Conflict Between Vitalism and Mechanism for Behavioral Research 209

14
The Inductive Basis of Comparative Ethology 213

15
Animal Keeping as a Research Method 221

16
The Origin of the Comparative Phylogenetic Approach in Behavioral Research 235

17
The First Steps in the Nomothetic Stage 241

18
The Research Personalities of Whitman and Heinroth and Their Findings 245

19
The Discovery of "Appetitive Behavior" by Wallace Craig *259*

20
My Own Contribution to an Understanding of the Instinctive Motor Pattern *271*

21
Erich von Holst's Discovery of Automatic Stimulus Production in the Central Nervous System *287*

22
Implications for the Analysis of Related Phenomena *303*

23
Conclusions *313*

Bibliography *317*
Index *325*

Editor's Foreword

This is not a final book written by my father Konrad Lorenz, who died in 1989; it is, in fact, his first book. To be more precise, it is the initial, introductory part of a synthetic treatise on comparative ethology that he had planned prior to the end of the Second World War—"the first attempt to provide a cohesive account of a very young branch of biological research. The focus of this research is that most vivid of all living phenomena: the *behavior* of organisms." (p. xxv) My father completed part one in a Russian prisoner-of-war camp near Yerevan, Armenia. The other three parts that had been planned, the organization of which is described in detail on p. xxix et seq., were never written in their originally intended form, as my father decided very soon after his return from captivity in 1948 that he would write a new version of the book. Confronted with many new developments in his field of research, he intended to take them into account and set about this task straightaway. Between 1948 and 1950, he produced two manuscripts for the first chapter and some additional sections. In the late summer of 1950, he sent the second version to Erich von Holst and Gustav Kramer (both of whom responded). It was presumably Kramer who encouraged my father to write not *one* book but *two* and in fact this is what he did many years later. *Die Rückseite des Spiegels* (translation published as *Behind the Mirror* in 1977)—alluded to here on pp. 13, 168, and 257 as "the other side of the mirror"—was brought out in 1973 by Piper Verlag in Munich. In 1978, *Vergleichende Verhaltensforschung—Grundlagen der Ethologie* (translation published as *The Foundations of Ethology* in 1981) was brought out by Springer-Verlag in Vienna. The original contract must have been some 20 years old by the time this second book appeared.

It seems obvious that, during the initial years after his return, my father simply did not have the time and peace of mind needed for his great undertaking. He had no appointment and hence no income. There was, indeed, a Station for Comparative Ethology in Altenberg, near Vienna, which had been founded in 1949 thanks to a nonrenewable grant from the English author J. B. Priestley and which later came under the aegis of the Austrian Academy of Sciences. But the resources of this station were extremely limited. No salaries were available for my father or for his co-workers Wolfgang Schleidt and Ilse and Heinz Prechtl.

Thanks to the foresight and energy of my self-sufficient mother, our family came through the last years of the war and the immediate postwar period quite well. After the end of the war, my mother helped to run a small farm belonging to her family. In addition, in 1941 she had exchanged a leased field for a profitable orchard and had planted a large vegetable garden. But cash was always in short supply. In order to help out in this respect, my father followed the advice of a female publisher of his acquaintance and wrote two books of "animal stories" (*King Solomon's Ring* and *Man Meets Dog,* both originally published in German by Borotha-Schoeler in Vienna in 1949 and 1950, respectively). These sold very well and led to a number of profitable lecture tours in Switzerland. My father also traveled a great deal for other reasons, partly to take part in congresses and partly in search of new sources of financial support. Then, in the fall of 1950, it transpired that von Holst and Kramer had succeeded in their attempts to secure the support of the Max Planck Foundation for my father, who would as a result have access to good working conditions in Buldern (Westphalia). Baron Gisbert von Romberg made available a large tract of land with a small lake for a goose colony, a glasshouse for an aquarium, and living quarters for the members of a research team. With the help of the researchers from Altenberg (joined by Irenäus Eibl-Eibesfeldt in 1951), it only remained to set up the Research Station for Behavioral Physiology of the Max Planck Institute for Marine Biology, Wilhelmshaven, which was officially opened in the spring of 1951.

During this period, my father used the manuscript that he had written in Russia as a basis for short publications, talks, and lectures. He continued to do this even after moving to the newly founded, far larger Max Planck Institute for Behavioral Physiology in Seewiesen, near Starnberg (Bavaria), in 1955. The titles of some of my father's publications from this period are taken from chapter headings or the text of the Russian manuscript. In these new publications, as in the two books mentioned above, many of his formulations and even some of his analogies and illustrations are the same. At some time in 1963 or 1964 (I cannot exactly remember when), the manuscript went astray, presumably after a lengthy period in which it had not been consulted. I suppose that it had either been mislaid or put away somewhere. I was a direct witness to this, as I served as my father's secretary, occasionally in 1954 and then on a full-time basis from 1955 to 1961. I often fetched the manuscript and then put it away again, and after its disappearance I took part in the extensive but unsuccessful searches for it. My father repeatedly spoke about the manuscript thereafter. Before his move from Seewiesen to Altenberg, following his retirement in 1973, he told me that if the manuscript failed to resurface, then he would have to assume that it was lost forever. In fact, the manuscript did eventually resurface, in December 1990. It was buried deep beneath old proofs and such in a remote corner of his library in Altenberg, packed in a Munich newspaper dated 1973, and labeled "*Russ. Manuskr.*" in an unfamiliar handwriting. The manuscript consists of about 750 pages slightly larger than A5 [5⁷/₈ × 8¹/₄ inches] in size and written in ink on one side. The ink text is apparently the fair copy of a draft written in pencil that can be seen on the backs of some of the pages. In addition, there are about a hundred unnumbered pages covered with crowded, tiny writing in pencil. All of the samples of these that have so far been deciphered are concerned with the themes of the book that was planned.

The approximately 750 pages of the manuscript thus represent the initial section of my father's *first book,* written (as already noted) while he was a prisoner of war. Some pages are written on rough Russian notepaper of the quality to be expected at that time, while others are pieces cut to size from paper cement sacks. The text is written mainly with diluted ink or potassium permanganate, mostly using steel nibs but occasionally with birds' quills. I do not know how he came by the writing material. He presumably obtained some of it by bartering food, but that is unlikely to have brought much, as rations were surely very restricted, especially during the first winter: In the winter of 1944–1945, when his health was at its lowest ebb as a result of a scarlet fever infection contracted after he had been wounded during his capture near Vitebsk, he weighed only 120 lb. (During this illness, incidentally, while he was living in an unheated building in which drafts made icicles grow horizontally out of the wall, he tried to tame young rats that tried to warm themselves against him, but that were repeatedly summoned back by the mother.) I assume that my father's Russian colleagues often helped him to obtain writing material. He served as the camp medical doctor and from the very beginning had no problem in communicating with the Russians in Latin, Greek, or French. Quite apart from the scarcity of writing material, my father had to write under extremely difficult circumstances: without a desk, under poor lighting conditions, and in extremely low temperatures during the winters. His library was confined to a copy of Goethe's *Faust I,* always a favorite source of quotations. But he wrote with great energy and enthusiasm, with innumerable exclamation marks and underlinings (also a fond habit). I am convinced that his enthusiasm for his subject, his overriding interest for his work, helped him—a man who could be regarded as having been pampered throughout his life—to survive unharmed the harsh times in the prisoner-of-war camp. His duty to care for his fellow prisoners as medical doctor and psychiatrist (following only 2 years of admittedly demanding practice) undoubtedly also played a major role.

My father must have expended considerable amounts of energy in those days. In addition to his activity as physician and the work on his book, he gave lectures, organized a performance of *Faust I,* and arranged variety shows for which he also wrote parts and in which he frequently appeared himself. He also tried to provide guidance to the best possible eating habits under the prevailing circumstances. This was particularly the case during the latter period in Yerevan, because in this camp—which was far from Moscow—the prisoners of war lived under much less strict conditions, especially the doctors. Some of the so-called working platoons were housed at the worksite, where the camp physicians visited them for examination and care. In this way, my father came into contact with civilians for whom he sometimes provided advice or treatment, as a result of which he was invited for a meal in return. Thus it was that he learned, for example, about the culinary customs of gypsies. Among other things, he ate hedgehog baked in clay (although I cannot imagine that he would ever have killed a hedgehog unless he was really in danger of starving, and that was not the case in Armenia), and he learned to cook snakes on a spit. During his cross-country travels, he had the opportunity to catch fish and to collect edible snails, which he tried to purge with sawdust before cooking them. But only a few of his camp companions (presumably mainly the other doctors) could be persuaded to avail themselves of such protein-rich food.

Thanks to his relative freedom of movement in Armenia, my father developed another new talent: he became an excellent recycler of refuse. He recounted how he "naturally" never passed by a garbage dump without taking a close look. (After his return to Austria, he found it hard to resist the temptation presented by the local dumps, which were doubtless equally luxuriant. Today, I am ashamed that we laughed about this at the time.) He used his finds from Armenian garbage dumps to barter for most of the (presumably stolen) electric cable from which he extracted copper wire to weave a large cage for his crested lark and a small carrying cage for a hand-reared starling. In fact, he eventually used these same cages to bring the birds home with him.

In Armenia, my father received the first reliable news that he might be released from the prisoner-of-war camp and permitted to return to Austria. As is indicated by a few sentences on the back of one of the manuscript pages, he had originally toyed with the idea of publishing his book from the camp as a prisoner of war—with permission. Instead, he now engaged in correspondence to obtain official permission to take his book home with him (as is also indicated by a few sentences on the backs of pages). Indeed, he was then allowed to travel alone to Moscow and to present his manuscript there for examination. He often spoke about the doubts that plagued him as his camp companions boarded the transport trains for the West, while he headed off alone toward Moscow, though admittedly in far greater comfort in a personal compartment (for the first time in 4 years). Close to Moscow, at a camp in Krasnogorsk, he had to make a copy of the manuscript (recently rediscovered) which he was told to deposit there after giving his word of honor that the content of the copy was identical to that of the original.

When my father arrived back at Altenberg on February 18, 1948—driven to the door by a truck, courtesy of the local government of lower Austria, which welcomed returning nationals in the Vienna Neustadt—the manuscript was his most important item of luggage. Apart from that, in addition to the two birds (one cage in each hand), he had only a pipe that he had made from a corncob, a metal spoon, and a basic minimum of toilet articles. We knew that, along with thousands of other liberated prisoners of war, he had been waiting in Máramossziget (Romania) since December 1947 for the reconstruction of a railway bridge that had been demolished by the crush of ice. The news of his impending return had been transmitted by word of mouth and in letters. Since the end of the war, my mother had been corresponding with Kramer, Wolfgang Köhler, Alfred Seitz, Käthe Heinroth, Paul Leyhausen, and Edvard Baumgarten. At that time, there was an entire generation of young ethologists, particularly in Otto Koenig's circle at the Wilhelminenberg station. A number of former students who had heard my father lecture in the period 1937–1940 had also kept up with the subject. Accordingly, my father was awaited impatiently not only by his family and friends but also by these future disciples. To all of them, he often read (accompanied by elderberry tea at Altenberg) extracts from his *Lehrbuch* (textbook), as he then liked to call it. The lectures that he gave at the Viennese Institute for Science and Art, thanks to the encouragement and support of Wilhelm Marinelli, were probably organized in line with his plans for the second (special) and third (general) parts of the book, at least at the beginning. The above-mentioned *unnumbered* pages include a list in which the initial points correspond to these lectures, for which there is an accompanying plan with keywords.

But my father did not want to publish the "Russian manuscript" in the form that it then had and still has today. When he wrote the text that follows, he knew of less than a fourth of the publications subsequently cited in the two books mentioned above (p. ix). The manuscript is now almost 50 years old and it is part of the history of behavioral research. I very much hope that my father, were he alive today, would not object to its publication.

This explains how the present book came to be published. I transcribed part of the manuscript myself and dictated the rest so that it could be transcribed for me. I have carefully checked the copy against the original. Practically nothing has been changed. Small errors have been corrected and inaccurate citations—my father was, after all, relying on his memory—have been discussed in footnotes, as have a few passages that were not completely legible because of bleaching, damage, or other reason. These passages are mentioned at the corresponding places in the text. In some cases, my material editorial comments have been included in the text itself in brackets. Original asides notated by the author have been printed as footnotes at the relevant place. The manuscript includes twelve hand-drawn illustrations, which are reprinted here as facsimiles together with the surrounding text. The sometimes mediocre quality of their reproduction is explained by the state of the original manuscript. The bibliography (p. 317 et seq.) includes publications by some authors who are not mentioned by name in the text, although they did in fact serve as source material.

My father had apparently not decided on a definitive title for this book, as several alternatives are mentioned in the text. The publishers chose one of these as the title.

I owe thanks to the many people who helped to transcribe the manuscript and clarify the factual content. First and foremost, thanks are due to the board members and employees of the Konrad Lorenz Institute for Research into Evolution and Cognition: Rupert Riedl, Erhard Oeser, and Wolfgang Schleidt; my cousin Monika Kickert; and Erika Hörzer. In addition to my sister-in-law Beatrice Lorenz, Wolfgang Schleidt kindly offered to read the finished typescript and pick up any mistakes that I might have missed. He spent a great deal of time on this task and answered many questions. The Department for Paper Restoration at the Viennese Academy for Pictorial Arts devoted great care to improving the legibility of bleached pages of the manuscript and also produced many photographs at various magnifications and degrees of contrast that greatly assisted the process of deciphering the text. As in all other stages of the undertaking, but particularly in this respect, Monika Kickert was extremely helpful in her capacity as an experienced former secretary for my father and as an expert in reading his—in fact, very disciplined and legible—handwriting. Special thanks are due to the staff of Piper Verlag for their cooperation and assistance.

Finally, I want to thank the unknown person who so carefully packed the Russian manuscript, along with the two revisions begun in 1948 and 1950 and my father's lecture notes from that period, and then transferred the whole bundle safely to his library in Altenberg.

Agnes von Cranach
Bern, Spring 1992

Translator's Foreword

I was delighted to receive the invitation from The MIT Press to translate *The Russian Manuscript* for two reasons. In the first place, over the years between the completion of my Ph.D. studies in 1966 and the initial period of my present appointment in Zurich, I had already translated four of Konrad Lorenz's books. I started out with translations of many of his key scientific papers under the title *Studies in Animal and Human Behavior* (volume 1: 1971; volume 2: 1972), progressing on through *The Year of the Greylag Goose* (1979) to *Here Am I—Where Are You?* (1991). Sadly, it seemed that this intermittent, challenging occupation had come to an end with *Here Am I—Where Are You?*, as this was the last book that Lorenz managed to write before his death in 1989 at the age of 86. I was, in fact, just planning to visit him in Altenberg to talk over old times when the news of his death reached me. Which brings me to the second reason why I was very pleased to be invited to translate *The Russian Manuscript.* This translation of the *first* book that Lorenz ever wrote has now given me a welcome opportunity to pay a final tribute to one of the founding fathers of modern ethology, who played a significant part in the development of my own career.

My contact with Konrad Lorenz began in 1964. I had just completed my degree in Zoology at the University of Oxford, where my interests had been fired by lectures on animal behavior given by Niko Tinbergen. It was obvious that Konrad Lorenz was a central figure in the development of this subject and that I would have to learn German to become acquainted with much of the key literature. At my request, Dr. Mike Cullen very kindly set up the initial contact for me to apply to do postgraduate work at the Max Planck Institute for Comparative Behavioural Physiology at Seewiesen in Bavaria. Konrad Lorenz was then director of this institute, and it was arranged that I would work in the laboratory of Dr. Irenäus Eibl-Eibesfeldt, conducting behavioral observations on tree shrews. Tree shrews were then widely believed to be the most primitive surviving members of the order

Primates, and it seemed that a behavioral study of tree shrews might yield valuable clues to the early evolution of primates. This amounted to an extension in a novel direction of the comparative approach that Lorenz had particularly championed. The German Academic Exchange Service generously provided me with a grant and also arranged for an intensive crash course in German prior to my arrival in Seewiesen. Thanks to this support, I was able to embark on my project and to do so while communicating in German from the first day that I set foot in the Institute.

The two years that I spent at Seewiesen (1964–1966) had a major and lasting effect on my subsequent academic development. During this period, I came to know Konrad well, for his openness and approachability to students counted among his many positive qualities. He frequently invited me into his office on the spur of the moment to talk about all kinds of ethological topics, and these were undoubtedly the most stimulating one-on-one tutorials that I ever had. Seewiesen was a special place at that time, packed with young investigators and technicians enthused by ethological research, and the Wednesday afternoon seminars were both lively and informative. Many of the world's leading ethologists passed through the Mecca of Seewiesen and gave seminars while I was there, and ever since I have regarded a thriving seminar series as the life blood of any academic institution. Eventually, just as my time at Seewiesen was coming to an end, Konrad asked me whether I would be willing to translate *Studies in Animal and Human Behavior.* I saw this as a great honor and unhesitatingly agreed.

Thus it was that I began to translate Lorenz's key papers while writing up my doctoral thesis back in Oxford. In fact, I had come to the conclusion—initially suggested by the behavioral evidence that I had collected—that tree shrews are not related to primates after all. Defense of this conclusion in my thesis in 1967 marked the beginning of a personal quest to reconstruct the early origins of primates that has now lasted almost 30 years. Throughout that time, I have never forgotten the debt that I owe to Konrad and to the team of ethologists gathered around him at Seewiesen in the 1960s. Indeed, I partly owe my present appointment to the period that I spent in Seewiesen, as I would surely not have ventured to take on this challenge without the fluency in the German language that I had acquired there some 20 years previously.

While I was in Seewiesen, probably in 1965, Konrad had told me about a book manuscript that he had written with rudimentary materials during his time as a prisoner of war in a Russian camp. I was fascinated by his story and urged him to publish the book as soon as possible. At that time, texts in ethology were still rare, and I found it astonishing that he had been able to produce a manuscript under such conditions. Subsequently, I often wondered why the book had never been published. It was only when I read Agnes von Cranach's preface to the German original of *The Russian Manuscript* that I realized that the original manuscript had gone astray soon after I left Seewiesen, eventually resurfacing just a few years ago.

My initial task of translating *Studies in Animal and Human Behavior* was not without its trials and tribulations. From an early stage, Konrad had been keenly interested in philosophical aspects of behavioral research. Indeed, from September 1940 until he was drafted into war service, he had briefly shared with Eduard Baumgarten the Professorship of Psychology at the Albertus University of Königsberg, a post previously held by Immanuel Kant. For some time thereafter Konrad's writings were seemingly heavily influenced by Kant, at least with respect to textual complexity. Thus it was that, while struggling to write my doctoral thesis, I was confronted with the translation of complex sentences sometimes up to half a page in length. Taking full advantage of the rules governing the construction of German sentences, Lorenz would repeatedly begin with an initial statement and thereafter progressively restrict its meaning with a whole suite of subordinate clauses, rounding the whole thing off with a separable prefix at the end! To deal with this, I had to develop a form of literary surgery that involved holding the entire German original of each sentence in my head while dissecting out shorter sentences that would retain the original meaning. Although I was quite desperate at first, I eventually developed a skill for this. I subsequently realized that I had not survived the experience unscathed. My doctoral thesis was written in Lorenzian English! In due course, however, I received a double reward for my efforts from Lorenz himself. He first of all assured me, with a huge grin, that the translated version of *Studies in Animal and Human Behavior* was "better than the original" and then followed this up by insisting that we should henceforth use the familiar "Du" in our correspondence.

Thus it was that I eventually came to produce the English translation of *The Russian Manuscript.* The fact that the German original was written at all under the demanding conditions of a prisoner-of-war camp represents a triumph of the human spirit over adversity. It is quite staggering to think that Lorenz wrote the text entirely from memory (including several quotations from his favorite source, Goethe), making only trivial errors in the process. At the same time, this achievement bears eloquent testimony to Lorenz's personal commitment to the discipline of comparative behavioral research, then in its infancy. Indeed, the text has a clarity and intensity surpassing that of any of his later publications, the roots of which can clearly be traced back to this "common ancestor." It is particularly important to note that *The Russian Manuscript* (unfortunately only the first part of the major work that was originally planned) is explicitly concerned with potential applications to human behavior. Perhaps in response to the trauma of the war years, Lorenz firmly believed that something had gone seriously wrong with human behavior and that only ethology offered any real hope of generating a solution. Unfortunately, that hope has not been fulfilled, but recent events in various parts of the world have shown that these acute problems with human behavior remain. It is currently unfashionable to suggest that ethology has any major contribution to make to the solution of large-scale human problems such as armed conflict, but it was this belief that motivated Konrad more than anything else.

For the first time, I have had to translate one of Konrad's books without being able to check any queries with him. I have tried to intrude as little as possible and to maintain the integrity and flavor of the original. For this reason, I have retained the author's liberal use of exclamation marks and italics. Although this is less customary in English texts, it is part of Lorenz's bombastic style, as indicated by Agnes von Cranach in her preface to the German original. Similarly, use of the royal "we" has been directly translated in various places, as it is a reflection of Lorenz's clear belief that he was acting as a spokesman for comparative behavioral research. Language that might now be regarded as sexist has been left in its original form. Lorenz always wrote in terms of "he" and "his," so the translation is linguistically correct. It probably never occurred to him that his exclusive use of the masculine form might one day be regarded as unfortunate. Certain specific decisions also need some comment. I have generally avoided using the word "ethology," as it was not established at the time when Lorenz was writing. Instead, I have used the term "comparative behavioral research," a direct translation which more accurately reflects the special nature of Lorenz's approach.

Special problems arise in cases where interpretations have radically altered over the 50 years that have elapsed since Lorenz wrote the original text. A prime example is provided by evolutionary explanations of behavior. Lorenz repeatedly refers to *preservation of the species,* referring to the then widespread belief that selection in some way operates to promote the survival of species. Such terminology is incompatible with the dominant modern view that selection acts essentially on individuals. Appropriate terms had to be found to maintain the original sense in translation, as this genuinely reflects Lorenz's conception of the evolution of behavior. The translated text therefore refers to "species-preserving adaptations" and the like despite my full recognition of the fact that there is no known mechanism by which selection might favor the preservation of a species rather than of individuals.

Particular problems arose because parts of the text are specifically concerned with philosophical issues, with which I am only tangentially acquainted. I have tried to combine accuracy with simplicity in the translation. The term *Ganzheit* deserves particular comment, as it appears very frequently in the original. I have translated this as "entity," although a more conventional rendering would be "totality." I have several reasons for this personal choice, but one of them is that *Unterganzheit* can then be translated as "subentity," whereas the alternative "subtotality" is redolent more of accountancy than of philosophy. A number of unexpected and unusual tasks also came my way. For example, it proved necessary to translate several quotations in verse from Goethe, Wilhelm Busch, and others. As a habitually nonliterary scientist, I have done my best; I trust that the reader will accept this as a special form of "poetic license." At the end of the day, one can only hope that the sense of the original has been retained as faithfully as possible. I am only too aware of the way in which subtle mistranslations can distort meaning, as is borne out by a sentence in a letter recently received from an Austrian academic: "Given the themes and the purpose mentioned above it is useless to say that your presence will greatly contribute to the success of this operation."

Translating *The Russian Manuscript* was a very rewarding experience, not least because it called for reactivation of my surgical skills for processing Kantian sentences and provided a graphic reminder of Lorenz's particular talents. He had, for example, a special gift for choosing simple everyday parallels to illustrate complex phenomena. This was repeatedly shown in his contributions to discussions following those seminars in Seewiesen and it was this gift (among others) that enabled him to write several successful popular books about animal behavior (an economic necessity following his return from Russia). *The Russian Manuscript* contains several good examples of this talent. One is provided by the analogy he uses to illustrate the tandem operation of a physiological process and its experiential accompaniment. He suggests that we picture a flat metal plate, held such that just one side is viewed by each eye. Neither eye can encompass both sides at once. Equally graphic is his summary of a key experiment in which nerve connections were severed in the middle of an eel to test whether wave-like contractions are transmitted from segment to segment. Subsequently, contractions were seen to occur up to the operated section and then to continue beyond it, after an appropriate time lag, showing that the contraction sequence is an inherent property of the spinal cord and not a result of direct physical stimulation from segment to segment. Lorenz refers to the unseen continuation of the contraction wave through the paralyzed middle section as being "like a train passing through a tunnel."

I was also pleasantly surprised to see how deeply Lorenz had already penetrated into the intricacies of phylogenetic reconstruction. His discussion of its complexities halfway through the text could still be read with profit by investigators in any field of biology. A good example is provided by his demonstration of the need to separate primitive from derived features, as only the latter indicate specific relationships between species. This point is now widely recognized, but this was certainly not the case at the time when Lorenz was writing. In fact, one of Lorenz's insights with respect to interpretation of behavior is of equal importance for phylogenetic reconstruction (although he himself does not draw this conclusion). He notes that the important phenomenon of Gestalt perception has a drawback in that it tends to give "incorrigible" impressions. It is hence vital to recognize the need to curb the operation of Gestalt perception in the recognition of scientific principles. In fact, this same process is responsible for various "incorrigible" perceptions of phylogenetic relationships. If two species (such as a tree shrew and a primate) share a Gestalt that is in fact based on retained primitive features rather than on derived characters, the observer may find it difficult to eradicate the erroneous impression that there is some phylogenetic link between them.

One of my greatest worries in conducting this challenging but rewarding translation arose from my relative unfamiliarity with various philosophical terms. This worry was eventually dissipated by the superb work done by the copyeditor who checked the translation, to whom I owe a debt of gratitude. Combining a full command of the German language and a special interest in philosophy, the copyeditor, Bill Helfrich, played an essential role, far beyond the call of duty. On the one hand, I was relieved to see that I had not made as many errors (notably in philosophical terminology) as I had feared. On the other, I am extremely grateful for certain corrections that saved me from potential embarrassment. I would also like to thank Dr. Gustl Anzenberger for helping me with the translation of some particularly tricky terms in the German original. Gustl, like me, is an old hand from the Lorenz days in Seewiesen and shares my admiration for Konrad's achievements.

Agnes von Cranach ended her preface to the German original with the following invocation: "I very much hope that my father, were he alive today, would not have objected to its publication." In turn, I would like to end this foreword to the English version by expressing the hope that my teacher and friend, were he alive today, would have approved of this translation. This book, more than any other, has brought home to me just how difficult it is to ensure accurate translation. Indeed, the word "translation" itself has several meanings. In a speech on the subject of translation given toward the end of his life, Anthony Burgess cited Snug speaking to Bottom in *A Midsummer Night's Dream:* "Bless thee Bottom, thou art translated." I trust that the present translation has been achieved with rather less disastrous consequences.

R. D. Martin
Zurich
April 1995

What Is Our Goal?

Instincts and organs are to be studied from the common viewpoint of phyletic descent.
—*Charles Otis Whitman (1898)**

This book is the first attempt to provide a cohesive account of a very young branch of biological research. The focus of this research is that most lively of all living phenomena: the *behavior* of organisms. However, the special nature of this new line of research resides not in the subject matter, which has also occupied human and animal psychologists, behaviorists and reflexologists, but in the *method of study.* The theoretical and methodological approach underlying comparative behavioral research is fundamentally determined by recognition of one basic principle. This is that all living organisms, including humans, owe not only the characters of their external bodily structures but also their complete array of *mental* and *physical behavior* to a unique evolutionary history extending over a period of millions of years during which complex "higher" forms developed from simpler "lower" forms. The structure of any higher form of behavior is thus fundamentally and exclusively understandable only in terms of the simpler forerunners that were a prerequisite for its emergence. All other biological sciences have long since drawn this elementary and self-evident conclusion from the phylogenetic history of higher organisms. It is only in the investigation of the behavior of organisms, be it from the mental or physical side, that it has just recently become recognized that the only practicable route to an understanding of many important functions of higher animals, including humans themselves, is through investigation of their phylogenetic origins. Methodologically

**The complete citation reads:* Instinct and structure are to be studied from the common standpoint of phyletic descent, and not the less because we are seldom, if ever, able to trace the whole development of an instinct.

precise progress along this route defines the nature of the research direction that concerns us here: *comparative behavioral research.*

The subject of this book is therefore not *animal psychology,* although it is very largely concerned with animals, but essentially with *humans themselves!* Many historical vestiges from the time of prehuman animal ancestry are retained in the structure of human emotion, thought, and action, and they are indispensable for an understanding of extremely important psychological and, above all, sociological phenomena. Our interest in these phenomena is by no means dictated only by theoretical or methodological considerations. The features that persist in the physical and mental makeup of modern humans as a heritage from our animal ancestors should *not* be investigated and portrayed merely because an understanding of them is a precondition for the analysis of higher functions! Certain ancient species-specific human response patterns—"instincts" as they were previously called—have lost their original value for the survival of the species in the course of the headlong changes in human social life. They have become genuine rudiments in the phylogenetic sense. Yet these redundant response patterns, with the tenacious ineradicability and animal obstinacy of all "instincts," continue to influence human behavior. Indeed, they do so in a way that is obviously *detrimental* to life and society, inhibiting any progressive development. So much so that naive religious belief sees the influence of an evil foe in our own "rudimentary instincts" and psychoanalysis has been led to the assumption that there are specific, regressive death drives that are seen as opposing the creative principle of the platonic Eros. As will be carefully demonstrated below, certain species-specific response patterns that have persisted from the time of prehuman ancestors and have lost their original function are now among the greatest and most forbidding obstacles to the organization of a rational human social system appropriate to modern living conditions! Worse still: They doubtless represent a *danger* that should be taken very seriously, not only for progressive and creative further development but quite simply for the continued existence of humanity. As will be discussed later, human society has in any case reached a critical phase in its development. The only means of making these obstacles and dangers *accessible to attack* lies in causal analytical *research.* This must therefore be an urgent concern for anyone who values the existence and further development of humanity! The "animal within" is not only a topic of theoretical interest with respect to development and psychophysiology; it is one of the most relevant and pressing problems of the present day.

On the other hand, a knowledge of retained ancient, prehuman structures in human behavior is indispensable for an understanding of all higher mental processes that are built upon them—it is *basic* in the truest sense of the word. Anyone who tries to understand human behavior without knowledge of prehuman organisms will build an edifice without a foundation. *The route to an understanding of humans leads just as surely through an understanding of animals, as the evolutionary pathway of humans has led through animal precursors.* Therefore, in complete analogy to the established procedure of comparative morphology, our method involves the inference of historical, phylogenetic relationships from similarities and differences among the characters of living organisms. In what follows, it will be seen that a high degree of correspondence between structural characters and behavioral features emerges from phylogenetic comparisons.

A research procedure leading *from* animals *to* humans is no more a "disparagement of human dignity" than is recognition of the theory of evolution itself. It is in the nature of the organic process of creation that completely *novel* and *higher* features are developed that *were completely lacking* in the precursor from which the development took place! Least of all does our research procedure, dictated by the fact of evolution, imply any *underestimation of the differences* that separate human beings from the highest animals. Quite to the contrary, we maintain that it is precisely through comparative research that we can recognize the *special features* of humans that are phylogenetically *new,* that were definitely completely absent from the preceding series of animal ancestors, and that thus sharply distinguish us from the highest living animals. It is our claim that we can identify with great clarity *intrinsic human features* in that they stand out particularly well against the background of ancient, historical features that even today are shared by humans and the higher animals. We will attempt to demonstrate how, in this particular context, comparative behavioral research can build a bridge that is at present often sorely lacking in reflexological studies of the physiology of the central nervous system. With such studies, there is often considerable doubt about the degree to which results from animal experiments can be extrapolated to the human species.

The vital phenomena that we wish to portray are extremely complex *entities,* whose component parts or "subentities" are so intimately interwoven and alloyed that it is impossible to understand one without the other. This makes it extremely difficult to portray them, for we are not sure how and where to begin so that the reader can grasp at the outset the conceptual thread that holds the entity together. Accordingly, by its very nature our portrayal must of course be didactic; this book is first and foremost quite clearly a *textbook.* At the same time, however, it must be generally understandable for the following reason: comparative behavioral research is a science that is concerned more directly than any other with *human beings.* As such, even on the restricted basis of the results achieved to date, it has already acquired considerable significance for philosophical anthropology. For this reason, our portrayal of the subject must be written just as much for the philosopher as for the student of biology who wishes to play an active part in comparative behavioral research. As the prior knowledge of philosophical matters possessed by the student of biology is scarcely superior to the complete biological ignorance of most philosophers, this book cannot make any great demands on the reader with respect to either philosophical or biological knowledge. Quite apart from this, however, I feel—perhaps because of a certain overvaluation of my own field of interest—that the subject matter is of such general human interest that I would very much like to think that this book will serve as a source not only of *technical information* for the scientist but also of *general information* for the educated reader! This desire has shaped the planning and construction of this account.

In textbooks for all biological sciences it has become customary to *begin* with a "general" section containing the basic principles of the particular field concerned and to follow this with a"special" section that presents all of the detailed evidence that is fundamental to that field. Regardless of whether one picks up a book dealing with zoology, botany, internal medicine, pathological anatomy, or any other biological subject, the same structure is universal. Such a structure doubtless has a major advantage: for the uninitiated reader, it is easier and more effective to present the general principles, the *outcome* of the research, rather than the wealth of individual details from which they were extracted. But it also has a major disadvantage: the reader is led in *the opposite direction* to that originally followed by the research concerned. Research leads to the recognition of natural laws through the abstraction of governing principles from a great number of individual, concrete facts that must first be observed and accepted. By contrast, a textbook first presents the reader with the general principle and subsequently supports it with "examples," that is, with suitably selected facts that clearly back up that particular abstraction. This process suppresses the vast sea of trivialities through which research had to navigate, the complications, and usually—"for the sake of instructive simplicity"—even the many "exceptions" that call for supplementary hypotheses. All of this is educationally inappropriate in that the learner is presented and even trained with something that is the deadliest sin in research, namely that of *first* formulating a hypothesis and *only then* searching for examples to support it! Critical and willful students show a very healthy response to this form of

presentation. They revolt against this kind of tuition and sim
special section first after skipping over the predigested example
eral section. Of course, there is a great deal that they do not u
a result, so that after a subsequent reading of the general secti
to read the special section again. Although this may conflict with the intentions of the pedagogical/pedantic author of the book, such circular reading of textbooks is virtually unavoidable.

In theory, with a critical student who wants to find out not only *what* we know but also *how* we know it, it would surely be better to follow as accurately as possible the actual pathway followed by research. In all natural sciences, this pathway leads from concrete observable facts to abstract natural laws, from the special to the general! However, for a beginner who is unfamiliar with the overall field, it is initially quite impossible to perceive the goal of the undertaking from the confusion of apparently unconnected facts that would have to be presented in following that route. In the case of the present book, for example, it would certainly be a mystery how the behavior of the slipper animalcule (*Paramecium*), with which the special section will begin, has anything to contribute to our understanding of the inner life of human beings. If the student obediently learns all of the individual facts—although they will inevitably be extremely boring because of the initial lack of connection between them!—then sooner or later the "intellectual thread" will become apparent. But at this point it would be particularly necessary to reread all that had gone before. Hence, even this approach does not permit us to avoid the circular kind of reading that makes the crucial difference between *learning* and *reading* a book.

Following my wish to produce a *readable* book, after extensive and carefully considered reshuffling of the existing parts of this text, I eventually decided to construct it to some extent in a circular manner. It is in this way that the following overall plan for the text arose as best fulfilling the requirement of presenting the material in a *form appropriate to the entity (totality):*

Part One, Introduction to Comparative Behavioral Research, is intended in effect to be the most general part of the book. The first section thereof, entitled Philosophical Prolegomena, is designed to acquaint the reader with the elementary theoretical (epistemological) bases of all research in the natural sciences. It is aimed to show the historical and contextual relationship to philosophy and above all to provide a description of the nature of *induction* as the basis of all cognition in the natural sciences. In addition, it is intended to lead to an understanding of the characteristic relationships of a *consistent hierarchical system* within which all inductively operating areas of research are linked to one another. It is also important to understand not only the relationship that *should* exist in this respect between psychology and general biology but also the position occupied by comparative behavioral research, as a combined psychological and physiological undertaking, in the context of the other biological sciences.

The Philosophical Prolegomena will themselves sketch out for the reader a philosophical *world image.* This is actually not the *precondition* but rather the *outcome* of our overall inductive investigation of nature. As such, it should really come at the end of the book and, expressed in other words and at other levels, this will indeed be the case.

This initial sketch for a *world image* already contains an indication of certain ultimate goals of this book. As already indicated in the title, the aim is to define such a *world image,* and this is an undertaking that doubtless seems very ambitious and even arrogant. In reality, however, this is not an attempt to construct a so-called philosophical system. Instead, it is aimed at the much simpler and more promising goal of spelling out the fundamental orientation that *already exists* in the modern natural sciences with respect to the theory of knowledge (epistemology). Quite literally, the aim is to present our ideas concerning the *world image,* our conception of the manner in which the world is represented in subjective experience. The most ambitious part of this enterprise lies in my belief that I can extract from the factual basis of comparative behavioral research new and precise *justifications* for these ideas. All inductive natural science is based on the *assumption of a quite specific relationship between extrasubjective reality and its representation in the experience of the perceiving subject.* Without this, all research in natural science would be futile. Accordingly, this assumption is necessarily and fundamentally the same for all inductively operating researchers, regardless of whether (depending on the degree of their interest in philosophy) they are aware of this or not. This also explains why, to my surprise, the "phylogenetic theory of knowledge" to be presented here was received with immediate agreement by a wide spectrum of research workers, including among others Max Planck and Werner Heisenberg (both physicists), Hermann Rein (a physiologist), Alfred Kühn (a biologist), and Viktor von Weizsäcker (a psychiatrist). This positive response followed the first preliminary publications ("Kant's Theory of the *a priori* in the Light of Modern Biology." *Zeitschrift für deutsche Philosophie,* 1941; "The Innate Forms of Possible Experience." *Zeitschrift für Tierpsychologie,* 1942*).

**The corrected original titles and dates are:* Lorenz, K. (1941) '*Kants Lehre vom Apriorischen im Lichte gegenwärtiger Biologie*'; and (1943) '*Die angeborenen Formen möglicher Erfahrung,*' respectively.

Without a shadow of a doubt, in modern natural sciences a quite definite and unanimously accepted set of ideas regarding philosophy and the theory of knowledge *already exists.* As no other natural scientist has so far attempted to present a synthetic account of this conceptual framework, I feel (as it were) empowered to do so. After all, comparative behavioral research is concerned more directly than any other natural science with the human species as such and is therefore called upon more directly than any other to occupy itself with the science of the human mind. It is a perfectly legitimate task for this discipline to serve as the link between the natural sciences and the humanities. The presentation of the *world image* of a natural science is definitely one of the functions of this book. We shall need to supply the concrete factual material that serves as a basis for all of our views of the human being as a cognitive subject and for the world that is mirrored in this cognition. Setting out from these facts, we shall sketch out a deliberately and literally "one-sided" *world image,* in other words, one that is based on the inductive foundation of *just one single* natural science. In this book we shall, so to speak, interview only one witness in order to determine what this source can tell us about human beings and the world. We shall set out to see how the *new* facts established by comparative behavioral research compel us to see the "a priori" human patterns of thought and interpretation as *organs* of an *organism.* These patterns have developed in this particular form through a historical process of interaction between a real, material living organism and a real, material environment. We shall see how this realization has certain inevitable consequences that completely dispense with the old, idealistic dualism between the subject and the outside world. Our book will need to show how one can proceed from the factual basis of a quite specific, comparative phylogenetic contemplation of mankind to the conviction that there exists *one single* real, material world. The human being, as the experiencing organism, belongs just as surely to this world as everything that is mirrored in human cognition. This conviction is by no means new. It corresponds to the *world image* that, consciously or subconsciously and explicitly or implicitly, provides the basis for all true natural scientific research. It was clearly formulated long ago in the philosophy of dialectical materialism. The value of our representation lies in the new, concrete facts that we can supply to back up this *world image* and in the new pathways that we have followed in order to reach it. As already suggested, it is this that constitutes the value of a new witness who has seen from a new and previously unknown angle a set of facts already seen by others. This, no more and no less, is the philosophical and ideological mission of this book.

Over and above this, however, the first sketch of a *world image* outlined in the Philosophical Prolegomena will include the proposition of another, far more demanding task. It is not the job of this book, nor even of comparative ethology as a science, to master this task. Here, our research breaches a border zone with two other disciplines, one in the natural sciences and the other in the humanities: sociology and ethics. For methodological reasons, with which we shall become more closely acquainted in the chapter on the consistent hierarchical system of the natural sciences [chapter 3], there is always an awkward state of affairs when research from a more general, more basic area of knowledge enters one that is more specialized. This is precisely the case when our biologically oriented research enters the territory of the humanities. It is both desirable and methodologically legitimate when the *inverse procedure* occurs. In this, the more specialized area of research gradually extends its analysis toward the *more general* principles of the neighboring, more basic discipline in the hierarchy and explains the phenomena investigated in terms of those principles. Any *explanation* is nothing more than such a reduction of a more specialized and complex principle to an already recognized, simple, and more general principle.

An entire chapter is devoted to this axiomatic rule of the consistent hierarchical system of the natural sciences and to its implications. The following implications of that rule suffice for an understanding of the methodological situation that arises as a result of such contact between comparative behavioral research and sociology. In order to be able to explain the phenomena at the focus of attention of his discipline, any researcher must indeed possess a very sound knowledge of the neighboring, more basic discipline which encompasses his own. But, as a matter of principle, it is *unnecessary* for him to be concerned with the narrower, more specialized, and always more complex matter of any neighboring discipline at the next level in the hierarchy. Nevertheless, if he has some grasp of the higher-level problems that are the focus of research in such a neighboring discipline, this can in fact be of some significance for his own research. This is because such knowledge will steer the *choice of his subject* of research toward an understanding of phenomena that *might* be of value as a basis for attempts to provide explanations in the more specialized discipline. The classic example of this process is provided by the research conducted by Pavlov. Although he fundamentally avoided any psychological approach, his choice of subject for purely physiological investigations was determined by the endeavor to provide a factual basis for an explanation of more complex psychophysiological phenomena. However, whether such a choice of subject will turn out to be so fortunate in other cases depends on the "flair" of the investigator—and on luck! In fact the inductive basis available to a researcher working in a more general discipline cannot provide him with a reasonably comprehensible statement about the *requirements* of the research in the more complex neighboring discipline. Rapprochement between more basic and more complex research always smacks of *dilettantism.*

As a more basic science, comparative behavioral research has neither the obligation nor the competence to deal with the more specialized fields of sociology and ethics. All that comparative behavioral research can do is *offer* these disciplines concrete facts and principles that possibly, or probably, play an important part in the more complex phenomena that represent the research terrain of those more specialized research areas at the next level in the hierarchy. If the field of philosophical ethics, which is by its very nature always idealistic, pronounces that the principles it has erected are absolute and immutable, the basis for our criticism of such a pronouncement is entirely different from any criticism that we may make of laws governing social morality that sociologists have derived from their entirely real factual evidence. It is different to the extent that, when confronting philosophical ethics, we do not have to apply such methodological precautions and reticence as is the case with sociology. When dealing with sociological principles, because of the laws of induction, we are obliged to take into account their inductive-scientific nature.

Our comparative phylogenetic investigations of innate, species-specific behavior patterns of animals and humans have generated a broad basis of concrete facts. This leads to the conviction that many features of human social behavior regarded in philosophical ethics—because of erroneous secondary rationalization—as an expression of rational responsibility are in reality the products of species-typical response patterns. These are far more primitive than generally assumed and, as such, are not even specifically human. The same inductive basis leads to our claim that modern scientific sociology has mistakenly interpreted a whole range of characteristics, especially *disruptions,* of human social behavior as the *exclusive* product of a morality that is determined by society and class. Here, a *contributory* role is surely played by functions, and especially malfunctions, of the same ancient, species-typical patterns of behavior and response.

Comparative research has demonstrated the presence of extremely complex systems of innate (i.e., unconditioned reflexive and endogenous-automatic) behavior patterns in the social behavior not only of animals but also of humans. Because of their individual constancy, these behavior patterns provide, as it were, a broad species-specific skeletal framework for human social organization. Just like organs, such normative behavioral responses show only limited variability from individual to individual, and just like organs they can only become adapted to changed environmental conditions at the slow pace that is followed by phylogenetic modification of species. They share with all other fixed structures of the organism the dual property of *providing support* on the one hand, while on the other hand *ensuring inflexibility.* It is precisely this inflexibility of innate patterns of human social behavior that in many cases leads to conflict with the demands that are made on the individual by a social order that is changing at a vastly greater pace. It leads to a yawning chasm between the "inclination" of the inherited drive and the norm imposed by culture. The dual function of all "instincts" in providing support and inflexibility also gives rise to the momentous double effect exerted by *domestication* of human beings on their capacity for culture. On the one hand, the inherited *deficits* in certain species-typical behavior patterns brought about by domestication are the indispensable precondition for particular degrees of freedom that are virtually constitutive for mankind. On the other hand, comparable hereditary modifications brought about by the same processes (operating according to the principle of blind chance that applies to all mutations) can just as easily affect response patterns that are essential supporting elements of social behavior even under the conditions of modern civilization. This can lead to severe disruption and the deficit can become a *lethal factor* in the strict sense that applies in modern genetics. Finally, the same can also be said for domestication-induced processes leading to spilling over, or *hypertrophy,* of certain endogenous, automatic behavior patterns and the drives that they generate (craving for recognition, emotional poverty, value-blindness).

These functions and dysfunctions of unconditioned-reflexive and endogenous-automatic behavior patterns in human social life are *only* accessible to investigation from the inductive basis of comparative phylogenetic research. For this reason neither scientific sociology nor philosophical ethics was able to recognize the phenomena concerned or to appreciate the antisocial and inhibitory influence that can be exerted by the dysfunction of innate, species-typical behavior.

This claim by no means amounts to an "unacceptable biologization" of human sociology, whose special laws are in no way questioned. After all, we are merely claiming that sociology, *as a natural science,* has a duty *to take account of* the incontestably real processes that we have discovered. We regard this as especially important *in practical terms* because malfunctions of innate social behavior do not depend on the environment in the same way as the disturbances in human social life that are otherwise investigated by sociologists. As a result, the elimination of failings in the modern social order *will not simply dispense with such malfunctions.* Instead, special corrective measures will be required. How sociology will deal with these problems is, in fact, its own *special* task! Our task is simply that of *drawing the attention* of sociology to problems at the basal levels of human inner life that are doubtless of great and immediate importance. We must bring the concrete facts underlying our interpretations to the attention of those concerned with more specialized research and, through the investigation of the causes of certain phenomena, establish a basis on which they may be tackled.

The relationship between comparative behavioral research and philosophical ethics is rather different in nature. Investigation of the complex systems of unconditioned-reflexive processes that we refer to as *innate releasing schemata* (to which a major part of the general section of this book must be devoted) has brought to light a fact that compels us to come to terms with the philosophy of *value.* Our subjective assessments of value turn out to be the experiential side of a quite definitely innate, species-typical response. This operates independently of any prior experience in relation to specific environmental situations that can be abstracted with remarkable simplicity, with the constancy and predictability of the unconditioned reflex. It displays a close resemblance to genuine innate schemata.

Our response is quite clearly built upon and dependent on such schemata and it is, indeed, quite likely that there is direct identity between them. This is also revealed by the manner in which pathological *deficits* in our response tend to occur in connection with other domestication-induced mutational deficiencies. The existence of this close bond between human assessments of value and unequivocal unconditioned reflex processes must appear highly paradoxical to those who are ensnared by the old idealistic prejudice that anything that is natural must be lacking in value or be at least value-neutral. This is not so, however, for anybody who has realized from other facts of comparative behavioral research just how directly human *emotional responses* reflect the experiential side of innate, species-specific behavior patterns. It comes as no surprise that the subjective, emotional phenomenon of value also falls within the purview of the inductive scientific approach. In principle, it is doubtless too early for a public presentation of the results of this approach, but I would justify such a step with an old and absolutely valid aphorism: It is always better to say something that is incomplete than not to say it at all. It seems to be advisable to make some statement now about this incompletely and perhaps quite imperfectly known subject, for the following reasons: Humanity is at present faced with enormous problems and dangers that stem from a particular critical stage in our biological, cultural, and social *development.* In this troubled situation, humanity has been left in the lurch by a support that is more necessary now than ever before. The most serious problems and dangers are of a decidedly *ethical* nature, in the face of which the traditional ethical systems of the great world religions and idealistic philosophy have *completely failed us.* There are several reasons for this failure, but the most important among them has been correctly identified by scientific sociology. It resides essentially in the inflexible *permanence* of all religious and philosophical ethical systems. These can never take into account the continuously flowing development of human society and instead attempt to define absolute social-moral laws for a given time and a given class that are then pronounced as eternal. But the reasons for the failure also include the above-mentioned malfunctions of inherited, innate behavior patterns. Such dysfunction of "instincts" becomes increasingly threatening as society becomes more complex, and traditional ethics and modern sociology have not made its dangerous effects apparent. Still less have they identified the nature of the problem.

The duty to make some public pronouncement also stems from another consideration. The causes of the ever-widening ethical failure of civilized humanity also include a fact that the scientific thinker is bound to recognize: It was, in fact, scientific-materialistic thinking that robbed the traditional laws of religious and philosophical ethical systems of their binding, obligatory character as categorical imperatives. Of course, this was unavoidable in the interest of the intellectual advance of humanity. But, in the life of organic systems, there is always a *dangerous* transitional phase when fixed structures have to be dismantled before a new support is available. Whether it be a crustacean in the process of molting, a human being shifting in personality structure from a boy to a man in the course of puberty, or the whole of humanity shifting from one social order to the next stage of increased complexity, it is always necessary to overcome periods of chaotic breakdown during which the life-preserving harmony of the organic entity is unprotected against serious danger. Given that scientific-materialistic thinking has been such a major factor in the present upheaval of human society, any convinced materialist and convinced scientist must feel a certain degree of responsibility for this process of "ethical molting." It is *our* contemplation and research that has brought this about and we therefore have a duty to *replace* the extremely and undeniably important sociological factors that have been eliminated by such thinking. The need for such a replacement amounts to nothing more and nothing less than the duty to create a new human ethical system. But neither comparative behavioral research nor sociology, nor indeed any other inductive natural science, is up to this task. The task could only be accomplished by a new philosophy arising from a genuine synthesis between the traditional humanities and their antithesis—the young field of inductive natural science. The last chapter of the Philosophical Prolegomena [chapter 4] is accordingly concerned with this notion and with the necessity for such a synthesis.

Section II, of the introduction, entitled Biological Prolegomena, is intended to give an idea of the particularities of subject, approach, and method that distinguish *biology* from other natural sciences. The aim is to establish the necessary basis for an understanding of the phylogenetic origins of organisms and of the conceptual and practical methods that we use to tackle investigations of this process. At the same time, the aim is to introduce the reader to the basic essentials of the conceptual and practical methods required for *holistically appropriate* analysis. This involves the principle of the methodological procedure that we refer to as *analysis on a broad front,* which is required whenever we are confronted with organic entities. This section is intended to show the logical and methodological relationship between causal and final effects, which even today are subjected to unbelievable confusion by vitalistic animal psychology. In closing, Biological Prolegomena will acquaint the reader with the fundamental relationship that exists between the investigation of the experiential side of psychophysical processes and that of the organic-physiological side. The aim is to show how comparative behavioral research attempts to approach the "mind-body problem." This is accomplished by trying to investigate a single psychophysical process while *simultaneously* trying to achieve a clear separation between the two incommensurate physiological and psychological domains.

Following Philosophical Prolegomena and Biological Prolegomena, we shall quite literally comply with the theoretical requirement of leading the learner as accurately as possible along our own cognitive pathway. We shall do this by devoting Section III of the introduction to a chronological account of the *genesis* and development of comparative behavioral research. We shall follow the remarkable developmental process of a *human* natural science whose origin *apparently* lay in the simple pleasure of observing animals! This developmental history of our field of research is essentially identical to the history of its most significant discovery. This surely lies in recognition of the fact that, in animal and human behavior, *there is a second elementary function of the central nervous system, alongside and in addition to the reflex, that plays an equally fundamental role.* This second elementary function is the *automatic, rhythmic generation of stimuli,* previously documented only for the pacemaker center of the heart. All of those behavior patterns that were referred to by earlier comparative behavioral investigators as "instincts"

(Whitman) or as "species-specific drive-governed activities" (Otto Heinroth) and that are now called *instinctive motor patterns* are not, as was previously believed, composed of chains of unconditioned reflexes. Instead, they depend on endogenous, automatic, rhythmic processes of stimulus production in the central nervous system itself. This discovery of the physiological peculiarity of instinctive motor patterns has major, far-reaching implications in several respects. First of all, it provides a precise *physiological* explanation for the phenomenon of *spontaneity* in animal and human behavior. Because this simply cannot be explained through the principle of the *reflex,* it has always provided support for vitalistic, teleological explanations of instincts. It did so by yielding welcome arguments against any proposed mechanistic interpretations, which were previously based entirely on the chain-reflex theory. The discovery of spontaneous generation of stimuli in the central nervous system thus dislodges vitalism from a commanding position that was previously defended with vigor and success. This opens up to causal analysis an entire field that was previously just a playground for fruitless philosophical speculation.* From the point of view of psychology, the discovery of automatic stimulus production presents tremendous opportunities for coming to terms with the laws governing the subjective phenomena accompanying drives, purpose, and the entire phenomenon of pleasure and aversion. It is precisely this that permits us to subject an undoubtedly singular physical and mental process to precise, dual physiological *and* psychological investigation.

A simple, chronological account of the individual cognitive steps followed by comparative behavioral research in arriving at an understanding of automatic stimulus production and its varied effects is, in itself, a very logical and consistent framework. As a result, the history of the development of our field of research provides a very good overview of the entire discipline. For this reason, it can very adequately fulfill a second theoretical requirement concerning the *holistic suitability* of didactic representation of organic systems. Wherever, as in this case, the nature and complexity of the subject matter compels us to divide entities into *pieces* for presentation to the student, it is a methodological necessity to provide a *sketch of the entity.* This provides a preliminary framework into which the dissociated pieces can then be incorporated. Our account of the history of the development of comparative behavioral research can easily achieve this goal while serv-

*"Lorenz's approach opens up to causal analysis a field that was previously just a playground for fruitless philosophical speculation." (*Max Hartmann in discussion following a lecture given by Konrad Lorenz in 1936 at the Harnackhaus in Berlin.*)

ing its primary function. In addition, the reader will in the process become acquainted with the methods and central issues of our research. This will suffice for the reader to understand why and in which respect individual observational facts are of interest, despite the great variety of details of animal behavior that are presented in the following "special part" (Part Two).

As a result, the special part can be placed where it really belongs from a methodological point of view, namely, before the general, nomothetic part. The special part will present the reader with a series of widely different living organisms whose behavior patterns are known in such fine detail that we can venture to outline the complete picture. There are not many animal species for which such a demand on the thoroughness of observation can be met. I do not feel that the scientific respectability of our presentation is in any way diminished if these animals and their behavior are described as graphically and as vividly as possible. That is to say, they will be described precisely as a good but scientifically untrained observer would see them. I shall also not hesitate to recount the life stories of individual higher animals exactly as I myself experience them, even in cases where some readers might be given a journalistic impression by such "animal stories." What I have to say about ducks and geese, jackdaws, and some fish largely goes back to my youth and even childhood. Wherever the special part deals with animals that I have myself observed intensively, I have tried to keep to the naive expressions of my diaries and publications of the time. This is because they do not involve the modern scientific jargon based on abstract concepts that the reader will not encounter until the general part. In this way, the reader will automatically and quite naturally relive our own progress toward these concepts. Only a few of them, most notably the concepts of the instinctive motor pattern and the innate schemata, cover such general and widespread principles that they cannot be dispensed with in the special part. To the extent that these principles are not known to the reader from the developmental history of comparative behavioral research, they must of course be explained, at least in outline form, in those places and for those animals where they first appear. This necessarily amounts to an anticipation of the general part. I have not shrunk from such anticipation, as the major principles of animal behavior should and must of course be contained in the observations that are reported, and it is doubtless helpful for the reader to present them.

I have, nevertheless, taken pains to keep such aids to a minimum and to allow the reader as far as possible to hit upon the abstract reason for the concrete fact. As everything is interconnected in the entity constituted by any organic system, any description that does reasonable justice to that entity is bound to contain repetitions because any given aspect must be discussed in several different contexts. For this reason, no attempt has been made to avoid such repetition. To the contrary, precisely those observations that the reader has come to know through the graphic descriptions in the special part have been used as examples for the phenomena discussed in the general part. As a reflection of this open recognition of the didactic necessity for repetitions, these have been indicated throughout with corresponding page references.

The general part (Part Three) contains a detailed presentation of the *principles* that can be extracted from our observations and experiments. These are primarily the physiological and psychological principles governing the endogenous-automatic, centrally coordinated motor sequences that we refer to as *instinctive motor patterns,* the *orienting responses* or *taxes,* and the *innate releasing schemata.* Since it is standard practice that we only conduct experiments on animals whose entire action system (behavioral repertoire) is already known in detail, the above-mentioned repetitions are particularly necessary in the experimental part. Here, we need to reexamine behavior patterns previously discussed in the special part, but this time from a precise theoretical angle.

This book cannot tackle the task of providing exhaustive accounts of the many sister disciplines that interact with the field of comparative behavioral research. We shall restrict ourselves to the bare necessities even for two sister fields that have yielded certain information that is indispensable for an understanding of the phenomena that we are investigating. These two fields involve research into the *conditioned reflex* and the psychological and physiological study of *Gestalt perception.*

In principle, *domestication*—with all of its accompanying notable modifications and deficits in innate behavior—is a phenomenon that borders on the *pathological* and is therefore organically and didactically distinct from the functional entity constituted by innate and acquired behavior patterns that is to be presented in the general part. A certain amount will, in fact, be said about domestication in the special part when discussing the behavior of the mallard, the graylag goose, and the domestic dog. In fact, I have devoted a special section at the end of the general part to domestication because a better understanding of the associated phenomena and principles is an absolute prerequisite for the subsequent chapter on *humans* in the fourth and last part of the book.

I have devoted a special section of this book to humans, rather than including them "after the dog" in the series containing other living organisms. This is not because I was concerned to avoid any insult to human dignity, but simply because the situation for our species is so complicated that it is only possible to present it in a reasonably lucid fashion if prior knowledge of the entire general part can be taken for granted. I have left the part dealing with human beings essentially in the form of a series of seven lectures that I gave several times to a group of doctors who were fellow prisoners in a Russian prisoner-of-war camp.

It is by no means our belief that comparative behavioral research in its present virtually embryonic state is in a position to provide a satisfactory, let alone definitive, picture of the human species. All that we can hope to do is to present in an easily understandable form the few *new* statements that comparative behavioral research has to make concerning human nature and the preconditions for human origins. It will be seen just how much and how little this is. It is quite a lot with respect to the original, deep emotional layers of human inner life and also quite a lot with respect to the actual process of human evolution, particularly as far as the preconditions are concerned. By contrast, as is only to be expected, it is relatively little with respect to phylogenetically completely new, highly advanced capabilities—such as language—which have no counterparts (homologues) among animals. Nevertheless, with respect to the nature of such capacities, *negative* conclusions are far from unimportant. We shall, in fact, be attaching considerable weight to the lack of certain capacities in animals, even in the highest evolved relatives of mankind!

Comparative behavioral research does have a considerable contribution to make to our understanding of the highest intellectual capacities of human beings in *one* respect: All of the forms of possible experience that are independent of prior experience, namely those forms of thought and intuition that Kant gathered together under the concepts of *a priori schematizations,* are exposed in an entirely new and to some extent much brighter light by physiological and phylogenetic investigation of the so-called *innate releasing mechanism.* The cognitive considerations that arise from this lead us back to the starting point of the most general subsection of the Philosophical Prolegomena. We shall encounter once again the interaction between the perceiving subject and the world of realities that exists independently of any perception of its existence.

But the conclusions for the humanities that emerge from the results of research into innate schemata also lead us back to deliberations concerning the *philosophy of value* that are already outlined in the Philosophical Prolegomena. The most important contribution that comparative behavioral research has to make to the foundation of the new ethical system, called for there as a synthesis between the natural sciences and the humanities, lies in recognition of the fact that the subjective value perceptions of human beings are linked to the functioning of innate releasing schemata. Acceptance of this attribute of value perceptions obliges us to draw the same epistemological conclusions as those derived from the organ-like character of the innate schemata and all other innate forms of possible experience. The innate form of experience can be seen as a *receptor* that has evolved in the course of phylogeny to serve a special survival-promoting function as the outcome of interaction between the organism and its environment. In other words, it responds to a quite specific, real condition. Demonstration of the existence of such a receptor organ inevitably leads to the conclusion that there must be something to receive from the extrasubjective real world. A real, objective factor in the concrete world must also correspond to anything that we are able to perceive as a value through some innate structure in our experience. But such a factor is nothing other than the ectropic development of organic creation. Investigation of human value perceptions reveals that these purely subjective responses, like the parallel behavior patterns that can be objectively and "behavioristically" recorded, occur in a systematic fashion *only* when a *quite specific process* is perceived in extrasubjective reality and penetrates to the conscious level. Wherever natural developmental processes, particularly of an organic kind, lead from simpler to more complex conditions of organization, from lesser to greater *improbability*, our perceptual experience registers "higher" with all the inevitability of genuine innate schemata! Everything that we see as *values* on the basis of this "a priori" structure of our experience and perception is not rational but *innate*. It is, without exception, the outcome of those laws governing all organic matter that determine a creative development from the simpler to the more complex, from the generally probable to the less probable. Thus, the "lower" and "higher" that are common even to obstensibly "value-free" science—for example, in the self-evident zoological concepts of lower and higher animals—can be defined *without* recourse to subjective human perceptions of value! The organic process through which the more complex can arise from the simpler, with the "lower" precursor giving rise to the "higher" successor as something entirely new, is far from being some product of human imagination. The laws that govern this process of development are natural laws entirely comparable to those applying to physics and chemistry and they belong to the material, concrete world just as much as they do. We are led to the conviction that whatever is concealed behind our subjective experience of values is also something real. Any phenomenon perceived by an intact human not suffering from value-blindness as the highest of all, indeed, as holy, is without exception an isolated case of something which has always taken place in the organismal world and which took place for a very long while before the origin of value-perceiving organisms.

This brings us back to our starting point in the Philosophical Prolegomena and the circle is closed. I have attempted to present an organic entity as a totality to the extent that this is permitted by mere words. This, admittedly, is not very much, as "words strive in vain to summon shapes creatively" (Goethe). I nevertheless hope that I have been able to "present a whole" to the extent that the reader should find it difficult to pluck it to pieces. For the reader can begin at any point. One can start with the graphic animal profiles in the special part (which is, for example, what I would probably do), or one can begin with the abstract principles of the general part. Indeed, as far as I am concerned, the reader could even begin with the discussion of human beings,* although this would require frequent consultation of the parts dealing with the innate schemata and with domestication. In any event, the reader will complete the entire circle, as long as the subject remains of interest. However, if I may be permitted to make a recommendation, I would suggest that the reader read the outline presented in Historical Origins and Methods of Comparative Behavioral Research [section III] before tackling the later sections of this book.

**The author refers, for instance, on p. xxiv et seq. and on p. xxxiv et seq. to the original four-part structure that was planned for this book: introduction, special part, general part, human beings. The present book contains only the first part from the original plan.*

Part One

Introduction to Comparative Behavioral Research

I

Philosophical Prolegomena

1

Natural Science and Idealistic Philosophy

In earlier times, there was only one field of scholarship. This was philosophy. Nowadays, we are often told—mainly by practitioners of the humanities—that there has recently been a split, a schism in scholarship between natural science and the humanities. This is, however, a fallacy stemming from a basic misunderstanding of the essence of natural science. The few probing attempts at genuine natural science that we know from classical antiquity initially came to nothing and a real natural science in the narrower modern sense has existed only since Galileo's time. It simply did not exist during any of the previous stages of civilization. Yet each of these stages possessed its own, in some cases very highly developed, philosophy. *Sages* have always existed, but the first scientist documented by recorded history was Galileo. For the history of emergence of human cognition, it was more than a Copernican revolution when humanity first began to *observe* rather than just *meditate* as in the past. In their endeavor to gain knowledge of themselves and the surrounding environment, human beings first learned to *ruminate* and only much later did they learn to *scrutinize.* This undeniable fact of intellectual history is extremely curious because, in both developmental and evolutionary terms, simple observation is equally undeniably a far older and more primitive capacity than reflection. Even subhuman organisms use their sense organs to perceive external reality, so if a philosophizing human being *neglects* this source of information in such a conspicuous fashion, some special *explanation* is required. We believe that we can provide such an explanation.

A Greek sage regards it as the "birth of a philosophy" when mankind first begins to *wonder* about something that was previously self-evident. We are all familiar with the strange experience whereby the most well-known phenomenon can suddenly appear in a new light, as if we had never seen it before. And our ancient Greek philosopher is by no means wrong in attaching so much significance to this! Philosophical θαυμάζειν, the experience of sudden astonishment, led human beings without a shadow of doubt to one of the most important discoveries ever made, the discovery of self. In the view of many philosophers, recognition of self, and hence *reflection,* is the most important feature of mankind. Marveling about that most obvious of everyday phenomena, the existence of the self, does indeed represent the root of all reflection about mankind and the surrounding environment. It has become the source of a large part of human striving after knowledge.

Over and above this, there is a second, far more ancient root for human striving after knowledge. This second root is not even specifically human, but it nevertheless has great significance for human nature and, indeed, for the *origin* of mankind. It is nothing other than simple *curiosity,* which characterizes specific prehuman creatures. Anyone who knows how to observe can spot this even in a suckling infant. It governs the experimental play of 18-month-old children and is clearly expressed in the tinkering of a growing boy. In some of us, naked elementary curiosity develops through a smooth transition, without the slightest alteration in its intrinsic nature, into the purposive striving for knowledge that we refer to as *research.* This far more ancient search for knowledge is fundamentally different from the philosophical quest for knowledge. In its higher stages of development, it is not always *free* of reflection, but it is always *independent* of it. This *must* be the case because it is incommensurably older than reflective philosophy. Later on, we shall consider how a particular form of curiosity characterized prehuman beings and how it came to be a *precondition* for the emergence of mankind. In contrast to the philosophical quest for knowledge rooted in reflection, curiosity is essentially directed *outward.* Even if it contains an element of wonderment, it is not the same as philosophical wondering about well-known phenomena that abruptly seem to be *new.* Instead, it is a far more naive, I would say home-baked, astonishment about *really* new phenomena. Above all, however, curiosity represents a striving after *education* through new things. Curiosity is literally a lust for novelty. As we shall see, it is plainly a goal-oriented form of behavior allied to orienting responses and not to drives. It is a "positive taxis" directed toward a stimulus situation in which the acquisition of a new *Gestalt* will occur. Anybody who has become really familiar with the phenomenon of curiosity through detailed observation of animals and children and has personally experienced the smooth transition from the quite unreflective curiosity of childhood to adult investigation will know that all inductive natural science has developed in the same way. For such a person, there will never be any doubt that the deepest roots of natural science make it something fundamentally different from any reflective philosophy.

Hence, philosophy and natural science are *distinct.* They differ from one another in their ultimate roots, in the *motives* behind their striving for knowledge. It was essential to establish this point. But how can we now explain why natural science—despite its origin from a far more ancient, prehuman root—developed so much later in human history than philosophy born of reflection? The reason is that, immediately after the discovery of self, a philosophizing humanity became blocked *in* this self as if in a blind alley. This strange process perhaps took place in a very similar fashion during other stages of human civilization in which genuine natural science failed to emerge. In our Western civilization, this process of intellectual development, which is at once both paradoxical and self-evident, quite definitely had the effect that for about two millenia (from Plato to Galileo) people were prevented from using their senses for the purpose of acquiring understanding. It is now incumbent upon us to portray this in more detail.

As is to be expected from everything that has been said above, a completely naive, primitive human being is initially more a natural scientist than a philosopher. Just like a small child, he at first directs his search for knowledge toward the outside world. Neither has yet discovered the existence of self. Such a prephilosophical human being is necessarily always a *naive realist.* Because he is not yet aware of the existence of the self, he is also unaware that he is *perceiving* objects in his environment and must therefore believe that those objects must be exactly what he perceives. His perception reveals objects to him in such a convincing fashion, so *tangibly,* that he is quite unable to hit on the idea that they may not in fact exist. I do not see the sun at the back of my eye but where it is *high up in the sky.* I see writing paper in front of me on the table, occupying a *different position* from that of the sun and certainly not at the back of my eyeball. My perception does *not* tell me that both objects form images at the *same* place, on the central areas of my two retinas. Still less do I perceive anything of the operation of the extremely complex central nervous organ system that processes the retinal images, the orientation of the visual axes, and other factors to construct perception of the sun and of the writing paper to *project* them into the corresponding position in space! Such a projection process is, in fact, incorporated in our sensory perception to permit us to perceive objects in particular positions in space. It allows us to see, hear, and touch them *outside* our bodies, despite the fact that all of the nervous processes involved in such perception take place *within* our bodies. It is precisely this *spatial* certainty of perception that prevents a naive human being from ever doubting that an objective world exists outside the self, independently of our existence or nonexistence.

But then humans begin to look inside themselves, to reflect. It is at this point that they discover for the first time that it is not objects themselves that are perceived but processes that take place internally. This discovery is, in itself, absolutely correct and it represents one of the greatest discoveries of human intellectual history. However, the ancient philosophers that made this discovery did not know that these internal processes are *responses* of an organism to perfectly real effects exerted by objects in the external world. Thus they saw that perception represents only an *image* of an object, but not that it is a *real* image of a real object. It was therefore inevitable that, in its early stages, philosophy at once drew a conclusion from the first great discovery of the human internal world that turned out to be one of the most disastrous fallacies of human intellectual history. This is a fallacy from which, right up to modern times, Western civilization has been unable to free itself entirely. That fallacy is *doubt about the reality of the external environment.*

The discovery of the fact that we exclusively experience internal processes inevitably led to doubts about whether external objects corresponding to our perception exist at all. It was unavoidable that perception of external objects should be compared with perceptual processes that are similar in form and content but do *not* correspond to anything existing in external reality, such as ideas, memories, fantasies, and dreams. One could not fail to conclude that variegated, colorful external reality perhaps represents nothing more than dreams. In the hands of a philosopher whose gaze is directed inward, the only reliable truth that remains when faced with an external world dissolving into dreams and emptiness is Descartes' pithy statement *Cogito, ergo sum* (I think, therefore I am). It was therefore only to be expected that in the human quest for knowledge this "only reliable" internal anchor should have been taken as the starting point for any attempt to explain the "unreliable" phenomena of the external world. My dog is lying in front of my desk. It is an undeniable fact that a picture, an idea of a dog—in short, something doglike—is present in my perception. But I cannot maintain with certainty that the dog itself exists. After all, only yesterday I dreamed of my dog and when I woke up the dog was not there! [Gustav von] Bunge wrote: "The nature of idealism resides in the belief that the only correct pathway to knowledge requires us to set out from the known, internal world in order to explain the unknown, external world."

In fact, the idealistic belief that it is only the "idea of a dog" that can be real (Morgenstern), and that all of the dogs running around in the world are only transitory, incomplete expressions of this permanent and complete idea, also has another, deep root in an understandable but very naive anthropomorphization of the process of creation. Whenever a human being *creates* something, for example, when a carpenter makes a table, the idea of a table is present in a more-or-less complete form in the brain of the creator before the realization of that concrete, individual table (though not before the existence of concrete tables in general). The idea is, indeed, more perfect than the table that is actually made, with its unavoidable chance defects, and it may also be longer-lasting than the concrete piece of furniture. The idea continues to exist when the table falls apart and can be realized if the old table is repaired or if a new one is made. *We* by no means forget the fact that this idea is a real thing in a real brain, but the ancient idealists did not see this at all. For them, the idea was some preexisting thing, independent of the presence of any brain. To put it crudely, for them the idea of a radio or of an electron microscope existed *prior* to the creation of the real world, as with any other thing. As human beings who believed in anthropomorphic gods, they were necessarily drawn to interpret the creation of the world in analogy to the process of human creation outlined above. Thus it was that Plato taught that the idea of anything that was created was the original, sole reality and all of Western civilization believed him. "All that is transitory is no more than a simile."

Most of us are not even aware of the extent to which all of European culture is interwoven with and permeated by the doctrine of idealism in its literature, in its philosophy, in every religion, and in all of its everyday habits of thought, right down to slang expressions. One is tempted to say that it is impregnated with this doctrine. This is true to such a degree, particularly in German intellectual life, that the word "idealist" has a highly positive moral significance, whereas the term "materialist" has taken on a pejorative connotation. It is as if a human being who holds the world of ideas for true reality must also live out his ideals. By contrast, it is more or less implied that anyone who regards ideas as a product of the brain can only be concerned with extreme material values. In its disastrous superficiality, colloquial language conflates the concepts of idea and ideal in the one word *idealist* and uses this word as if anyone who has a materialistic approach and regards the outside world as something real must be ipso facto lacking in ideals.* Of course, in reality, the ideals that guide a man's lifework are in no way bereft of their purposive function if we divorce ourselves radically from the heritage of Platonic idealism. We know that ideas are not

* *The following sentences, up to* "but an archenemy of ideism" *were written in rough form on a scrap of paper that was found tucked between the leaves a few pages further on.*

preexisting plans of real objects but images cast in the human brain. Yet even if ideas have the character of images they are nevertheless important and significant. And the ideas that are most significant for the entire progress of humanity are those that we refer to as "ideals." We define as an ideal the idea of some thing or condition in the real world which *does not yet exist but is fundamentally possible* and which carries with it, for anyone who has fully understood, the moral *obligation* to work toward the *realization* of that thing or condition. As a research worker deeply committed to the materialistic way of thinking, I would hold out most energetically against not being an "idealist" in this sense! In order to escape from the disastrous superficiality with which colloquial language confuses the concepts of idea and ideal in the terms *idealist* and *idealism,* in what follows we shall refer not to idealistic philosophy but instead, in quite proper linguistic form, to ideistic philosophy.

A true inductive natural scientist is an archidealist but an archenemy of ideism. We are all so used to encountering the maxims of ideism in a wide variety of formulations that we have completely lost any feeling for the extremely paradoxical fashion in which ideism turns facts upside down. The image of an object that is formed in the human brain is erroneously taken for true reality while the real object is explained as an image of something that is itself the image. If this sentence seems to be very confused, the blame lies not with some failing in my expository style but with its quite accurate reflection of the incredibly confused factual context that is apparent to the impartial observer. In all seriousness, idealism has completely reversed the relationship between extrasubjective reality and our perceptual image of it. As a result, considerable harm has been done to philosophy and, still more, to empirical research. In classical antiquity, promising beginnings of a materialistic philosophy were indeed present, but the immense proliferation of idealistic interpretations completely suppressed them. Western philosophy was so captivated by the relatively recent influence of Platonic idealism that the interpretations of Heraclitus, which strike us today as unbelievably modern, had to be virtually rediscovered. It is no exaggeration to claim that this state of affairs simply *blocked* the emergence of real natural science, or rather the further development of its preideistic roots in classical antiquity, up to the time of the Renaissance.

Who, then, would have wished to be a clown concerned with the laborious task of observing and describing the endless small details of the external world if this was no more than a void, a dream, the investigation of which would not advance our understanding of the world and its inner workings by one iota? Who would have wished to take the trouble to untangle the infinitely complex relationships of the external world when this, at best, could lead only to the recognition of the principles underlying a mirage? How could this compete with the opposing approach of looking inward to gain immediate revelation of ultimate, fundamental truths? It is only on the basis of such deeply rooted thought processes that we can understand such excesses. How else can we explain, for example, the fact that in Aristotle's time men who were neither mad nor in the business of reveling in bad jokes seriously *discussed* the question whether houseflies have six legs or four, rather than simply catching a fly and counting? To put it quite crudely: If these Platonic ideists had concluded that a majority among them possessed the idea of a four-legged fly, would they then have explained all real flies as an imperfect expression of the ideal fly? It is plain to see how fundamentally an ideology of this kind could have been an obstacle to the development of a science that would derive the great laws that govern the universe, abstracting them from the multitude of concrete details identified by our perception from the objective world. Because all of philosophy, until the modern era, was dominated by an ideistic train of thought, it was impossible for inductive natural science to arise *from* this source. It arose *despite* philosophy, bursting forth as something entirely new during the Renaissance. Like water in a test tube that suddenly begins to vaporize when the boiling point is reached, it arose with the abruptness that is peculiar to ideas that are actually long overdue and whose birth is held back only by external circumstances. Even much later, when the humanities slowly—how slowly!—found their way back to a belief in the reality of the objective world, doubtless in response to the pressure exerted on their way of thinking by the successes achieved by research in the natural sciences, they had definitively lost the ability to utilize perceptual evidence as a source of knowledge. No experiments are conducted in the humanities, not even in cases where they would be entirely compatible with the underlying principles on the basis of empirical or realistic viewpoints. Even today, for an ideistic thinker, natural science is the "culmination of dogmatic narrowmindedness," to quote that most implacably consistent of all Neo-Kantians, [Kurt] Leider. For a natural scientist, by contrast, all ideists are "individuals who are *incurably obstructed* by epistemological considerations from using their senses as a source of scientific knowledge," to cite an equally candid appraisal by the Gestalt psychologist [Wolfgang] Metzger.

Because philosophy retained its ideistic nature until the modern era, it could not lead into natural science. Quite to the contrary, such philosophy blocked such a development for centuries, perhaps even for millenia. For its part, however, natural science eventually led along a long, tortuous, and thorny path to a particular form of philosophy. The widespread belief that the natural sciences and the humanities are offshoots from a previously unified human quest for knowledge, and that a regrettable schism led them to drift apart over the course of the past few centuries, is incorrect in two senses. In fact, the natural sciences and the humanities are derived from different sources and they are growing together, such that they have quite recently come into contact and reciprocal understanding. To the extent that the humanities are returning to belief in the reality of the external world, they are ceasing to scorn the natural sciences and beginning to see the necessity of including their findings in a philosophical conception of the world. The natural sciences, on the other hand, with an approach that is purely directed toward observation of the real external world, could not fail sooner or later to encounter human beings among the living organisms that inhabit the earth and thus to include humans as one object of their investigations. To the extent that the natural sciences turn their attention to the organic functions of this most important of all "objects" and hence come to include the functioning of the human central nervous system within the scope of their research—that is, learning to regard *experience* and *knowledge* as organic functions—they begin to encounter the humanities on their most elemental terrain.

This encounter perhaps represents a major turning point in human intellectual history. Superficially, this does not seem to be anything very special at first sight. On the contrary, it would seem to be quite obvious that curiosity-driven research that is directed purely outward in an objective search for knowledge should eventually include humans within its scope, like any other natural phenomenon. In reality, however, this gives rise to a point of view that is just as new in kind as the discovery of the self in prehistoric times was a new standpoint for the beginnings of reflection. Until now, there have been *two* ways of observing the universe, one leading inward and the other leading outward. But now these two diametrically opposed approaches have been joined by a third that is entirely different in kind. The human being itself, the perceiving *subject,* has become the *object* of objectifying, outwardly directed scientific research. Still more, *perception* itself—the "camera that takes a picture of the world"—has become the object of an investigation which is by its very nature directed at the real external world and which sees this "world-image apparatus" as an *object in that world* and not as the diametric opposite of that world. An entirely new kind of reflection about reflection has emerged, a completely original way of observing *the inner world from outside.* A naive, prephilosophical human being is aware only of the external world and not of the fact that it is being perceived. But until now reflective philosophy was also unaware that the phenomena that were revealed by introspection can also be seen *from another side,* namely, from the outside. Recognition of the remarkable and unusual *duality* of all living processes that are accompanied by experience was the first world-shattering fruit of the newly acquired point of view. Introspective philosophers never hit upon the idea that the mental processes that they were studying could also have a physical side. It never occurred to them to regard the human brain as an *organ.* Naive realism admittedly failed to recognize that the perceptual world is just a mirror image of reality. But equally, idealism, even Kant's idealism, was not aware of the undeniable fact that the mirror that forms this somewhat hazy and simplified image of the objective world is *itself also a thing,* a real material thing of just the same kind as the other material things that it depicts. The prephilosophical human being looks only outward and is not aware of being a mirror. The idealist looks only *into* the mirror and turns his back on the outside world. From his angle of view, he is quite unable to see that the mirror has a material, *nonreflecting* side, a side that places it in the same category as all other *reflected* objects. But research is now beginning gradually to take the mirror itself as an object for research. It is beginning to regard our world-image camera as a human *organ.* Above all, research is now beginning to investigate the mirror *simultaneously* from its reflecting side and its nonreflecting side, by using exact methods to investigate simultaneously psychological and physiological functions of the nervous system that are accompanied by perception, without mixing or confusing them. This book will present some aspects of the way in which *one* science, that of comparative behavioral research, tackles these problems.

Because the human world-image apparatus came to be a focus for scientific research in this way, a sharply defined and above all *very well-founded* notion has emerged concerning the *relationship between the real and the phenomenal world.* In a number of very important points, this notion is in close agreement with concepts developed by Kant in his theory of a priori forms of possible experience. But in other, even more fundamental, points, it diverges from Kant's purely idealistic doctrine. Understanding of the goals, conceptions, and research methods of our own discipline can be promoted by briefly outlining these points of accord and discord.

Biological research, a small special area of which is to be portrayed in this book, has led to the conviction that human beings are organisms that—like all other living organisms inhabiting our planet—developed from other, simpler forms of life in the course of an interminable struggle for survival. If this is indeed the case, every human organ has acquired its particular form and function through millions of years of exposure to the circumstances and laws of the external world, *adapting to* these scarcely changing circumstances and laws. The human brain is also an organ of this kind and there is no justification whatsoever for seeing the innate forms of its functioning in any light other than that in which we see any other organic functions determined by the form and structure of the organs concerned.

Kant, the great discoverer of our innate patterns of thought and intuition, did *not* quite see these as organic functions determined by structure. He saw something that was not seen by the English empiricists, the first philosophers to regain confidence in the evidence provided by the sense organs. Bacon, Locke, Berkeley, and Hume were all of the opinion that the fundamental structures of human perception of the external world, such as interpretations of space and time and conceptions of causality, substantiality, quantity, etc., all arose from the individual experience of every individual human being. They regarded them as *abstractions* that human beings derive "a posteriori" from the information about external reality provided by the sense organs. Their perfectly correct guiding principle was: *Nihil est in intellectu, quod non ante fuerat in sensu*—"The intellect contains nothing that was not previously contained in sensory perception." Although they were opponents of idealism and although they quite rightly took account of the sense organs and their specific structures, they were still "idealists enough" to overlook completely the brain as an organ that can also possess structures—and what structures! Although Kant was even less inclined to interpret the forms and structures of human reason as organic functions determined by structures, he was the first to see something that had not been seen previously. He saw that quite specific *forms of human thought and intuition* are possessed by human beings that are independent of any experience and that, conversely, experience is only permitted by the *existence* of these given "a priori" forms of thought and intuition. He recognized with great clarity something that the "*sensualistic*" tenet of the empiricists completely omitted, namely, that the form in which the external world is depicted in our perception is determined not only by the structure and function of our eyes, ears, and other sense organs but, far more extensively, by *internal* laws inherent in human perception itself. Like no other before him, he thoroughly examined and dissected these laws. With overwhelming clarity, he analyzed the image generated by our perception of the objective world, and with sheer superhuman acuity he grasped the structures of this image that are to be derived not from the nature of the object depicted but from the laws governing human thought and intuition. Kant not only discovered our "world-image apparatus" but also analyzed it far more than anyone after him. For this reason, it seems almost like a satire of human intellectual history that the idealist Kant largely did this in criticism of the empiricist Hume!

As an idealist, however, Kant saw his human subject and human perception exclusively *from the inside.* It never occurred to him to regard the perceiving subject as a part, as a *member* of the objective world. Instead, that subject was seen as an opposing counterpart to the objective world, standing in diametrical contrast. As a result, for him all preexisting a priori laws remained completely outside the objective world and the laws governing that world. Kant saw *two* worlds where we see only one. For this reason, he failed to notice the close functional analogy that links the world-imaging *structures* of human categories of thought and forms of intuition with those of the likewise world-imaging sense organs and even more with the structures underlying *perception.* For him, only the latter belonged to the objective world and only they were accessible to scientific investigation and explanation. Perceptual physiology did not yet exist to reveal to him that the functions of the internal structures of the central nervous system linking the sense organ to experience in fact represent an intermediate level between sensory function and internal forms of intuition. There was as yet no theory of evolution to compel him to tackle the question of the origin of human reason. Hence, the "a priori" structures of human thought and experience remained for him things that had *not developed* but were *necessarily* as they were, things that belonged not to the external objective world but to another, supranatural "*intelligible*" world, things that were not organic but simply *divine.*

For Kant, there was no comprehensible, logically tangible relationship between the objective world and the intelligible world comprising only those *a priori* schemata that create images of objects in our perceptual world. For this reason, there could be no comprehensible connection between the image of things and the things as they really are. Although Kant assumes the existence of extrasubjective reality, of *things in their own right,* for him this reality has no connection whatsoever with the manner in which these things "influence our senses" and emerge as a phenomenon in our experience. For him, a thing itself necessarily remains forever unknowable. This is the point at which our interpretations fundamentally diverge from those of Kantian idealism. The Kantian inference concerning the absolute *unknowability* of objects would be unassailably correct if the underlying premise were correct. This is that the perceiving subject and the thing reflected in its perception belong to two separate worlds with absolutely no relationship between them, such that there is really no connection between things as they are and the manner in which they influence our senses. I would like to draw an analogy here: Let us assume that a given person is unknowingly suffering from mild poisoning with an exotic poison of which he is completely unaware and with which neither he nor the species *Homo sapiens* has previously had any contact. The poisoned person will doubtless experience the poison in some way. His ears might buzz, his hands might tremble, and so on. But because for this person the nature of the poison has no comprehensible relationship with the manner in which it influences his senses, his perception cannot tell him anything about the properties belonging to the poison as such. His perception is not an *image* of the poison and has no analogous relationship whatsoever with the properties of the responsible chemicals *as such.* This analogy reflects reasonably exactly the relationship between a phenomenon and a thing itself as represented in Kantian idealism. But, in our eyes, the premise that there is no relationship between the two worlds is erroneous. The human being, along with his perceptual structure, belongs to the *same* objective world as the things that are depicted in his perception. For us, the innate forms of thought and intuition are determined by organic structures in exactly the same way as the peculiarities of seeing and hearing are determined by the architecture of our eyes and ears. The structure of our sense organs is also given prior to, and is independent of, any individual experience and *must* be given for perception to be possible at all. In this respect, the definition that Kant himself provided for the a priori also applies fully to organic structures. Modern perceptual physiology and psychology have revealed extremely complex but nevertheless apparently entirely reflexive processes in the central nervous system that construct from sensory data perceptions of space, form, and size with respect to objects. These processes, which are undoubtedly associated with specific *structures,* surely provide the precondition for and—as we shall see—the phylogenetic precursors of our interpretation of spatial relationships. In the chapters dealing with the innate releasing mechanism and the conditioned reflex, we shall see the great extent to which a similar conclusion applies to the relationship between these likewise extensively analyzed physiological processes and our interpretations of time and categories of thought. We therefore have a quite firm scientific *basis* for the assumption that the a priori forms of human thought and intuition, which are predetermined in the absence of any experience, are nothing other than *organic functions* whose form is determined by their connection with physical structures of the relevant organ.

If we accept, however, that the predetermined forms of our experience, of our world image, are in this precise manner predicated upon the structures of the organ responsible for that experience and upon the specific structure of our "world-image camera," it is necessary to consider the following: As has already been noted, it is the reciprocal interaction between real things that has conferred, over the course of millions of years, a particular form on every organ of all currently living organisms. Certainly, in the course of this interaction living organisms themselves have in turn had a very limited influence on the inorganic world, bringing about small changes. As a general rule, however, the organic material that covers our planet as a thin "layer of mold" has been engaged in a hard battle with the pitiless laws of the inorganic world. It has had to obey laws that are infinitely older than organic creation and that apply regardless of the existence or nonexistence of organic material. These laws govern organic and inorganic material alike and all individual organisms have had to adapt to them, developing forms that gave them the *capacity* to participate in the struggle for existence. Seas made waves like those we see today millions of years before there were any fins to carve through them. The sun shone as it does now for billions of years before there were any eyes to capture its rays. If fins have evolved with a form permitting them to slice through water, this form has been determined by properties that water has always possessed and always will possess regardless of whether fins exist to interact with those properties. If eyes have evolved with a structure permitting them to concentrate rays of sunlight on their retinas, that structure was determined by the laws of optics that have always governed light rays and that will always exist regardless of whether eyes exist to capture them or not. A real, heterogeneous, colorful world, a world composed of indestructible matter and indestructible energy, some kind of world behind our perception of space, some kind of world behind our perception of time, existed for an infinite period of time before an organismal brain with its narrowly limited capacities ever attempted to generate a crude image of this world. This world existed before any brain was able to understand information about space and time, substance and causality in "coded," analogous form. And when such a brain evolved with a structure permitting it to do this and to construct a model-like, simplified representation of external reality and its inputs, it evolved in a form that *was determined by these inputs from the outside world.* Every one of our innate forms of thought and intuition is a receptor that has developed in a particular way through interaction with and adaptation to a specific input from external reality. The reality of this interaction between the internal receptor and the objective world is witnessed by the *success* of its functioning. Our perception of space and time and the categories of causality and substantiality are in fact nothing less than a functionless *luxus naturae!* Indeed, we *live* by them exactly as a fish lives by the performance of its fins and as every other organism depends on the functions of its organs. For sure, we can only perceive phenomena for which our species has developed the corresponding receptors. From our perception, we can only derive phenomena that can be transcribed in our "code writing" into the *symbols* of space, time, causality, and so on. But on the other hand we may draw the following conclusion: The naked fact of our *existence* proves that, for all of the inputs from the eternal, extrasubjective world *that are relevant to us* and which directly influence our lives, we possess *adequate* receiving devices. If we did not have these, we would not exist.

In one respect, Kant correctly presents our interpretation of a priori structures for human perception of the world as organic structures with a phylogenetic history. This entire "world-image apparatus" that Kant discovered certainly does not develop anew in every human being as a result of individual experience, contrary to the sensualistic explanation advanced by the empiricists. It is an ancient phylogenetic inheritance, like the form of the sense organs, and like them it is not the result but the *precondition* for any kind of perception. As seen purely from the standpoint of the *individual,* it is indeed something that exists a priori, and Kant of course was not familiar with any other viewpoint. Nevertheless, the empiricists were also right to a certain degree. Like the sense organs, the perceptual apparatus has not existed in its particular form forever, but has been *acquired* like them through organic creation in the course of a phylogenetic history. It is a historically developed *acquisition* of the organic world, representing an adaptation to external conditions, such that its form has been *dictated* by those conditions. Thus, as seen from a phylogenetic perspective, the innate structures of experience arose a posteriori. However, they arose not through experience but as a result of a phylogenetic process that determined their form in a manner that *fitted* them to those inputs from external reality whose internal representation constitutes their function for preservation of the species. Hence, although it happened a long time ago, the form of perception has indeed not been shaped by experience as such but it has been shaped by the form of perceived reality, as the empiricists maintained. It has not existed necessarily and forever as an immovable, divinely determined given as Kant assumed.

Our entirely well-founded views about the nature of human forms of thought and intuition have far-reaching and quite revolutionary consequences with respect to the question of the *knowability of a thing as such.* The manner in which we perceive the outside world is based on the function of our "receptors," which have developed in a quite specific way in adaptation to anything that is to be perceived through a highly material interaction between one reality and another. As a result, there *exists* a material and therefore *fundamentally investigable* relationship between the phenomenal and the real world. Kant's premise that there is no relationship between these two worlds is incorrect. It is relatively simple and easy to grasp the general nature of this relationship. In a certain sense, even an organ whose adaptive value resides *not* in the representation of real things but in a purely mechanical interaction with a specific thing becomes, as a result of its functional adaptation, an image of that thing. In the words of Jakob von Uexküll, it becomes its "counterpart." The form of an organ is, so to speak, the negative impression formed in the plastic matrix of organic substance by invariable features of the inorganic outside world. Thus, in its form and still more clearly in its movement, a fin is an image of a wave. A horse's hoof is just as much an impression of the steppe substrate as is the real footprint that it leaves behind, which is *its* impression in a virtually literal sense. The same applies, perhaps even more directly, to the sense organs, whose adaptive function resides in the representation of particular features of extrasubjective reality. The analogy here between an organ adapted for the specific function of representation and the reality that it represents goes even further. "If the eye were not sunlike, it would never be able to see the sun." Goethe undoubtedly believed in some kind of "prestabilized harmony" between the receptor organ and the input, but he nevertheless correctly *saw* that the spherical shape, the radiating structure of the iris along with the functions of the lens and the retina together constitute a small "antisun," a contrapuntal impression of the eternal laws obeyed by light rays. But that which applies to the organ also applies to its function. The world-imaging structures of our central nervous system doubtless have a much closer and more detailed *analogous relationship* to the nature of that which is represented than does the eye to the sun. This itself attests to the relationship that exists between our perceptual world and the real things underlying our perception. This is quite simply the relationship that, in a more-or-less fitting analogy, exists *between any image and reality.* Our perceptual world is a *model* of extrasubjective reality that is generated by receptors in our central nervous system. These receptors belong to our sense organs just as much as the nervous structures that form perception from sensory data and as much as the most central structures of the brain that determine "a priori" forms and categories of intuition.

For the assessment of our own perceptual capacities, the recognition of a model-like or analogous relationship between our phenomenal world and the reality concealed behind the phenomenon renders us, in a most peculiar way, at once modest and excessively ambitious. We must accept that we are fundamentally incapable of knowing how far the analogy extends between the perceived phenomenon and the thing itself, that is to say, of knowing the *relative accuracy* of our world model. The number of details and the nature of the real features that this model can represent are basically dependent upon the tightly limited number and the phylogenetically determined construction of our external and internal receptors. Our present world model is quite definitely no more than very approximate and *just sufficient* for the practical representation of extrasubjective reality! The infinite quest for knowledge that drives the natural sciences fundamentally prevents us from making any voluntary concessions to agnostics. We simply do not believe that there is anything in the universe that is *fundamentally* unknowable! Yet insight into the structural dependency of all of our perceptual functions obliges us of course to accept that we can never know objects as they really are. Even if, in the course of our further evolution, the central nervous system and hence our perceptual capacities should develop into something unimaginably more complex and more advanced, both will always depend upon a *finite* number of neural elements and receptors. The external reality that is represented, by contrast, is most probably infinite. Practically, there will always be at least *one* real datum in the universe that will be beyond the possibility of complete knowledge even for the most advanced rational beings imaginable, namely those rational beings themselves. The development of the universe, which includes its most advanced rational beings as an integral part, will and must always be and remain a step ahead of the development of perception. An organismal brain, however advanced it may be, will never be able to know how many features exist in the extrasubjective world which are not represented in its phenomenal world because it is blind and deaf to them in the absence of receptors to respond to them and central nervous structures that can think them.

On the other hand, however, our recognition of the model relationship between our phenomenal world and the real world provides a sound basis for confidence in the (albeit limited) information yielded by our perception of extrasubjective reality. We can be quite confident that the fraction of the objective world *that is* represented in our phenomenal world *really corresponds* to something analogous in extrasubjective reality. This is comparable to a poor photograph in which every detail that is just barely recognizable "adequately corresponds" to a matching detail of the scene recorded. Our world-image apparatus, which has proved itself over the course of millions of years of practical struggle for survival, is not afflicted by false information. What it records is reliable enough. The drawback for a scientist striving for knowledge is that it provides a representation of reality that is too approximate and too superficial, yielding insufficient detail. In a much later section of this book, we shall conduct a precise comparison between the world images of higher and lower organisms. This comparison shows, in an extremely significant and instructive fashion, that even the world models constituted by the simplest systems differ from that of humans only in that they reproduce *fewer details.* The few features that they reproduce are represented in the same way as in the far more complex human, scientific world image. When subjected to the critical gaze of the "better" human apparatus, the simpler world images emerge not as "distorted" or less correct but merely as *far more simplified* models of that which really exists!

However coarse and incomplete the model-like analogy may be in our world picture of objective reality, recognition of its existence shatters the rigid boundary that Kant erected between things that lie *within* and *beyond* possible experience. As soon as the phenomenon is seen to show no more than an analogy to the thing itself, any advance, any improvement in the analogy, represents an *approximation* of the perception of that which really exists and hence a displacement of the boundary between the "immanent" and the "transcendental." Not only every small step in the evolution of the brain but also every perceptual advance, every improvement in our way of thinking, and, above all, every new *instrument* can bring some previously unrecognizable feature of extrasubjective reality into the scope of possible experience. Every single advance in research represents an approximation of the reality of the objective world. Human knowledge of the world is fundamentally incapable of attaining things themselves, but it approximates them along an asymptotic curve, *with which we cannot predict how close it may eventually come to the things themselves.*

To many natural scientists, these observations will appear to be commonplace. All those who devote their lives to research naturally believe in the reality of the subjects of their research and they are convinced that their results will lead to approximations to reality. Even with scientists, not so few in number, who are disciples of idealistic philosophy, I cannot quite *believe* their credo concerning the unreality or unknowability of the objective world. It is impossible for a human being to squander his life's work on the investigation of something that he regards either as unreal (Platonic idealists) or fundamentally unknowable (Kantians). Every scientist believes in his heart of hearts that his research will lead to a better understanding of "the innermost coherence of the world!" Yet even today many natural sciences and many scientists are still at the stage of *pre*philosophical, unreflectively naive or "animal" realism. For such people in particular, our arguments concerning the world image and world-representing organs will seem to be self-evident and superfluous, along with any other epistemological considerations of this kind. Is this attitude correct? Can any natural science completely dispense with any consideration of the perceiving subject and of the limitations of its "apparatus?"* In the initial stages, all natural sciences have certainly done precisely this, and achieved—as shown by the results—great things. All natural science initially set out from the naive-realistic belief that things are simply as we perceive and experience them. As has already been shown in some detail, this belief is far from justifiable. But it suffices as a working hypothesis for a superficial and sketchy investigation. This is because the model of things represented by our perception is accurate enough, *detailed* enough, to reveal the elementary facts and approximate relationships that are of interest in the early stages of research. At the "intermediate level" of our world, the analogies in the model constructed by our world-image apparatus actually hold up quite well and its inadequacies first become obvious when we approach the limits of its capacities. As long as a research worker using a *microscope* restricts himself to the use of low and intermediate magnifications, he can dispense with any knowledge of the laws that constrain the capacities of the instrument. He will not be misled into making any major errors if he assumes that the observed objects are as they appear in the microscope image. But as his research becomes more detailed, requiring the use of ever-increasing magnification, he will come closer and closer to the maximum capacity of his instrument. He will be obliged to include this aspect in the range of topics considered and hence to become acquainted with the origins and limits of the instrument's performance. Otherwise, he will be in serious danger of attributing to the observed objects properties which they do not really possess but which represent artifacts generated within the range in which the instrument is no longer reliable. In just the same way, any research will sooner or later reach the limits of the human cognitive apparatus. *In the cause of further objective investigation* of the topic of interest, it will then

**At this point, the original text reads as follows:* Is this attitude correct? Can one in any natural science dismiss consideration of the perceiving subject and of the limitations of its "apparatus?"

be necessary to include the perceiving *subject* itself in the quest for knowledge. It was precisely this process that I meant above in stating that all natural science, after a long detour, eventually leads to a particular form of philosophy.

At the highest levels of its knowledge of nature, research will always be obliged to engage systematically in "theory of knowledge as apparatus technology." Even naive-realistic natural science follows a similar course in a far more primitive fashion that does not always involve conscious reflection. The entire quest for *objectivity* that characterizes the natural sciences even at their naive-realistic stage of development has its source in nothing other than recognition of the fact that the observing subject, with its specific patterns of perception, is inclined to attribute to things properties that they do not actually possess. The more *specific* the schemata that respond in human perception, the less that perception will have the character of an image adequately corresponding to the object concerned. This is particularly the case where perception is based on the response of innate schemata. This can be illustrated quite crudely as follows: What we perceive of the movement of a pendulum or the expansion of a metal rod when heated is, without any conscious participation, a far more objective picture than our perception of the behavior of an attractive woman. Even great thinkers may respond to the latter in a manner that seems quite incomprehensible to anybody who is not subjectively affected in the same way.

In the discussion on the innate releasing schemata of human beings, we shall examine more closely how such schemata—which exist for the comprehension of specifically human properties and display patterns—can respond *erroneously* to other conditions, to animals, and even to inanimate matter with an accompanying "physiognomic" perception of these objects. The outcome is that *human* properties and activities are perceived where none exist. The closer an organism is to humans, the greater becomes the danger that properties are "projected" onto it as perceptual extensions of human schemata. As a result of the blind, reflexive nature of the schemata, such perception is incorrigible. It cannot be prevented; it can only be compensated and corrected through conscious knowledge of the operative principles. Understandably, comparative ethology, which is most particularly interested in animal behavior patterns that are *really* analogous or even homologous to those of humans, has had to conduct a particularly arduous battle against these "reading glasses" of anthropomorphic schemata. A considerable part of our daily occupation with objectifying research is devoted to careful correction of the resulting "faulty images," which can all too easily be *overcorrected!* But we can now claim that we have had considerable practice with such correction. For the first time, precisely where the sources of error in subjective experience were most disruptive, a proven technique for compensation has been developed. To return to our comparison with the microscope: Even as a schoolboy, I never believed that paramecia and other transparent organisms *really* have rainbow-colored margins, despite the fact that my ancient small microscope clearly indicated this to be the case. I equally quickly found out that an owl is not really wise, that a crane is not proud, and that an eagle is not fearless, although my physiognomic perception of these animals attributed these qualities to these birds with great assurance.

But "epistemology as apparatus technology" can achieve much more than the simple exclusion of "colorful margins." Just as a knowledge of the capacities of a microscope often permits us to apply its principles to draw conclusions about things that are *no longer* visible in the image itself, a critique of human forms of perception can sometimes open up new perspectives on things that actually exist but are not amenable to the usual forms of thought and intuition. "Sometimes" is a rather audacious generalization, as it has so far only happened once. In the fields of quantum theory and wave mechanics, *physics* reached a frontier at which the thought categories of substantiality and causality broke down.* In other words, it had encountered entirely real processes which the human brain is unable to represent in the thought forms of these familiar categories and which we simply cannot think of or imagine in an illustrative form. These processes can, however, be determined using the category of quantity in the form of calculations of statistical probability and they can be *predicted* in a manner that permits them to be confirmed by our final adjudicator, the experiment. Max Planck's great achievement was precisely this *replacement* of more specialized, more graphic, *more anthropomorphic* forms of thought and intuition with the more general, more abstract, *more objective* category of quantity. This represents a breakthrough with respect to the frontier that Kant erected between the immanent and the transcendental, a breakthrough without parallel in human intellectual history. This would never have been possible without a *critique* of the capacities of the human perceptual apparatus and without recognizing sensory perception as an arbiter operating in an experimental context that *calls for* a critique of those human forms of

***Notes in the margin of the handwritten manuscript are reproduced in this printed version as (appropriately indicated) footnotes. The marginal note in this case was:* Transcendental numbers, infinitesimal calculus.

intuition and categories that are claimed by the transcendental idealist to be divinely determined and necessary. I freely admit that my confidence in my own epistemological conclusions was boosted by a critique made by Planck himself of my 1941 paper "Kant's Theory of the *a priori* in the Light of Modern Biology."* In this critique, he noted that, from an entirely different direction, he himself had arrived at identical ideas concerning more specialized, more anthropomorphic vs. more general, more objective ways of thinking concerning a model-like, analogous relationship between the phenomenal and real worlds. He had drawn the very same conclusions as those which we, on the basis of the modest factual evidence accumulated by our young science, had reached, perhaps more rapidly than was actually warranted.

Comparative behavioral research itself arrives at an encounter with philosophy by another, more direct path than that portrayed here in a generalized fashion as applying to all inductive natural sciences. In fact, comparative behavioral research is more closely related to the humanities than any other genuinely inductive science in that the theory of knowledge is applied not just as a *means* to the end of advancing the objectification of the outside world. Over and above this, comparative behavioral research has set itself the *direct* goal of understanding human beings themselves and hence their capacities as perceiving subjects.

It is the task of this book to portray the way in which comparative behavioral research is attempting to reach this goal. Because this attempt is founded on the conceptual and investigative methods of induction, we must begin by setting out the essential aspects of this approach.

See footnote to p. xxx.—Ed.*

2

Induction

In the preceding text, we have developed particular notions concerning the *roots* and *motives* of natural science, which are distinct and independent from those of philosophy. These notions are fully confirmed by the *developmental pathway* that has been repeatedly followed by the natural sciences, not only as an entire outcome of the collective quest for knowledge conducted by all participants but also during the individual development of every individual research worker. Max Planck himself, in a paper published in the journal *Naturwissenschaften* in 1941 [1942], demonstrated quite convincingly that the collective quest for knowledge in any natural science is not fundamentally different from that of the completely naive, outwardly directed search for knowledge conducted by *any* single human being. He showed that the development of the natural scientific world image takes place in a manner that is fundamentally the same as that followed by any individual human being in forming his more or less complete world image in the course of ontogenetic development.

For any young child, *everything* is new at the outset and the construction of that individual's world image starts with restless, indiscriminate, and avid incorporation of new images. In just the same way, in the initial stage of its development a young natural science indiscriminately *collects* facts, following nothing more than the simple motto: "Look at all there is to discover." For the natural sciences, just as for a young child, this initial, preparatory recording of everything that exists is closely linked to the function of *nomenclature.* Exactly as a 1-year-old child will proudly point its finger at any objects in a picture book that are already familiar and stammer out their names, natural science always begins with the creation of names for the individual, concrete phenomena that it has discovered. But the child's grasp and *bestowal of recognizability* on what it has seen, which is achieved quite simply by *noticing* objects, can only be achieved by means of *description* in research. Because of this primary function of describing individual, concrete phenomena, this developmental stage of inductive natural science is referred to as *idiographic* (from the Greek εἴδωλον, picture, and γράφειν, to write), after Wilhelm Windelband.

This idiographic stage gives rise, in a smooth transition that begins early, to the *systematic stage.* Once again, as with the development of the child's world image, the concrete facts are *put in order.* The same is grouped with the same and the similar is allied to the similar, such that an ordered and *manageable* system is generated from the jumble of individual phenomena. Our *Gestalt perception* plays a decisive role in this function of arranging concrete individual phenomena into a systematic entity. It permits us to identify in a direct fashion similarities and dissimilarities in the great diversity of phenomena. The major part played by perception of the forms of things both in the recognition of concrete details and in their ordering according to degree of similarity makes the boundary between the idiographic and systematic stages rather less sharp than is implied by Windelband's concepts. Some children with particularly well-developed Gestalt perception first apply names not to individual concrete facts but to identifiers for similarities of form. One particularly gifted 10-month-old child of my acquaintance uttered "ball" as his first word, using it to refer in a blanket fashion to all objects with circular contours, including spheres, circles, and the ends of cylinders. He was a passionate *collector* of circular objects and crowed with joy whenever he discovered a new concrete case of circularity in the pattern of a carpet, in a knob, or in a lampshade. In a quite analogous manner, even in the early idiographic development of a science, the collection of concrete facts is steered along particular paths that in principle represent a systematic arrangement of phenomena. The process of collection of phenomena cannot be sharply separated from their systematic organization for the very reason that the Gestalt perception of similarities is itself one of the strongest *motives* for collection. Wolfgang Köhler, in a profound observation, stated that any organism can only recognize a given conditioning feature if it is able to perceive it as a Gestalt. As will be discussed in more detail in the section on "inquisitive creatures," the tendency to learn something new is identical with an innate drive to acquire through Gestalt perception all those conditioning features that can be recognized. In objective terms, curiosity—as active, purposive striving for *acquaintainceship* with the outside world—is a process of education by everything that lies within an organism's capacity for Gestalt perception. In physiological terms, this striving is not a drive but, as has been emphasized earlier (p. 6), an *orienting response,* a "positive taxis" directed at everything that contains features accessible to the organism's capacity for Gestalt perception. In all evidence, it is precisely the inherent property of "beauty" in any phenomenon, that is, any regular or rhythmic features that are amenable to Gestalt perception, that renders it *conspicuous*, in such a way that it arouses and attracts curiosity. Subjectively speaking, it is the pleasure attaching to any form accessible to Gestalt perception that is the driving principle behind all learning processes based on curiosity. The same curiosity as that which was shown so splendidly by our young collector of circular forms underlies the tremendous learning capacity of human beings in their early years and is also the root of all scientific *collection*. In research, too, the pleasure associated with recognition of form leads not only to collection and accumulation but also to the simultaneous systematic ordering of the tremendous wealth of concrete facts that represents the most indispensable basis of any natural science. In many branches of natural science, this collection of knowledge about the forms of things is combined with actual collection of the things themselves. This reveals with particular clarity the elementary pleasure derived from the object and thus allies the scientific quest for knowledge even more closely with its childish counterpart, so that the boundary between them virtually disappears. Zoology, botany, paleontology, mineralogy, and other sciences arose from the pleasure of collecting, initiating their research activities with a systematic collection of the relevant objects that in its initial stages barely contained any *conscious* striving for knowledge. Almost every person who becomes a researcher in one of these fields relives in his own career the developmental history of his science. The best zoologists and botanists are precisely those who became collectors in their youth because of the pleasure they derived from the objects concerned and were thus led into the subject of their later research. I must admit to a mild suspicion that every biologist who has *not* personally experienced this developmental stage of his science is a so-called armchair biologist. He lacks the intimate bond to the object of his research that alone can lead to the growth of a *personal* store of knowledge of concrete details. Thus, in science, the idiographic and systematic stages develop from the *single root of pleasure derived from the Gestalt*, as in the development of the naive world image of any individual human being. Accordingly, these stages may be conceptually separable, but their development is not distinct.

The *third* stage in the development of the world picture identified by Windelband is more sharply separated. This is the stage of *formulation of laws*, referred to as the *nomothetic stage* (from the Greek νόμοζ, law, and τίθημι, to compose). In this case, too, the development of science follows very similar paths and is based on functions similar to those involved in the prescientific world picture of every individual human being. The formulation of a law is a direct outcome of the discovery of *order.* Any form of order is generally *improbable.* Only chaos is generally probable, and in the philosophy of classical antiquity it occupied a highly symbolic place at the beginning of all existence. Any regularity requires an explanation and any order must have a particular cause. For this reason, the complex and harmonic ordered states recognized all around in the systematic stage oblige the researcher to ask questions about the *causes* that generate these regularities *in a lawful manner* and about the *laws* that govern them. The route toward answering this question lies in the conceptual attempt to extract from the concrete individual facts through *abstraction* the regularity that governs them. This process of abstraction grows out of systematic ordering in a smooth transition, since both depend on closely related functions of Gestalt perception. Recognition of a similarity between two concrete things in itself represents perception of some principle that controls both of them. It is a *perceptual performance* of the same fundamental kind as that involved when a child recognizes the general Gestalt quality of a dog in a pug, a poodle, or a dachshund or when a researcher achieves a "breakthrough in understanding" by abruptly *seeing* a great natural principle governing a wealth of individual facts! With a picture puzzle, the sought-after Gestalt separates out quite suddenly from the background of confusing, immaterial, and superfluous lines to emerge with convincing self-evidence. In just the same way, this highest function of Gestalt perception also leads to recognition of *the governing principle* in the confusing chaos of individual concrete details with all the suddenness and conviction of a revelation. We shall learn more about this function of complex Gestalt perception, which is commonly referred to as *intuition,* in the next chapter and in the discussion of Gestalt perception in the general part. For present purposes, suffice it to say that it is functionally very close to conscious abstraction and is without a shadow of a doubt a *precondition* for this achievement, which takes place on a higher plane. The decisive part that Gestalt perception thus plays in the emergence of the nomothetic stage from the systematic stage means that in this case, too, the boundary between the two stages is not completely clear.

The broad outline of the three developmental stages of natural sciences that has been sketched here is sufficient to show the *direction* of the route toward knowledge. This route leads from the concrete individual fact recorded by our sensory perception to the abstract natural law, from the particular to the general. Mathematics and logic operate in precisely the opposite sense, proceeding from *axioms,* that is to say, from statements of *general, inherent* validity, and then "deducing" particular cases from the general principles by *derivation.* In the absence of a German word for the thought process that operates in the opposite direction to this form of deduction, it can be referred to as *induction.*

The great power of the inductive method resides in the fact that the *degree of probability* of the correctness of any identified governing principle is known. The *greater* the *number* of concrete individual facts for which a given principle is shown to apply, the smaller becomes the probability that this is due to chance. The greater the diversity, the complexity, and the regularity of the array of individual facts that are encompassed and systematized by a given explanatory principle, the more *improbable* becomes the assumption that an alternative explanation might exist. This *statistical* nature of all true induction dictates the requirement that results are valid only if *all* the relevant facts are taken into account *impartially.* This includes not only facts that conform with an identified explanatory principle but also those *that do not do so.* Unfortunately, however, a human being who has discovered an explanatory principle is greatly inclined to offend against this requirement for impartial treatment of conforming and nonconforming facts. Indeed, this inclination is rooted in the physiological nature of Gestalt perception, which has the effect that facts that are *not* governed by a previously identified principle are simply not seen. It is precisely this limitation on human objectivity, which cannot be avoided even with the most penetrating self-criticism, which makes it desirable to postpone the search for explanations as long as possible. It is better to remain as long as possible at the level of idiographic-systematic collection of individual concrete facts and to rein in one's own intuitive recognition of the governing principles, keeping it, so to speak, in suspension. The exactitude of inductive natural science is heavily dependent upon this curbing of the researcher's own Gestalt perception, at least in biological disciplines. Every premature formulation of a hypothesis unavoidably leads to the result that one sees concrete facts *supporting* the hypothesis far more readily than those that do not fit the explanation. But this itself invalidates the statistical basis that underlies any induction. What has been said so far has adequately demonstrated the great significance of unbiased idiographic and systematic work for the establishment of a solid foundation of concrete individual phenomena. The breadth and impartiality of this foundation determines the validity of any inductively derived conclusions.

Yet the following procedure, which is often still regarded as permissible and "scientific" in behavioral research, is founded on a complete misunderstanding of the nature of all inductive methods: A hypothesis is developed on the basis of just a few facts and there is a *subsequent search for examples* that appear to be suitable for support of that hypothesis. For anyone who in this way overlooks the statistical nature of all inductive research, let it be said that any hypothesis, however absurd and meaningless, can be easily supported with "examples" if facts are selected in such a biased manner. Organic creation is so rich and diverse that any nonsense can be supported by an appropriate selection of examples. Crude examples of this least scientific of all methods can be found in various attempts by [Edgar] Dacqué, [Max] Westenhöfer, and others to deny the theory of evolution, or at least to circumvent the evolution of humans from apes.

The validity of the results of all natural science stands or falls on the foundation of individual concrete facts from which those results were derived. The *inductive basis* must be statistically unquestionable. Any "slant" in this basis invalidates the result and the basis must be "broad," as its breadth, in a mathematically testable fashion, is geometrically proportional to the probability that the result is correct.

The predominance of induction as the most important conceptual and practical procedure in natural science by no means implies that deductive thought processes are completely unnecessary. To the contrary, the latter play a decisive part in the testing of results that have been obtained by induction. Let us assume that we have observed a large number of cases in which solid bodies have fallen to the ground under the influence of gravity. By an inductive procedure, we have derived from the concrete individual events the generally applicable laws governing free fall, including (for example) the law that states that the velocity at any given point is equal to half the acceleration due to gravity multiplied by the square of the time elapsed ($q/2 \cdot t^2 = V$). What means are available to test the correctness of this law that we have identified? In the first place, we can increase the probability of the validity of this principle by following the old-established route of increasing the breadth of the inductive basis. We simply continue to observe further concrete cases and check whether they obey the law that has been identified. In addition, however, there is a far more elegant, new route involving the application of the deductive method. With a small number of additional observations, perhaps only one, we can demonstrate the validity of the law that we have identified. On the basis of the law, we can deduce a single case of our choice: After x seconds, the velocity of a falling body should take a particular value. That is to say, we *predict* a single case that we have *not* yet observed! And then we take a decisive step that nobody had consciously taken prior to Galileo: *we deliberately set up the single case that we have predicted.* We allow a solid body to fall, measure its velocity after the time interval that has been selected, and compare the result with our prediction. The deliberate contrivance of such a concrete single case in order to confirm the validity of a law that we have identified or inferred is referred to as an *experiment.* An inferred law that can be used for the construction of experiments is termed a *working hypothesis.* The formulation of working hypotheses and their confirmation by means of experiments is the most important tool permitting inductive natural science to engage in *voluntary* broadening of its inductive basis. In this procedure, deduction serves the relatively modest role of formulating the question that is addressed to the outside world. The last word always and necessarily belongs to the concrete observed fact that represents the answer given by external reality to the question posed by the experiment.

Accordingly, all inductive natural science stands or falls on the assumption that the information about extrasubjective reality provided by our perception has an adequate counterpart in that reality. The preceding chapter dealing with philosophy and natural science was placed at the beginning for the sole reason that the epistemological justification and confirmation of this assumption represents a precondition for understanding the productivity of induction. Above all, it yields the insight that the concrete observed fact is unconditionally predominant over any result of deduction, however correct and conclusive it may be. Especially if he is an idealist, anybody who regards human forms of thought—even the laws of logic and mathematics—as unconditional truths will be inclined to mistrust his perception rather than his deduction in cases where there is a *mismatch* between the information from his senses and the conclusions from deduction. As [Christian] Morgenstern has so aptly put it: "Because, as he so trenchantly concludes, a thing cannot be if it may not be." In the light of all that has been said in the previous chapter about the forms of thought and interpretation, these laws of thought and cognition that have been developed in the course of human evolution for the purpose of interaction with the outside world are far from being infallible "truths." In their nature and their function, they are much closer to something closely related to a working hypothesis that a human being has developed through his individual interaction with extrasubjective reality. They are hypotheses that are to a certain extent innate and, above all, have proved to be *successful* over the course of hundreds of thousands of years of operation. There is a far from insignificant analogy between them and the laws, existing in their own right, that they serve to identify. As the "measure of all things," which they undoubtedly represent for us, they correspond quite adequately to that which is to be measured, at least in the *central range of measurement.* This—as has already been demonstrated—is, of course, simple proof of our *capacity to exist.* On the other hand, however, we should certainly avoid regarding the range of their validity as absolute, let alone infinite. Like any hypothesis formulated by a human being, they are *certainly simplifications* of the existing laws that they encapsulate. Wherever our research develops beyond the "central range of measurement" of our world-image apparatus, we must seriously and practically expect to encounter entirely real things with which our "innate working hypotheses" cannot cope. Such things cannot be expressed in the symbols of our innate forms of thought and intuition. We are confronted with prime examples of this in quantum physics and wave mechanics. If, in a similar fashion, a hypothesis generated by human beings reaches the limits of its validity, we attempt to formulate a *new* one. Alternatively—and more commonly in the case of such *successful* hypotheses—we attempt to extend the original hypothesis, which has failed to fit the facts *because it was oversimplified,* by adding a *supplementary hypothesis.* Neither of these developments is easily permissible with the innate working hypotheses of our forms of thought and intuition, for our intellect is trapped in the straightjacket of the schemata imposed by brain structure like a lobster in its carapace or a knight in his armor. Just as these rigidly structured creatures can only perform certain movements that are permitted by specific joints in their armor, a human being can only think and—as a general rule!—experience that which is allowed to take place within the limits set by his mental structures. These structures are a quite marvellous "construction" brought about by evolution. Like the parts of a well-designed machine, the fixed components and the freedom of their movement that is permitted are mutually adjusted so that no *internal* conflicts or incompatibility can arise between them. The same applies to the relationship between the individual parts of a well-considered working hypothesis or theory formulated by a scientist. *In themselves,* human forms of thought and intuition fit together in perfect harmony and mutual "intelligibility." But the smooth and unfaltering operation of an idling machine is only the *precondition* and not a *guarantee* for adequate, not to say perfect, *practical functioning.* Even the best machine has its deficiencies that are not apparent while it is idling and that only emerge when it is in operation. The *internal* consistency of a theory is merely a precondition and by no means a warranty of its validity. Even the best theory has its imperfections, which first emerge when it is *applied* and *can* only emerge then!

Now, the great question as to whether everything that we deduce from the inner certainties of a priori axioms is *really true* can fundamentally only be answered in exactly the same way as the question of the validity of a theory formulated by the human mind. In other words, it can only be tested according to the concordance or nonconcordance between the deductively generated prediction and new, concrete facts. "Innate working hypotheses" are also testable only in their operation, for the adjective "real" is linked to the verb "to realize." Wherever *nonconcordance* occurs, we must always be prepared to grant more credence to concrete facts than to the most evident internal truths. Because of the ancient tradition of the idealistic way of thinking, when faced with the frequent small discrepancies between necessity of thought and reality, we are all too inclined to regard both as "true." We do this, as it were, by allocating each to a different world or even by attributing the lack of concordance between thought and reality to the "imperfection" of the latter. Instead, we should always clearly recognize that thought is simply a box that is very approximately, but in practice satisfactorily, adapted for that which it is "constructed" to receive. To take an example: Even the thought form of quantity, embracing the axioms of mathematics, constitutes no more than a box of this kind, despite the fact that it is of great utility for practical interaction with reality and is, indeed, the most useful of all in this respect. At first sight, the sentence "Two times two makes four" is perceived even by a nonidealist as an undeniable truth. But this "truth" is only the *internal* consistency of an *idling* machine, with which we have been equipped for the purpose of quantification of extrasubjective matter, for the counting of atoms, apples, and sheep. The wise German poet Wilhelm Busch expressed this marvellously in the following verse: "Two times two makes four, that's right./But sadly this is void and passive./I'd rather have it clear, not trite./Something that is full and massive." Every equation is basically no more than a tautology which states that an empty box of one is equal to itself. As soon as we *fill* any box of one with a real thing, the equation loses its character of absolute truth. The category of quantity then reveals itself for what it really is, namely, a quantifying machine that functions with high statistical accuracy. Two atoms, apples, or sheep added to two more of the same kind are in reality *never* exactly the same as four others, for the simple reason that there are not even *two* exactly identical atoms, apples, or sheep. In probabilistic terms, when applied to real individual things, the equation "two billion plus two billion makes four billion" shows a much *lower* percentage discrepancy, because with large numbers the differences between concrete individual things are statistically balanced. This particular character of a working hypothesis that is indeed of great practical utility but is actually quite coarse applies to *all* innate forms of thought. Any natural scientist should always be aware of this in order to grasp the full, scarcely to be overestimated significance of the unconditional precedence of the observed fact over the thought, of induction over deduction. The "intelligible world" is merely the reality of an idling receiving apparatus with self-evident internal consistency. The individual form of thought is just a box that fits more or less exactly to a specific thing that is to be received. Now comes the crucial point that must be understood in order to comprehend the peculiar cognitive function of induction: As long as our search for knowledge encounters phenomena which are adequately matched by the boxes of our forms of thought and intuition, as long as we remain within the central range of measurement of the measure of all things, the receiving machinery of our world-image apparatus will function without complaint, without knocking or banging. But what we perceive in the process is, in a way, nothing that is fundamentally new. We only perceive things that we can "imagine in any case." The things that we come to know may have great potential significance as the inductive basis for far-reaching understanding. But as long as everything operates smoothly, the process of their representation in our world image yields nothing that is substantially different from the idling of the receiving machinery. The latter *appears* to be complete and all of its axioms *appear* to be absolute truths. Induction operates only within the framework of that which is deductively accessible, and since no problem arises within that framework there is no cause, indeed no possibility, for exposing it to criticism. But as we have seen, all scientific research eventually moves beyond the central range of measurement of the measure of all things and encounters phenomena whose representation generates serious problems for our receiving machinery, which subsequently grinds to a halt. The machinery is exposed to things that become hopelessly embroiled with the "boxes" of our forms of thought and intuition and are simply beyond its grasp. At just this point, where deduction must admit to a hopeless "*ignorabimus*," induction reveals its unique perceptual capacity. For inductive scientific research, those "problems" and breakdowns of the human receiving machinery, the knocking and banging of real phenomena against the boxes of our forms of thought, are not something *regrettable* that should not actually exist. To the contrary: all of these discrepancies are nothing other than *the real world of things looking in at us between the joints of our lobster carapace.*

It is not concordance but *demonstration of the nonconcordance* of a hypothesis that is the starting point for further advances in knowledge. *Any* hypothesis is merely a greatly simplified model of an approximation to reality and a priori forms of thought are just the same. "Truth" is the working hypothesis that best keeps open the path to something better. Accordingly, for all inductive quests for knowledge, discrepancies in the previous hypothesis always constitute the point of attack for new research. The most revolutionary development in the history of human knowledge came when physicists treated the discrepancies in the thought form of causality in quantum physics and wave mechanics like discrepancies in a hypothesis created by the human mind. When faced with the "collision" between experimental facts and the category of causality, this proven, unquestionably rational and "necessary" form of thought was simply and unhesitatingly discarded like any obselete hypothesis and replaced with other forms of thought. However, this frankly stupendous epistemological achievement has by no means generated the widespread excitement in the humanities that one might justifiably have expected. More's the pity for the humanities! Yet for anybody who has fully grasped the organic nature of the human cognitive apparatus, Max Planck's breakthrough across the boundary between the immanent and the transcendental represents a victory of induction over the limitations of the human mind. It opens up incalculable perspectives into the realm hidden behind appearances and offers a view into infinity.

3

The Consistent Hierarchical System of the Natural Sciences

The conviction that there is a *single* real world, which simply encompasses *everything* that may be an object of research or knowledge, obliges natural scientists to draw an extremely far-reaching conclusion: Their results must be free of any *contradictions.* It is possible to have laws with a very general, extensive range of validity and laws with a very narrow, quite particular range of validity, *but there must be no contradictions between them.* The great laws of physics, such as the central principles, the law of inertia, and so on, apply to all matter whether it be organic or inorganic. They encounter no exceptions, not even in cases where other, more special laws apply *in addition.* The more special and narrower laws that exist in connection with the specific, complex structure and movement of particular forms of matter are fundamentally subject to the *framework* and the *foundation* provided by the great and general laws that govern all matter. To take a very coarse illustration involving the most general and simplest in conjunction with the most special and complex: The first principle, which states that two bodies cannot simultaneously occupy the same position in space, also applies to the highest functions of human thought, which are of course entirely dependent upon material processes. A brain that is packed to the limits of its capacity with detailed physical knowledge cannot simultaneously house the mind of a great politician and vice versa. The same applies to the principles concerning the indestructibility of matter, entropy, and suchlike.

The universe is governed by a *hierarchical framework* of laws in which the broader, simpler principles are *always incorporated* into the narrower ranges of validity of special, more complex principles, because they provide the foundation for them. Some of the most generally applicable laws stimulate no desire for further explanation in the human subject because they coincide with innate forms of thought and are hence "mental necessities." The first principle, for example, is not just an empirical fact. Over the course of millions of years of interaction with this fact, the structures of our perceptual apparatus have developed in such a way that our form of interpretation of space and of our thought of substantiality render it impossible for us to think of two material objects occupying the same position in space. A similar relationship exists between the principle of energy and the category of causality. There is undoubtedly a reason for the fact that two things cannot occupy the same position in space and this needs some explanation. But this explanation will eventually prove to be quite difficult, insofar as the latest concepts of modern wave mechanics, which explain matter in terms of wave processes, prove to be correct. Yet, because the simple fact of impenetrability of physical bodies is already "anticipated" as an *axiom* in our mental structures, it is "self-evident" for us and no further analysis seems necessary. In contrast, for any human being seeking knowledge, the demonstration of every narrower, more special, "higher" natural law arouses the need to *understand* the reasons and causes that explain *why* things behave in this particular lawful fashion. In other words, the attempt is made to trace them back to laws which are self-evident to us in the above-described manner. To *explain* something always means to trace the complex back to the simpler, the special back to the general, and the unknown back to the known. In the final analysis, everything is reduced down to the most general laws, which—as demonstrated above—coincide to a large extent with phylogenetically predetermined patterns of thought. To provide an inductive explanation for a special law means in effect to render it deducible from a more general law. In practice, this means that analysis can only be regarded as complete when it permits us to conduct a reverse synthesis from simpler elements.

Among the natural sciences, only physics has so far been able to push the explanation of individual laws (those of a relatively simple kind) to the limit of a priori forms of thought, genuinely reducing special principles to the most generally applicable natural laws. And only chemistry has achieved an immediate connection with physics at the level of processes taking place in atoms and molecules. It is for this reason that physics and chemistry are collectively labeled as *exact* natural sciences. All other branches of scientific research are concerned with objects whose structures and principles are so complex that their reduction to basic physical principles is not only a futuristic dream but must necessarily remain impossible for a number of reasons. In the first place, even with the objects of inorganic research—for example, in mineralogy or geology—the historical factor is so important in the causal chain that it must inevitably generate a residue of some size that cannot be rationalized. This applies even more forcefully to all areas of biological research. In addition, however, the sheer *systemic complexity* of even the simplest organism renders it practically impossible to reduce the processes taking place within it to elementary physicochemical events. Although this is in all probability fundamentally conceivable, a single human brain will never be able to grasp the multitude of processes and the unpredictable interactions among them. Accordingly, in all of the more specialized natural sciences the capacity to explain a natural law is limited to its reduction to the *next* broader and more general law, which may itself still be incalculably far removed from the elementary physical principles. But in all branches of natural science the *direction* in which the unknown is reduced to the known always remains the same. It is the direction from the special to the general, from the narrower to the broader law, as prescribed by the nature of induction.

As a result of progressive growth in known factual evidence, all natural sciences have been obliged to become more specialized and to divide up the work to be done. In an entirely appropriate manner, this division of labor has taken place not so much according to the different concrete objects of research involved but far more according to the *range of validity of natural laws.* Only purely systematic sciences, such as systematic mineralogy, botany, and zoology, are separated according to the objects with which they are concerned. All of the nomothetic fields of research, such as physics, chemistry, physiology, genetics, comparative behavioral research, and psychology, are separated according to the laws involved. In this way, relatively automatically and without the involvement of any conscious methodological planning or organization, a *consistent hierarchical system* of the natural sciences emerged to match the above-described hierarchy of natural laws. The global and most basic category is physics, which is concerned with the most fundamental laws that govern all matter. The principles and laws of physics apply without restriction to even the most special of the specialized sciences. They govern the events taking place in the mechanical functioning of our muscles, tendons, and joints just as much as the action currents in the neural elements of our brain. The next level in the hierarchy is undoubtedly occupied by chemistry. The relationship between chemistry and physics is precisely that which exists, or should exist, between any nomothetic science and the next most general discipline. For this reason, this particular example will be examined more closely. Research in chemistry *began* as the science of substances and of qualities of matter, quite independent of physics. But to the degree that chemistry succeeded in finding *explanations* for the laws that were discovered and learned to understand the processes taking place in atoms and molecules, *it became physics.* Atomic chemistry and atomic physics flow together as one. The highest level of understanding of chemical processes so far achieved is attributable to their reduction to the most generally valid laws of physics, namely, to those governing atomic processes. It is therefore perfectly correct and appropriate to maintain that all chemical processes are in essence physical processes taking place at the particular level of atomic physics. The observation that this statement does not apply in reverse seems to be superfluous here, as it is quite obvious that by no means all physical or even just atomic physical processes are simultaneously chemical in kind. But we shall later see why this needs some emphasis at this point. The fact that further progress in a particular science in this way leads it to *dissolve* into the next broader category in the hierarchy (after all, the Greek word *analyze* means nothing other than *dissolve*) is fundamentally illustrative of the relationship between *any* nomothetic science and the next most general neighboring field. However, the dissolution from a more specific to a more general level and hence the perfection of analysis has nowhere achieved anything like the success that is seen with chemistry.

Even the unique and most remarkable of all natural phenomena, that which we refer to as *life,* occurs thoroughly and without exception within the framework of the general laws of physics and chemistry. Just as we were able to state above that all chemical processes are "actually" or "basically" physical, we can also claim for all life processes that they are "actually" physicochemical processes. Living processes are undoubtedly very special and complex and we are today still far from understanding the basis for their special character. Nevertheless, we already know for sure *that the fundamental laws of the exact sciences do not encounter any exceptions in the realm of living organisms,* that life processes never under any circumstances *infringe* the energy principle or any other basic laws. In other words, we know that life, as a natural phenomenon, is not a "*miracle*" in the sense propagated by vitalists. The narrower hierarchical level of biology, the science of life, in its entirety rests on the broader "levels" of chemistry and physics.

Our interpretation of life as a *material* phenomenon by no means represents an underestimation or disparagement of life and its highest psychological and intellectual manifestations. Instead, it represents a proper *high* estimation of *matter* and its intrinsic potentials, which include not only the possibilities for the laws of lever action and gravity but also those for our mental laws and intellectual achievements. There may be those who nevertheless see superficialization, obfuscation, or abasement of the concept of life in our statement that it basically consists "only" of physicochemical processes and that all life processes are "actually" physicochemical processes. But such critics are above all revealing only that they have correctly recognized the *peculiarity* of life. They by no means have to be vitalists and are therefore not necessarily opponents of our view. To such critics, it must be admitted that use of the words "only" and "actually" is misleading. The fact is that life processes are not *only* governed by physicochemical laws. These *also* apply and they do not suffer any exception, but *in addition* there are countless special, new laws that also apply. These arise from the incalculably greater systemic complexity, from the special structure and function of organic matter, and they apply *only* to living substance and not to anything else! So much for our use of the word "only." As far as use of the word "actually" is concerned, it must be admitted that it could justifiably be *reversed.* It can well be said that life processes are basically physicochemical processes, but *actually,* that is to say, with respect to their own *actual* properties, they are something special and unique. They *actually* possess unique properties that are not covered by the customary concepts of physicochemical phenomena.

A completely analogous relationship to that existing between the narrower, "higher" laws of biology and the broader, more basic "level" of the exact natural sciences also exists between the science of *mental processes* and biology itself. A major, puzzling developmental leap similar to that separating life from mere physicochemical processes also separates all mental processes from the general processes of organic matter that are unaccompanied by psychological phenomena. Nowadays, no reasonable psychologist who is close to nature and life can doubt that all mental and psychological processes, without exception, *also* have a *material* aspect and, in particular, involve *neurophysiological processes.* All we need is a small effect of poisoning or injury to the brain and what becomes of the "immortal" soul and the spirit that reigns over all material things? One must have direct practical experience of the sinister and shocking personality changes that follow brain injury to know just how mortal the human soul is, far more mortal than the body!

The extensive analogies that exist between the great leap separating the organic from the inorganic and the great developmental leap separating psychological from nonpsychological processes has misled many philosophers into regarding them as identical and hence into equating life processes with mental processes. But closer acquaintance with psychological processes and their relationship to morphological and physiological processes shows that such an equation is quite false. There is nothing to lead us, far less oblige us, to assume that the purely automatic and reflexive life processes of lower animals, or even the growth processes of plants, are accompanied by any subjective form of consciousness or mental activity. Such accompanying features are lacking here just as they are from many of the adaptive life processes in our own bodies, such as the imperceptible processes of normal digestion, elimination of urine, beating of the heart, and so on. Yet all psychological processes in higher animals are *also* physicochemical processes, just as all life processes are *also* physicochemical and all chemical processes are also dependent on atomic physics. This sentence, however, is not *reversible.* Just as it would be wrong to regard *all* physical processes as chemical or *all* physicochemical processes as life processes, it is equally wrong to regard all life processes as being accompanied by mental processes. Only a very few, quite specific life processes are simultaneously mental processes. Recognition of this fact was the goal toward which this presentation of the hierarchical system of broader and narrower natural laws was aimed.

The preceding discussion is far from being an empty, purely theoretical presentation of epistemological considerations. On the contrary, it leads to very practical demands that must be fulfilled by any individual research worker if fruitful cooperation between the individual fields of scientific research is to be achieved. Such cooperation between individual fields is an inescapable precondition for the research conducted within them, with the exception of physics. As the most basic science embracing the most general laws, as the lowest "level," physics is independent of any other branch of natural science. All other natural sciences need broader, more basic fields of knowledge for their endeavor to reduce special phenomena to more general laws. As has already been said, no single human brain can ever survey the entire hierarchy of narrower and broader laws such as that, for example, extending from complex life processes down to the underlying processes in atomic physics. But it would be a great error to believe that the exactitude, the "scientific status," or even the *significance* of any field of scientific research should be measured according to the degree to which it succeeds in pushing the analysis of its subject matter down to the atomic level. Absolutely any phenomenon that we are able to observe can and must be an object for inductive scientific research, even if its resolution down to the level of atomic processes is, at the present state of our knowledge, absurdly utopian. This is, indeed, the case for life processes in general and even more so for the mental behavior patterns with which this book is concerned. The scientific character and exactitude of a field of research depend instead on the clarity, consistency, and self-discipline with which the rules of inductive procedure discussed in the previous chapter are applied. Above all, they depend on the purposive accuracy with which the *direction* of analysis, from the special to the next more general principle and to the next more basic science, is respected.

It is precisely the need to respect that direction that dictates the *practical* demands on any individual research worker that were mentioned above. As already defined, an *explanation* of a law means reducing it to a known, more general principle. Every research worker is hence faced with the immediate task of becoming acquainted with the next more general level of knowledge. In other words, every research worker must be closely familiar with the *next more general* discipline. Otherwise, he is confronted with the insoluble problem of acting *alone* to reduce the complex phenomena in his field of research down to the atomic level. He has no need to concern himself with the narrower, more specific levels that are higher in the hierarchy than his own discipline. It is perfectly possible for a physicist to conduct research without knowing much about chemistry and without being prevented by such ignorance from advancing scientific knowledge in his own field. By contrast, for a chemist it would be a major obstacle to understanding of his own field if he did not possess sufficient knowledge of physics to be able to comprehend the atomic processes that provide an explanation for the formation of all chemical bonds. In the same way, it is entirely possible and scientifically legitimate to conduct research in chemistry without possessing any knowledge of biology. On the other hand, it would be utterly inappropriate to try to conduct general research in biology without possessing the fundamental chemical knowledge that is necessary for an understanding of metabolic processes. For cooperation within the hierarchical system of the natural sciences, what matters is that every research worker *only needs to concern himself with the next more general discipline.* The metabolic physiologist in the example cited above only needs to understand a special part of organic chemistry. The question of how the chemical processes underlying the metabolic processes can in turn be explained in terms of atomic physics can be safely left to the chemist. Indeed, a specialized researcher in organic chemistry can leave this problem to somebody concerned with a more general field of research, namely inorganic chemistry. In order to establish the necessary link with the next more general law, with the underlying broader level in the hierarchy, it is essential for a scientist to know approximately where his own area of research is located within the framework of the general natural sciences. When Sutton and Boveri were striving to find an explanation for Mendel's laws of inheritance, it would not have been of the slightest help to them to possess knowledge, however exhaustive, of metabolic physiology or organic chemistry, much less atomic physics. What they needed to know were the principles governing the processes operating in mitotic cell division, in meiotic division, and in fertilization. *For the explanation of any law,* the intimate and complex hierarchical organization of laws renders it fundamentally impossible *to pass over the next broader, underlying law and to trace it directly back to an even broader law.*

Compliance with these fundamental principles behind all inductive scientific research is so self-evident for any practicing scientist that he would doubtless regard everything that has been discussed here as superficial and commonplace. But there are very good reasons for placing great emphasis on the clarification of these matters in this introduction to our presentation of comparative behavioral research. The methodological principles dictate that we should investigate the physiological and psychological aspects of all behavior accompanied by subjective phenomena. Accordingly, an essential feature of the nature of comparative behavioral research is that it is *also* psychology and is thus obliged to concern itself with psychology as a science. However, the fecundity of this interaction is unfortunately greatly impaired by the fact that modern human psychology is an offspring of the humanities. As such, it has almost completely failed to incorporate the theoretical and practical principles of scientific research. Because of this, the utility to us of results obtained from human psychology is extremely limited. Even where human psychology characterizes itself as a natural science, including the use of experiments, virtually no link has been established with the next more general fields of research and indeed this has generally not even been sought. In the few cases where an attempt has been made to establish links, it has occurred with the typical omission of the next-more-general law that has been described above as a methodological error. A classic example is presented later in the discussion of *association psychology,* in which the attempt has been made to reduce all psychological phenomena to very simple "elements," bypassing all the laws of biology in the process. In fact, modern branches of human psychology commit a basically identical error in a less naive fashion: many highly intelligent psychologists suffer from the fact that they have no idea of the fundamental laws of biology. Although human psychologists recognize in principle that all psychological processes are life processes that also have a physiological side, they fail to draw the conclusions from this that would at once be obvious to any inductive natural scientist. As a result, even today many investigations in human psychology take all kinds of directions other than that which would be necessary to ensure fruitful cooperation with the inductive natural sciences, namely, toward the next more general laws of psychology and biology. An exception is provided by perceptual psychology, which has established a highly appropriate link with perceptual and sensory physiology.

It is easy to understand in historical terms why human psychology has so far failed to connect up with the consistent hierarchical framework of the natural sciences. In principle, of course, a field of observational and experimental research such as psychology has a claim to recognition as a natural science. The statement made above, that any phenomenon in our universe, however complex it may be, can and must be made an object of inductive research, is of course entirely valid even for the whole realm of psychological phenomena. On the other hand, these are also subject to the above-described principles of inductive research, according to which the special must be reduced to the general and the unknown to the known. But in this respect human psychology was in a particularly unfavorable position. As a general rule in the history of the natural sciences—we may say "fortunately"—the broader, more general laws were in most cases discovered *before* those of neighboring, more specialized fields of research. Usually, as the analysis became more penetrating, any narrower field of research encountered a well-prepared basis of knowledge in the next more general field, whose broader laws could serve for reduction of the phenomena investigated. Thus it was that the advance made by physical chemistry into the atomic realm encountered an already established basis in atomic physics. As physiology began to advance into the realm of chemistry, organic chemistry had long since established a basis of known facts that the physiologist required. And so it goes.

The situation was different with psychology. The concept of psychology as a discipline goes back to the distant past and already existed in classical antiquity. It existed as an introspective branch of the humanities long before there was any natural science in our sense. As a natural science, psychology is relatively young, but it is still older than the next broader field of research. Even at a comparatively late stage in which psychology—doubtless heavily influenced by the Darwinian theory of evolution—began to regard humans as living organisms and psychological phenomena as life processes, there was still no sign of the next more general field of knowledge that would have provided a link with biology. At that time, virtually nothing was known about the psychological behavior of animals as such. One result of this was that the so-called discipline of scientific psychology founded by Wilhelm Wundt at the turn of the century embarked on an undertaking that was condemned to failure from the outset. The attempt was made to *bypass* the next more general field of knowledge and to reduce the psychological phenomena that were investigated *directly* to extremely simple, elementary laws.

Another, far more significant branch of psychological research that originated from medical psychiatry remains remarkably isolated and disconnected, although it deserves more than any other field of psychology to be labeled as scientific. I am referring here not to the medical psychology of the Kretschmer school, which notwithstanding important advances in knowledge still retains a strong tradition of purely idealistic philosophy and logic, but to *depth psychology*. However much we may reject the theoretical edifice constructed by Sigmund Freud and Carl Jung—particularly with respect to its ideological basis—there can be no disagreement that both of these depth psychologists are *observers,* indeed gifted observers, who saw *for the first time* certain *facts* that are irrevocable, inalienable components of collective human knowledge. Modern behavioristic and reflexive schools of behavioral research are very dismissive of the results of depth psychology. This is because the latter speak a different language than that of inductive psychology—although they set out from quite identical mechanistic basic concepts and envisage entirely material processes in all psychological phenomena. But this dismissive attitude incorporates something which we shall characterize as one of the worst infringements against the spirit of inductive research, namely, a *refusal of knowledge.* Although depth psychology today still lies outside the consistent hierarchical system of the natural sciences, this is exclusively due to a problem that affects psychology in general: the absence of a neighboring, more basic natural science to provide the simpler, broader laws to which extremely complex psychological phenomena can be reduced. It is precisely in this context that we identify an important task for comparative behavioral research, but one which is by no means easy to fulfill. The accumulation of collective knowledge in the natural sciences can be compared with the piling up of a cone-shaped heap of a loose material with a low stacking coefficient. The point of this comparison lies above all in the fact that it is impossible to increase the height of the pile even to a small extent without greatly enlarging the base. A further point that this comparison brings out very well is the following: At the beginning, when the base of the pile is being laid, it is not easy to predict where the tip will eventually lie. It is very difficult to determine in advance through conscious planning of the base that the tip will come to be located at a preselected position. Most natural scientists have not even attempted to do this. Instead, they have cheerfully and thankfully pushed forward in any direction that offered the possibility of expanding the base. As a result, even the most recent and most advanced results of research have virtually always been determined by analytical possibility rather than by planning. Thus, when extremely complex phenomena with no easily recognizable relationship to anything simpler exist in isolation, as is so decidedly the case in psychology, it is exceedingly difficult to find the broader and more general laws to which they can be reduced. In such cases, in announcing individual concrete facts, one must refer far back to much "lower," simpler, and more general levels in the hierarchy to establish the necessary inductive basis. One must not object to establishing a base that is very broad, perhaps broader than is absolutely necessary, as it can never be *too* broad. This task must be fulfilled without fail if psychology is ever to become an inductive natural science.

What, then, is the next broader "level" and what are the next more general laws within which all the phenomena of human behavior and mental life are contained? These must first of all be *identified* so that we create the possibility for *explaining* special aspects by reducing them to a known, more general basis. If we wish to pursue human psychology as an inductive natural science, we simply cannot avoid the fact that humans, *Homo sapiens,* are *living organisms* that have arisen from other, simpler living organisms through a historically unique process. The next broader, more general, *simpler* laws underlying the special, complex laws of human behavior and mental activity are therefore those *that govern the behavior of animals in general.* Quite apart from the various conclusions that emerge from the nature of inductive research, even the simple fact of evolution means that all of the more complex structures and functions of higher organisms are fundamentally understandable only in historical terms. That means that we must first understand the simpler evolutionary stages that gave rise to them. For this reason, the comparative phylogenetic study of the behavior of living organisms *including* humans is undeniably the next broader field of science on which the science of human behavior and mental activity is based.

These claims, which are *irrefutable* from the standpoint of natural science, have nothing to do with an impermissible biologization of the mental and, particularly, the *social* behavior of human beings. The relationships within the consistent hierarchical system of the natural sciences have been discussed in some detail precisely in order to avoid such a misunderstanding. With advances in analysis, the special laws governing a more complex and (in terms of human value judgments) *higher* organic system become *explicable* in terms of broader, simpler natural laws. But those special laws are not rescinded or abolished following the complete success of such analysis. Life processes and the laws governing them will always remain as something special and (according to our value judgment) higher, even if causal analysis eventually succeeds in completely reducing them to physicochemical processes and thereby explaining them. In the same way, the most special, most complex, and highest functions of the mental, moral, and social behavior of human beings will suffer no loss of their distinctiveness and unique laws if inductive causal analysis should ever succeed in explaining them completely on the basis of more general, broader laws governing living organisms or—in an as yet purely utopian manner—on the basis of chemical and physical laws. We therefore explicitly reject the conscious or unconscious interpretation made by many mechanists that the success of causal analysis provides proof of the fact that the more complex phenomena in the material world are "nothing more" than the simple processes invoked to explain them. As we have, it is hoped, convincingly demonstrated, they are something more. Any "biologization" of human behavior, particularly mental and social behavior, that attempts to apply the simpler laws of more general biology *directly* to these more complex and higher phenomena is therefore unacceptable. These phenomena possess their own, far more complex laws. Although these can indeed be *reduced* to the general laws governing living organisms, they are not *contained* within them. Human beings are not contained in apes, nor, to take the process further, are they in the chemical elements of which they are composed.

To summarize: The comparative phylogenetic study of animals is the next broader field of science within which all of human psychology *as an inductive science* is located. Comparative behavioral research is by no means expected to replace human psychology, let alone render it unnecessary. It can no more do this than physics can replace chemistry or organic chemistry can render metabolic physiology unnecessary. But psychology, and particularly depth psychology, *must be informed by comparative behavioral research for the same reasons that physics must inform chemistry and chemistry must inform metabolic physiology.* But at present the whole of psychology fails to fulfil this requirement for the simple reason that the young discipline of comparative behavioral research has so far remained virtually unknown to psychologists. Although this field is very young, although the number of scientists working within it is still very small, and although the results obtained so far may be regarded as relatively modest, it is nevertheless undeniable that it must serve a vital role in linking the other inductive natural sciences to human psychology. According to the laws of inductive scientific research, it provides incontestably the only legitimate access to and understanding of human behavior and mental activity. For this reason—at least provisionally—it is justifiable to regard it as the "natural science of the human species," as indicated in the title of this book.*

Insert†

Discussion of the question as to how far the next more basic discipline is authorized and obliged to engage in criticism of the narrower, more specialized, and more complex explanations of an independent and older-established, more specialized discipline, following the successful demonstration of additional laws. The question of economy of thought with respect

**See the comments on the title in the foreword (p. xiv).*

†Apparently a note on later thoughts about this topic (p. 108a of the manuscript, in different handwriting.

to a simple explanation (e.g., on the basis of the innate schemata) and a complicated parallel explanation (e.g., on a psychoanalytic basis). Analogous problems between research into instinctive behavior and sociology are more acute. Chemistry has to "put up with" the statement that the processes it investigates are in effect atomic physical processes. Expansion of a "lower" field of research into that "above" is scientifically legitimate. "It is none of your business" does not apply! It is also incorrect and illegitimate to say "You do not understand it," as the "lower" field can perfectly well become more specialized.

4

On the Possibility of a Synthesis Between the Natural Sciences and the Humanities

As we have seen in chapter 1, the conditions necessary for reciprocal understanding between the natural sciences and the humanities are present from *that* point at which scientific research turns to *human beings* as an object of its quest for knowledge and at which philosophy regains its belief in the reality and fundamental cognizability of the external world. Without a doubt, both of these requirements are being met to a rapidly increasing degree. The natural sciences and the humanities are growing toward one another and they will eventually link up in the foreseeable, although not immediate, future. The fact that such a link barely exists today, if at all, is probably due mainly to the major and purely external obstacle constituted by *communication problems.* The deepest roots of the quest for knowledge in the natural sciences and in philosophy are so different and the resulting mental and practical approaches so dissimilar that philosophers and scientists speak different languages. From my own experience, I am all too familiar with the futile manner in which they can talk at cross-purposes. This has been apparent in many discussions between scientists and philosophers, despite the fact that both disciplines have taken part in such discussions with the best will in the world! Impartial observation of such discussions reveals something that has actually already emerged from what was said in chapter 1 about the different roots of natural sciences and the humanities. This is, namely, the difference in *type* between the people who are attracted to one discipline or the other.

The aptitude, and doubtless also the inclination, to philosophize resides in the capacity to direct the gaze *inward.* It resides in the talent to perceive the harmonious Gestalt of a given entity and to grasp the most complex relationships in an intuitive fashion. A philosopher has a certain propensity for contemplative inactivity and, lacking the urge to engage actively in the course of events, shows a Diogenes-like *satisfaction* with his intuitive findings. By contrast, the natural scientist is characterized by a burning desire for new facts that is directed *outward.* He possesses an eagle eye for the smallest, seemingly unimportant *details* of colorful, intricate reality and the constant readiness to engage *actively* with the real, objective world. A natural scientist *must* be active and industrious and can never be satisfied with previous achievements. What characterizes the scientist above all else is the need to identify the *causes* of phenomena and, once they have been identified, to seek the causes of these causes. He is not satisfied with the recognition of a truth; he demands *proof.*

The twinned characteristics presented here as differences between philosophers and natural scientists represent four pairs of typical contrasts. Although there are intermediates and transitional states, in their most developed form these contrasts are indeed "*typical*" and mutually exclusive in their expression. This applies most particularly to the first pair of contrasts described. Inwardly and outwardly directed people, "introverts" and "extroverts" in Jaensch's terminology, represent two types that are particularly exclusive. But the capacity for intuitive perception of complex Gestalts and the talent for observation of fine details are also traits that are rarely found combined in the same person. It is in the nature of Gestalt perception that details dissolve into the total quality of the whole and thus disappear. For precisely this reason, people who possess a particularly well-developed ability to perceive entire Gestalts are commonly inclined to display a contempt for small details. They do not *need* any conscious evaluation of the details in order to recognize the important relationships. They are not aware of the fact that the details registered by the sense organs are not only present in the complex quality of the perceived Gestalt but are actually present as an integral component contributing to the *determination* of that quality. It is for this reason that they erroneously regard details as quite unimportant. We shall shortly demonstrate that intuition, the direct recognition of profound internal relationships, is in fact nothing other than a high-level function of Gestalt perception. It is therefore not surprising that almost all those who have a particular talent for intuition show a very dimissive attitude to the fine details of anything that they perceive. For a talented intuitive "seer," any recognized truth is *sufficient* by virtue of its power of conviction, elevated above all doubt and bearing the character of a *revelation.* Such a person does not *need* any causal-analytical proof and any scientist who categorically demands such proof is seen as a pedantic philistine who is blind to the important relationships. The classic example of a great seer misunderstanding and scorning an inductive, causal-analytical scientist is provided by Goethe's utter misconception of Newton's theory of colors. Goethe expressed real conviction in the following verse: "Still secretive in the light of day,/Nature will forever remain veiled,/And whatever she will not give away/Cannot be prised loose or un-nailed." In Goethe's eyes, any nonintuitive research worker who bases his opinion on induction from a wealth of individual details is the "poorest of all the sons of the Earth," granted immortality in the character of Wagner in *Faust.* This is the picture of detailed research painted in Goethe's head. "How hope somehow lives on within the mind/Of one who always to pale evidence clings/A greedy hand to search for treasure brings/Yet is pleased when mere earthworms are the find." At the time when we had our big breeding colony of cichlid fish (in Königsberg), my assistant [Alfred] Seitz and I had the habit of quoting this verse whenever we were out looking for an invaluable food source: earthworms!

On the other hand, for a scientist who has a talent for inductive and causal-analytical investigation but who has less of a gift for intuition, the directly perceived truths of the philosophical seer are utterly inaccessible. He cannot follow the process by means of which the philosopher arrived at his conviction. Since the latter is equally incapable of explaining *how* he arrived at his conclusions, the scientist sees him as a hollow windbag and dismisses him with disdain. There are quite a number of accomplished, indeed brilliant, scientists who deny that philosophy has any right to exist, and in particular reject any justification for referring to it as a *science.*

This greatly simplified and deliberately somewhat exaggerated presentation of the contrasts between natural science and the humanities suffices to show quite clearly that these contrasts are not at all fundamental in nature but are actually the relatively trivial effects of human mental barriers. Insofar as both the philosopher and the natural scientist are convinced of the existence of a single world of things that is independent of any cognition thereof, a complete agreement between them must be attainable. The second precondition for this seems to me to lie in insight into the function of the two so different and yet in many ways analogous pathways to knowledge: intuition and induction. We have already dealt with the most important aspects of induction. All that remains, therefore, is for us to discuss intuition, to examine its functional analogies to induction, and to assess the role that it plays in all inductive scientific research—although the individual researcher is often quite unaware of this. At first sight, intuition and induction seem to have absolutely nothing in common. It seems initially to be a quite different kind of truth when Heraclitus makes the statement πάντα ῥεῖ (everything is in flux) as opposed to when Galileo states: Acceleration due to gravity is 9.81 m/sec. Everyone must admit that both are truths and, insofar as we accept the *unity* of the extrasubjective world, both truths can only concern the *same* world. Much of what is seen by a poet observing nature similarly appears to be fundamentally true. Any natural scientist who doubts this should first of all read Goethe's *Faust* with an open mind!

The poet, like the philosopher, is satisfied with contemplation of recognized truth and regards any proof thereof as superfluous. By contrast, the scientist often believes that only proof can *furnish* truth. It is easy to demonstrate from the nature of intuition how erroneous both beliefs are. Any scientist who disdains intuitively recognized truth overlooks a fact that is already contained in the concept and even in the word *verify.* It is only possible to verify something that has already been suspected or claimed in some form *previously.* It has so far never occurred that *a truth has been demonstrated with the inductive approach of natural science without some previous recognition of its existence through the intuitive approach of philosophy!* Even Galileo first *saw* the law cited above before he provided the proof. If the immediate *overview* that is so often left unmentioned by natural scientists *did not exist,* all research would be nothing more than blind playing with dice without any idea of the likely outcome. Where would their results be under such circumstances? It is always and everywhere the case that natural laws are first intuitively seen and only afterward proved by induction. In *this* sense, the verses from Goethe quoted above are entirely correct: only after nature has unveiled something to our eyes can we prise loose or un-nail the elements of proof. Hence, even in scientific research itself, intuition *leads the way* and points out the direction for inductive analysis. This despite the fact that many research workers are by no means conscious of this fact or of the performance of their *own* intuition. Even in scientific research, inductive analysis is *dependent* on the lead provided by intuition. In the same way, philosophy could and should be the guide for the natural sciences as a whole. But how should philosophy proceed in order to achieve in the collective undertaking of all science conscious fulfillment of the function that it serves in any case (usually without conscious recognition) in analytical research? In order to answer this question, we must examine this function more closely.

Like "instinct," intuition has been and still is regarded by many as an *extranatural* factor which "reveals" unconditional truth to human beings in a manner that is not accessible to further analysis. This is, of course, nonsense. At the appropriate place, we shall need to take a closer look at the infringement against the theoretical principles of natural science that the introduction of such "factors" represents. As has already been said, the immediate recognition of complex relationships, which we are accustomed to labeling with the word "intuition," is a function of *perception* in general and of *Gestalt perception* in particular. It is accordingly an entirely natural physiological process.

A highly complex and extraordinary holistic organic apparatus in the central nervous system constructs completed perception of location, size, color, and form of objects from individual sensory inputs. The functions of that apparatus show remarkable parallels to *inferences.* Perception of the *distance* of a visual object, for example, is achieved in a completely unconscious fashion through the operation of a mechanism which records the *angle* formed by the visual axes of the two eyes (*convergence*). This mechanism then uses the "known" distance between the eyes to "calculate" the distance of the object seen, using exactly the same principle as the telestereoscope. These processes are functionally so close to inferences based on calculations that Helmholtz—who was the first to discover and investigate them—interpreted them as "unconscious inferences." Schopenhauer regarded these functions as a manifestation of *rational processes* and we shall see in the section of Part Four* dealing with *taxes* and *insight* to what extent this interpretation is correct. However, the fact that perceptions are definitely *not* unconscious inferences is evident from the presence of the same mechanisms in lower vertebrates, which are certainly not capable of trigonometric calculation. These perceptual achievements are definitely not due to rational processes and they are quite inaccessible not only to reasoning but even to introspection. The message transmitted from perception to conscious experience is always presented in a finished form without any indication of the way in which it was generated. If I hold a pencil 20 inches from my eyes, I simply see it in a specific position in space with self-evident clarity. I know exactly where it is. But I experience *nothing* of the basic facts from which this knowledge is *extracted.* However hard I may pursue introspection, the angle of convergence between my visual axes and the distance between my eyes remain inaccessible to me. If I arrange a situation in which my depth perception mechanism deceives me, I shall perceive only the distance that is determined by the experimental setup. For example, I can conduct a stereoscopic experiment in which two pictures are presented, one to each eye, in such a way that I shall see an object at a specific distance corresponding to the angle of convergence imposed on my eyes. I can *change nothing* in this perception, despite the fact that I am fully aware that the two objects I can see are very close to my eyes.

**The reader is referred here to the footnote on p. xliii. From this point on, cross-references made by the author to the other parts of the book that he had planned will not be given a special commentary in every case.*

The *color* and *size* of visualized objects are "calculated" by means of an analogous process. I see my writing paper as having the same "white" color, regardless of whether I look at it under electric light with a strong yellow hue or by daylight with a more pronounced blue content. Yet in *objective* terms the paper reflects light of an entirely different color in the two cases. I also see the sheet of paper as having the same size, regardless of whether it is 1 yard or 2 yards away from my eyes, although the size of the image formed on my retina is quite different at these two distances. Such *constancy phenomena* are also generated by mechanisms that are functionally comparable to unconscious inferences. In the perception of constant coloration of a given object, the illumination of *all* things seen simultaneously is "taken into account" and the dominant color of the impinging light is "determined." The *object's own* characteristics are "concluded" from the color of the general illumination and from the color of the light reflected by the object, which is redder, bluer, or yellower than that reflected by other objects and thus indicates its "color." The latter is then directly seen under quite different conditions as a constant quality intrinsic to the object. In a quite analogous fashion, for the generation of *size constancy* of an observed object, *its distance* is taken into account and its actual size is determined from the size of the image formed on the retina. This particular size is also seen directly as a constant feature intrinsic to and characteristic of the object.

These few examples suffice to demonstrate certain peculiarities that are just as inherent to general perceptual functions as they are to those very special and complex forms of Gestalt perception which we refer to as intuition and which concern us here. Firstly, *many elements* are always combined into *one entity.* In other words, many sensory inputs are integrated into a single fact by means of a process that demonstrably takes place entirely mechanically but is *functionally analogous to an inference.* Secondly, in every case, it is only the *result* of this "inference" that is experienced by the perceiving subject. The process through which this result is obtained takes place in the absence of any subjective phenomena and remains virtually inaccessible to introspection. Thirdly, the message that is delivered from perception to subjective experience always bears the character of a clear-cut, vivid, "*evident*" and undeniable "*truth.*" The German word *Wahrnehmung* for perception is, in fact, derived from the roots *wahr* (true) and *nehmen* (to take), meaning literally "to take as true." Fourthly, the perceptual picture *cannot be corrected* by any feat of reason, even in cases where the result of the "unconscious inference" is quite obviously due to an illusion and hence completely false. Fifthly, wherever perception communicates the *properties of objects,* its essential adaptive function resides in the fact that, by means of extremely complex and amazing processes, it records constant *characteristic properties of any object* as a consistent quality thereof. As in the examples provided above for perception of color and size, this is achieved *independent of fortuitous, inessential* accompanying circumstances.

The latter property of "ignoring inessential features" is particularly important and especially complex in *Gestalt perception.* Even the two relatively primitive examples of constancy phenomena provided reflect direct perception of a far from simple *law* governing the multitude of variable sensory inputs. Whenever depth perception makes the report "nearer," the retinal image reports that the object is "bigger" in a consistently lawful geometric relationship to the "nearer." A similar relationship applies to color constancy. But the basically analogous function of Gestalt perception is far more complex and far less easy to analyze by means of experiments. This is true even in its most simple and primitive expression, in which it is literally concerned with the perception of "Gestalt" (form), that is, with the external appearance of objects. The principle that governs the sensory data and is directly "seen" from them is immeasurably more complex than that involved in the constancy phenomena discussed above. The image formed on my retina of the pipe that I am smoking while writing this text has a great variety of different shapes as I turn it in different directions between my fingers. Nevertheless, the form of the pipe that I perceive remains constant in just the same way as its size remains constant as I move it toward or away from my eyes. The perceptual feat that permits us to see these shape changes in the retinal image as movements and not as changes in shape of the object, that *correctly* allows us to "*interpret*" them as movements and not as shape changes, is so familiar and self-evident to us all that we no longer marvel at it. Yet it represents one of the most wonderful achievements of our brain. Just imagine the number and complexity of the "unconscious inferences" that must be made in order to take the vast number of different combinations of sensory inputs that can generate the image of the pipe in all its different spatial locations and that nevertheless are able consistently to "infer" the same, constant form of the object!

As for all other constancy phenomena, the original significance, the species-preserving function, of form constancy doubtless resides in the fact that it permits the organism to perceive real objects in its environment as unmistakable individual entities that can be recognized under a great variety of different perceptual conditions. The function of constancy phenomena is *objectifying* in the truest sense of the word. By extracting the properties inherent *in objects* in the described manner, they are in fact entirely responsible for generating the entities in our phenomenal world that we refer to as *objects!* On the basis of comparative psychological evidence, it can be confidently concluded that form constancy, with its capacity for rendering objects recognizable as constant phenomena, is phylogenetically the most primitive form of the *Gestalt phenomenon.* The perceptual capacity to "*identify*" a governing principle from the unlimited wealth of sensory data quite definitely had its phylogenetic origin as a mechanism generating *object constancy.* But, as has happened in striking individual cases in the course of organic evolution, the organic apparatus in the central nervous system that was developed to serve this very special function proved surprisingly to be capable of other, more general feats. Even the perception of higher animals goes beyond the capacity of "seeing" the principles that emerge from the *identity* of visualized objects. A given principle is recognized as a "Gestalt"—as an "unmistakable experiential entity," in [Hans] Volkelt's terminology—even in cases where it is present in *different* individual concrete objects. A higher bird or mammal or a human child can go beyond the recognition of an individual visual object, for example, our dachshund Kathi. Here, constancy phenomena permit recognition of the same individual dachshund under all kinds of illumination, at a distance or close up, from the front or from behind. Over and above this, the capacity for Gestalt perception permits such higher organisms to recognize in this dachshund, in the neighbor's poodle, in the butcher's Alsatian, and in auntie's Pekingese a *supraindividual,* common quality, that of a *dog.* This capacity is certainly not based on a feat of *abstraction!* An 18-month-old child is definitely not capable of abstracting Linnaeus' zoological *identification key* for *Canis familiaris.* Still less can we attribute such an ability to a raven, a cat, or a mongoose lemur. Instead, the capacity of Gestalt perception to generate object constancy by extracting the essentials and excluding fortuitous features is here extended from *the individual to the generic level.* In this, a common governing principle is *directly seen as a Gestalt* in just the same way as in the original, more primitive function of recognizing the characteristic Gestalt of an individual object. [Christian von] Ehrenfels has emphasized the essential features of Gestalt perception as being its *independence of the elements involved, its transposability, and the primacy of the entity over its parts.* Accordingly, these are referred to as the "Ehrenfels criteria." All three of these criteria are already inherent to the primary function of Gestalt perception, which was concerned only with object constancy and with the recognition of individual visual objects. But all three are also decisive for the function of Gestalt perception at issue here, which consists of the extraction of a *principle* and which is *functionally similar to an abstraction.* Very good examples both of the Ehrenfels phenomena and of this special function are provided by auditory perception. A *melody* is an almost classic example of a Gestalt. It is recognized independently of the sound elements of which it is composed and it can be transposed to different keys and different pitches without suffering any change in its unmistakable individual quality. The primacy of the entity over its parts is particularly pronounced in this case. Anyone can recognize the melody more easily than the absolute pitch of the sound elements, the individual intervals, and so forth. Similar comments apply to *chords*, which are perceived even by the most untutored, unmusical ear as something special and *lawful*. Chords are sharply distinguished from uncontrolled dissonance. Yet the human subject is quite unable of determining through introspection the laws of vibration frequencies and harmony that are decisive for determining the quality of the unmistakable experience.

Very good examples of Gestalt perception of extremely complex laws, already leading into "intuition" and thus revealing its close relationship to Gestalt perception, are provided by *zoological systematics* with their arrangement of living animal species according to their degree of phylogenetic relationship. Here, we find all imaginable intermediates between simple feats of Gestalt perception, as in the recognition of a dog by a young child, and the highest achievements of scientific intuition, commonly referred to as "systematic finesse." At the age of 5 1/2 years, my daughter Agnes was able to recognize any bird belonging to the rail family after becoming familiar with just two representatives, albeit in some depth. Her achievement was remarkable for the following reason: Both of the species with which she was familiar—the moorhen (*Gallinula chloropus*) and the coot (*Fulica atra*)—are waterfowl that have adopted a generally ducklike form through adaptation to aquatic life. But the rail family (Rallidae) contains birds with a wide variety of different forms, including small steppe inhabitants with a pronounced chickenlike external appearance, such as the corncrake, and long-legged wading birds, such as Allen's gallinule, along with many others. Carefully avoiding any suggestive questions or involuntary guidance, I conducted a test on my young daughter using the comprehensive bird collection at the Zoological Garden in Schönbrunn, with which she was quite unfamiliar. She made no mistakes, either with the most chickenlike members of the rail family or with the most rail-like members of the chicken family, and in an enclosure that also contained herons and other wading birds she immediately picked out Allen's gallinule as a rail, despite the fact that this bird is a very unusual member of the family in external appearance because of its ultramarine blue coloration. In doing this, she solved a problem that had completely defeated Linnaeus, the founding father of zoological systematics. In his "natural system," the individual members of the rail family are classified into different groups of striding, swimming, or wading birds according to their external adaptations! How was it that this 5-year-old *saw* something that Linnaeus had failed to see? It was because she was closely familiar with the *living* bird, whereas Linnaeus had only examined the prepared specimen. In the general behavior and locomotion of the living bird, a wealth of fine details generate a *Gestalt.* This experiential entity incorporates *more quality-determining characters* than Linnaeus took into account in his identification key. When I asked my daughter *how* she could recognize members of the rail family as such, all that she could say was that they "are like a moorhen." The number of characters that determined the quality of my daughter's "rail Gestalt" for the birds with which she was familiar must have been so great that they completely overwhelmed the aquatic bird features shared by moorhens and coots. In spite of the somewhat distorted basis of her "inference," she was able to achieve a correct result through intuitive Gestalt perception. The most knowledgeable and eminent modern zoologists concerned with fine systematics exploit this feat of Gestalt perception quite consciously by taking observations of the living animal and of its "physiognomy" as indicators in their systematic investigations.

The function of intuition in directly identifying the law governing a wealth of individual concrete facts is quite definitely very closely related to the functions of Gestalt perception discussed above. The former is surely based on the latter and certainly requires them as a precondition, and it is indeed probable that they are simply identical. As with perception of a Gestalt, an intuitively recognized relationship emerges abruptly from the background of incidental features as a vivid and convincing result. This occurs *without* the person concerned initially being able to specify *which* concrete facts provide the basis for the certitude and power of conviction of this "revelation." He is no more able to do this than we are to specify the angle between our visual axes for our depth perception or a child that can recognize all dogs as such is able to state the zoological identification key for the domestic dog. Nevertheless, intuitively derived certitudes are just as *correct* as those that are based on more primitive perceptual functions. In other words, in general they are in most cases almost *completely correct,* but in a few awkward special cases they are *completely wrong!* But as the process of intuition is far more complex and unfathomable than the process leading to conscious inferences, yet yields results equivalent to and sometimes surpassing those of the highest achievements of conscious research and its conclusions, it strikes us as something bordering on the miraculous when it is actually *correct,* which is predominantly the case. Precisely because the perceptual processes on which intuition is based are inaccessible to introspection, it has the effect of a revelation, of something instilled from within. For this reason, many people—including those who do not admit it—believe in a supernatural *infallibility* of intuition. From what has been said, it is sufficiently obvious that this is a fundamental mistake that can be disastrous under certain circumstances. Of course, intuition exhibits the general accuracy and "somnabulistic" reliability that characterizes all perceptual processes. Nevertheless, just as the accuracy of any conclusion depends on the correctness of the premises on which it is based, the same applies to the accuracy of the "unconscious inferences" (Helmholtz) of all perceptual processes, including the most complex form of Gestalt perception that we refer to as "intuition." The most complex intuition is fundamentally open to deception, just like the simplest form of depth perception involved in our stereoscopic experiment. Further, once such deception has occurred, intuition clings to the "evident truth" of its results, remaining just as immovable and just as refractory to any rational objections. *Following any adulteration of the sensory data underlying perception, intuition delivers an incorrigibly erroneous message like any other form of perception.*

The following conclusion can be drawn: Just as with induction, decisions derived through intuition are based on a multitude of details recorded by the sense organs. As with induction, the probability that intuition will arrive at a correct result is proportional to the breadth and statistical accuracy of the mass of individual data processed. In these utterly essential and integral features, the functions of the two processes are *completely identical!* Despite the enormous differences in the neurophysiological processes and the psychological levels involved in the two processes, the function and the *value* of both intuition and induction are uniformly dependent on the derivation of *general* principles from *individual* concrete sensory impressions. (Although the preceding text has discussed in some detail the different ways in which induction and intuition serve these analogous functions, this was done in order to highlight the strengths and weaknesses of both processes and thus to show how they not only complement one another but also require reciprocal concessions. This is precisely what must happen with the natural sciences and philosophy if they are to arrive at a fertile synthesis.)

The *strength* of intuition resides above all in the fact that, as a genuine and immediate function of perception, it can *see* in the profoundest sense of the term. Things that would never occur to human conscious thought literally catch the eye of intuition. Intuitive vision is independent of any working hypotheses and can even see quite unexpected things, not just those that are significant for already existing questions. This wide-ranging scope of intuition is due to the fact that *it can simultaneously encompass a far greater number of individual facts to incorporate them as qualitative determinants in the total experience of derived knowledge* than is ever the case with conscious induction. Only intuition can achieve a holistic overview of a multitude of individual things in order to extract *one* law, *one* set of connections, and *one* harmonious relationship. It was intuition that led Heraclitus to the view that everything is in flux. It was intuition that led Copernicus to perceive in the strange, apparently convoluted movements of the planets the simple, splendid laws governing their true paths. And it was also intuition that led Darwin, from his observation of the confusing profusion of living and fossil plant and animal forms, to abrupt recognition of the Gestalt of the evolutionary tree of living organisms.

The *weakness* of intuition lies in the fact, already repeatedly emphasized, that the route that it follows to achieve its results is so inaccessible to our consciousness. Although these results are based on real perception of real things, such that there are in fact tangible *reasons* underlying the unconscious inferences, intuition itself cannot inform us about these reasons. Only through a relatively difficult process of introspection is it possible to uncover these to some extent. As a consequence, the result of intuition cannot be checked by conducting a critical "audit" of the route that was followed. This *unaccountability* of intuition is one of its great weaknesses; another is its *incorrigibility*. Every now and then, one of those awkward special cases arises in which Gestalt perception falls into a trap laid by a misleading set of sensory data. When this occurs, intuition clings to its revelation-like results with almost the same animal incorrigibility as that shown in the stereoscopic experiment, in which our depth perception clings to its illusory message regardless of any insight exercised by the perceiving subject. Even when we possess complete rational insight into the origins of the illusion, it retains its utterly convincing vividness. For this reason, those philosophers who have the best powers of intuition are also the most incorrigible once they have strayed into the cul-de-sac of a perceptual illusion.

The *strength* of induction resides in its accountability. With any result gained through induction, we can show any doubter the route that we followed in obtaining it. We can also use the breadth of the inductive basis to assess with genuine mathematical accuracy a probability value for the correctness of the result. Where that breadth is sufficient, the probability is so close to certainty that we can confidently equate the two. The strength of induction stems from the fact that its base is exactly known. It is therefore possible to identify the concrete facts that must be *sought* in order to extend that base, whereas this is never known with intuition. Therefore, only induction can increase the reliability of a result and only induction can *validate* a result that has been obtained exclusively through intuition!

The *weakness* of induction resides in its blindness or, to be more precise, its shortsightedness. The accumulation of inductively obtained knowledge resembles a conical pile of a loose material with a very low stacking coefficient. Not only the height but also the location of the peak is directly determined by the form of the base of the pile. If we use a really unadulterated inductive procedure to collect the facts that form this base, we can have no idea where we will end up. As a result, under certain circumstances the most advanced knowledge that emerges, represented in our metaphor by the peak of the pile, might come to lie in a quite uninteresting, unimportant place. The fact that this practically never occurs is due to the fact, already established above, that no natural science is really *free* of intuition.

This identification of the strengths and weaknesses of induction and intuition leads directly to the following major conclusion: As inductively operating fields of research, the natural sciences must *fundamentally recognize the leading role* that belongs to intuitively operating philosophy. Philosophy is really the "queen of all disciplines" and should remain so, or, to be more precise, it should *become* so. The natural sciences need such a queen more than ever because, with their rapidly increasing specialization, there is a growing danger that the individual research worker will lose the view of the great relationships that provide the rationale for his painstaking work on the details. But if the natural sciences are to accept the leading role to be played by philosophy, they must be allowed to make certain demands on philosophy. In order to formulate these demands, we shall now proceed to summarize all of the preconditions for a synthesis of philosophy and natural science that arise from what has been said above.

First of all, any philosophy that is nominated as the leader of the entire human quest for knowledge must of course be a *materialistic* philosophy that is convinced of the reality and unity of the world.* It is not our chosen task to try to convince the idealists, who regard the evidence of the senses as nothing and who ipso facto will not accept as such the facts that we may present. For this reason, our rejection of any form of idealism will be justified here with only two arguments which are *not* derived from the facts of inductive research and which the idealist must therefore also accept. The first is as follows: We see a quite unacceptable *lack of consistency* in all idealistic theories in that, while on the one hand they do indeed regard the evidence of the senses as without value as a source of knowledge, on the other hand

**Note in the margin here:* Umbrella discipline.

they accept as an unconditional certitude *the existence of fellow human beings,* which is known to us only through the testimony of our senses. It is undeniable that there is an amazing lack of consistency when Schopenhauer (for example) utterly rejects the existence and recognizability of an extrasubjective world and yet, of all things, drops his fundamental scepticism when it comes to the existence of fellow human beings. Further—in a purely dogmatic manner from his point of view—he assumes the presence of other perceiving subjects who supposedly possess a world constructed from imagination and willpower similar to that subjectively experienced by the philosopher himself.

Why should it be the *same* world? This question encapsulates the second fundamental point of our criticism of the idealistic or monadological view of the world. The high degree of correspondence between the phenomenal worlds experienced by so many different subjects *needs explanation!* All idealistic theories concede that this correspondence exists, but they fail to explain it. If the perceived world is only an illusion, a *dream,* then—by Sirius, as Plato would say—it is almost infinitely improbable that we would all dream *the same thing.* If we sit as a group philosophizing around a table and all without exception see six wineglasses standing there, then we are entitled to *demand* of the nonrealistic philosopher a logical and parsimonious explanation of this fact. Yet no nonrealistic philosopher is capable of providing such an explanation. As we have already seen, even for Kant—who fundamentally assumes the existence of extrasubjective reality—there is no logical (not to say causal) connection between the "thing-in-itself," which Kant only ever sees in the *singular,* and the manner in which it appears in our perceptual world. The assumption of the existence of "wineglasses themselves," which are *mirrored* in just this way in the perceptual worlds of the six participants in the debate, is a hair-raising absurdity for any Kantian. It has already been noted that we critical realists readily admit to the Kantian that we do not know and *cannot* completely know everything that is involved in the phenomenon that we perceive in the form of six wineglasses. But we are quite convinced that properties of this phenomenon *as such* are responsible for the *correspondence* between our perceptions, for example, determining the number *six,* which we have all counted. This correspondence between our individual worlds can only be understood if it is a property of *things* as such that for you, me, and the rest of mankind the sun shines, the wind blows, and the trees grow leaves in our environment. The lack of contradiction between individual perceptual worlds renders any *chance* correspondence, even in the simple case of our six wineglasses, extremely *improbable.* But this improbability becomes virtually infinite in view of the lack of contradiction in the natural sciences, which encompass such a multitude of details and yet fit together so well. The only idealistic philosopher immune to the charge of inconsistency is one who draws the final, bitter conclusion from his doubts about the reality of the external world and questions *everything,* including the existence of his own self. The solipsist is the only consistent idealist.

Consequently, the demand made by natural science of any philosophy for it to fulfill its calling as the leader of the entire human quest for knowledge is first and foremost the obvious requirement that it should be a kind of philosophy that recognizes the real existence of the subject matter that natural science strives to comprehend. All of the more specialized requirements follow from what has been said about the reciprocity between induction and intuition. Because intuition is *not a miracle* but a natural process that extracts general principles from concrete perception in basically the same way as induction, it similarly depends on the existence of a *base* of known individual facts. In order to perceive new and great relationships that point the way, relationships that are present among the confusing multitude of facts accumulated by the industrious individual activities of natural scientists, the philosopher must *know* these facts and possess an *overview* of them. If he succeeds in mastering this huge and extremely demanding task, he will achieve something that is beyond the reach of any single natural scientist. The latter, because of the significance that he must attach to individual *details,* always suffers a certain basic handicap in achieving an *overview.* As has been seen, it is precisely the special talent of the most analytically skilled, penetrating, and exacting research workers concerned with detailed investigation that renders them little inclined toward or suited for a synthetic overview. Quite apart from their characteristic special talents, they are too close to the frontline where the battle is raging to win new facts. Just as the commander cannot stand shoulder to shoulder with the combatants if he is to maintain his view of the overall situation, the philosopher who aims to identify new internal relationships in the collective knowledge of humanity *as pointers* cannot be engaged in the frontline of inductive scientific research. Nevertheless, like a painter who stands back from his easel after working on some detail in order to see its effect on the painting as a whole, any scientist should from time to time look at his work from a distance and examine its position within the framework of our overall picture of the world. This is all the more necessary because the individual scientist is not working alone on the painting concerned. But the philosopher, whose profession it is to survey the entirety of the human world image, should stand close enough to the natural sciences and should *know* enough about them to be able to identify the places to be occupied by individual results of research in the scientific world image and to estimate their significance. The philosopher's capacity to do this will be greatly enhanced if he has *himself* worked in some area of research, just as it is beneficial for a field commander to work his way up through the ranks. This is no empty theoretical requirement: The most significant philosophers of our time are such "commanders"; they are "painters" working on parts of the human world image who have taken a step back from the canvas in order to take a look at the painting as a whole. It is only through such *intentional* observation of the whole that a science can develop with the capacity to coordinate the individual research branches and to point the way forward. The natural sciences need such a coordinating and guiding authority more than ever before and only philosophy can provide this authority. This is the concession toward the humanities that is required of the natural sciences as a result of recognition of the fact that it is intuition which points the way for research. However, this concession brings with it the momentous requirement that an encyclopedic knowledge of the various fields of natural sciences must be combined with mastery of the entire heritage of the humanities.

These requirements arise from the need for leadership, so to speak, from a *weakness* of induction. Conversely, however, there is a whole batch of requirements arising from the *strength* of induction or from the weaknesses of intuition which the natural sciences must decisively present to any philosophy that is to be accepted as providing a lead. These requirements concern the treatment of individual concrete facts. The great demands with respect to the impartial evaluation of individual facts that are made of any research worker working within the consistent hierarchy of the natural sciences must also be addressed with undiminished vigor to the philosopher if a successful synthesis of the natural sciences and the humanities is to be achieved. Any research worker must always be ready to sacrifice the most beautiful theory, despite its apparent prior confirmation by a great number of facts, when confronted with a single contradictory observation. In just the same way, a proponent of the humanities must always be prepared to accept that an intuitively recognized "truth" is false when confronted with an inductive demonstration of its inconsistency. Even for an inductive researcher, a great deal of self-discipline is required even to *see* uncomfortable conflicting facts. Understandably, such self-correction is even more difficult for the unconsciously operating inferences involved in intuition. Gestalt perception is, in fact, constructed in such a way that it simply overlooks any details that do *not* fit into a perceived Gestalt and, under certain circumstances, virtually *falsifies* them.* The philosopher must remain completely aware of this fundamental weakness of intuition even when faced with the most directly convincing revelations of intuitive vision. He must never forget that even the deepest and most final knowledge available to the human intellect is only *approximate in character.* He must always be prepared to *improve* on this approximation as soon as a new fact provides some reason to do so. Research in the natural sciences tolerates no absolute truths, either for itself or for any philosophy that is to play a leading role. It tolerates only working hypotheses. Above all, it tolerates no "*personal opinion*"! In the natural sciences, a real conflict of opinions of the kind that is virtually the "father of all things" in philosophy is in principle not really possible. If *different* opinions exist with respect to a given class of phenomena, this simply shows that the factual evidence is still insufficient to permit proper formulation of an opinion. Only the working hypotheses can differ. Two research workers who happen to favor two opposing hypotheses will not try to decide between them on the basis of a *discussion,* which at the most can only serve for clarification of the differing viewpoints. They will engage in further interaction not with *one another* but with the facts, which will then show that one of the hypotheses is false, or perhaps that both are wrong.† The fundamental lack of decisiveness of any opinion is also very clearly expressed in the role played by the theory of even the greatest master in the research activity of his disciples. In philosophy, it is entirely permissible and legitimate simply to *accept* the theory of some great teacher, in other words simply to regard it as *true* without any obligation to engage in criticism or further refinement. In philosophy, "-isms" have been recognized and revered. Rationalism, empiricism, idealism, sensualism, and so on are all designations that are no more *derogatory* in their implication than labels of schools of thought derived from the *names* of the masters, such as Kantian, Hegelian, and the like. In the natural sciences, by contrast, the opinion of the teacher can of course only be a working hypothesis for the apprentice. If advances are made in research, the apprentice must be just as critical and sceptical toward the teacher's hypotheses as toward his own. The apprentice may not accept any word of the teacher unexamined. For this reason, names for schools of thought in the natural sciences, such as "Darwinism," "Lamarckism," and "Weissmannism," always indicate that the dynamically developing research of a great man has given rise to a static creed with a philosophical character.

**Note in the margin here:* Tachistoscope.

†*Note in the margin here:* Or both might be right, in which case the way of thinking is wrong = waves ↔ particles.

The latter has nothing left in common with the living, eternally forward-striving work of the natural scientist. The individual personality of the research worker pales into insignificance behind the facts that he has discovered. Indeed, the more important these facts are and the more rapidly they become a part of collective human knowledge, the more rapidly the researcher himself will be forgotten. The *more important* the new building block that the individual has added to the collective edifice of natural sciences—that is, the greater the possibilities that it opens up for *further construction*—the faster it will be covered over and concealed by that which is built upon it. In my lectures, I have often illustrated this phenomenon with an experiment. I first ask: "Who was Aristotle, Descartes, Hegel?" Most students with an average education know that these men were great philosophers, but only a tiny minority of them can state the merits and particulars of their theories. Then I ask: "What is the law of inertia?" "What is the thermal expansion of solid bodies?" "What is a reflex?" All of the students can provide an approximately correct answer to each of these three questions. But if one then asks who *discovered* these three widely known natural phenomena and established the corresponding laws, not one student knows the answer!

It is precisely this *collective* character of the treasure of knowledge accumulated by the consistent hierarchical system of the natural sciences, precisely this participation of so many human beings in the construction of a single great edifice that is permitted only by unconditional obeisance to facts. And it is this that we must demand with particular insistence from any philosophy for it to be in a position to coordinate such a cooperative effort and to point the way forward.

Unfortunately, however, it is this very demand on philosophy, the most important of all, that proves to be exceptionally difficult to fulfill in any *practical* attempt to achieve collaboration between the natural sciences and the humanities. Even philosophers of the subject who fully accept the justification of this demand in theoretical terms prove in practice to be still utterly permeated by the old idealistic conviction of the primacy of reflection over observation. One could say that they are so infused with this that it is simply impossible for them to discard completely a logically consistent theory, an intuitively recognized conviction, just because of a single small, troublesome fact. They are unable to allow a new theory to crystallize out in a fundamentally modified form in the manner demanded by a natural scientist, who is more accustomed to such rather painful but rejuvenating intellectual acrobatics.

But the thing that is most difficult to achieve with the humanities is insight into the *statistical* nature of inductive evaluation of factual evidence, into the unconditional requirement for *impartial* treatment of *all* pertinent facts. Even for realistic practitioners of the humanities, it is unfortunately even today still regarded as entirely acceptable and scientifically legitimate to construct a complex edifice of high-flying speculation on the basis of a tiny selection of biased facts. Indeed, for many philosophers, even the procedure that we have already denounced as a serious methodological transgression—namely, *searching* for examples to support a *preconceived* opinion and *selecting* them in a statistically inadmissible fashion—is far from being regarded in its true light as a highly unscientific fabrication. In the chapter on induction [chapter 2], we discussed the way in which premature formulation of a hypothesis can distort the inductive basis by tuning the investigator's eye to any facts that seem to support it and shielding it against any conflicting information. But, as we have also seen, the validity of philosophical intuition is just as dependent on the breadth and statistical reliability of its factual basis as is the validity of scientific induction. But in the case of intuition the danger of *distortion* of the basis is *far greater* for the very reason that the process of evaluation of the facts takes place unconsciously and uncontrollably. It must be demanded as a basic precondition of any philosopher, as a "specialist in intuition," that he should be intimately acquainted with the weaknesses of his greatest source of knowledge. He should show particular mastery of the "technique" discussed in chapter 2. This consists in putting a bridle on one's own intuition and suspending any perception of overarching relationships until a sufficiently broad basis of underlying, interwoven facts has accumulated. The suspended animation of an anticipated Gestalt, the deliberate bridling of its development, keeps the way open for the entry of additional and novel facts that can no longer be incorporated once the entity of the Gestalt has secluded itself. What we expect of any philosopher who is to serve as our leader is mastery of the great art of keeping Gestalt perception open and mastery of insight into the nature of these internal processes, which are so difficult to control and yet are fundamentally amenable to disciplinary power.

The greatest infringement against the laws governing the assessment of factual evidence—and one which we must repeatedly brand as unpardonable in contemporary work in the humanities—is the *denial of knowledge,* that is, the more or less *deliberate* overlooking of facts that *conflict* with particular preconceived opinions. Just as the deliberate evasion of decisive facts is shunned in the natural sciences, it should be equally shunned in the humanities. Even someone who is exclusively striving for the philosophical recognition of purely mental phenomena may not shut out new facts that are relevant to them on the grounds that they stem from a different scientific source. At this point, I am *not* referring to the requirement arising from the consistent hierarchical framework of the natural sciences, that anyone who sets out to investigate the human psyche must be equipped with a basic general knowledge of *life processes,* for the simple reason that the human psyche is just *one* of many different life processes. However justified this requirement may be, it is not the one that I presently have in mind. The "lower hierarchical levels" of physiology and biology have so far contributed virtually nothing to our understanding of purely mental processes and can therefore be dispensed with for the time being. But even here there are extremely important facts which were first *noticed* at a time when inductive scientific research was under way and which, as entirely *mental* phenomena, nevertheless demand incorporation and consideration in the humanities. *Even in the science of the human mind, new facts that conflict with old theories require a revision of those theories.* For example, if Darwin discovers the fact that human beings owe their existence not to a unique act of creation but to an extremely drawn-out process of evolution, this fact has important consequences for our contemplation of "a priori" forms of thought and intuition. Yet in the humanities epistemological theory responds to these inevitable consequences in the most indolent manner possible: it simply ignores them! In a further example, Max Planck discovers that there are cases in which the human mental categories of causality and substantiality cease to "correspond adequately" to reality (as Kant puts it), whereas the category of quantity in the form of statistical probability calculations continues unhindered to master the facts. This discovery is by no means just a *physical* fact that the Kantian can dismiss with the explanation that it "only" concerns our physical world image. The creation of "acausal" physics is primarily epistemological and is, so to speak, only secondarily a scientific act. It is far more revolutionary and exciting in epistemological than in physical terms. It could only be achieved by one of the greatest epistemologists and one of the best *Kant experts* of our time. Indeed, there is reason to doubt that it could have occurred in this way *without Kant,* without the prior existence of Kantian analysis of the human cognitive apparatus! But how do modern Kantian philosophers respond to this great achievement by a direct intellectual descendant of their master? One might expect—indeed any reasonable person would expect—that in their thoughts and words the practitioners of epistemology on a Kantian basis would be vigorously exploiting this powerful new development of *their own school.* But what happens in reality? Living Kantians *ignore* Planck because he offends against the absolute mental necessity and truth of a priori schemata, because he has dared to *extend* and therefore *change* the theories of the master, which have now become a matter of faith. There is nothing that can be done with this kind of philosophical school. It is a fossil which admittedly possesses a certain value as a museum exhibit but with which it is no longer possible to interact. In fact, however, Planck's results are in themselves already the *fruit* of a genuine synthesis between the natural sciences and the humanities, between highly specialized individual research and extremely general epistemology. As such, they bear witness to the fact that such a synthesis is really *possible*. At the same time, however, they provide the most convincing evidence of the unconditional necessity for unequivocal recognition of the primacy of the concrete individual fact over forms of thought, of induction over deduction. Those results were possible only because the experimentally determined fact was recognized as an authority fully justifying even criticism of human rational thought. They were permitted only by a truly unshakable conviction both of the reality of the extrasubjective world and of the fact that our cognition corresponds to this reality in an adequate fashion. But in this sense Planck's revolutionary achievements are also the fruit of a genuine, unconditional unity between the philosophical view of the universe and our factual interaction with that universe.

This brings us to the last and perhaps most fundamental requirement that the natural sciences must make of any philosophy that is to play a leading role. Conviction of the existence of *a single* world of realities revives the ancient and yet ever-topical requirement for absolute unity between the philosophical worldview and everyday actions, between theory and practice. Sages in ancient Greece not only made this demand, they also *fulfilled* it. The idealists of antiquity fully accepted the consequences of their belief in the futility of the phenomenal world: they renounced material possessions. In a metaphorical or even literal sense, they crept into a barrel in order to live out their ideas undisturbed. But with the vast majority of modern proponents of idealistic philosophy, even the most acute observer would be unable to see, or even suspect, from their behavior in an everyday practical context that they are less convinced of the reality of things than their naive-realistic contemporaries. Even Schopenhauer, the great sceptic and pessimist, led an entirely "realistic" life and never intended to apply suicide, which he valued so highly, to himself. It is my claim that, in the depths of their souls, all of these philosophers simply *do* believe in the reality of things and that they only cling to their doubts at the highest, but also most superficial, levels of their rational thought! This applies quite particularly to those far-from-rare scientists who devote their daily work to inductive scientific research but proclaim a belief in idealism in their philosophical worldview. Such a schism between philosophical conviction and daily activity is least damaging in cases where the person concerned leads a genuine double existence in two separate worlds that never come into contact with one another. For example, one of the most penetrating and exact physiologists of my acquaintance is a convinced Kantian, but he does not allow his philosophical belief in the absolute unrecognizability of the material world to distract him in the slightest from his eager investigation of that world. But this internal schism becomes far more threatening when a scientist attempts to combine the uncombinable. In this case, the logical rupture occurs *within* the scope of practical research, severely damaging natural science in the process. The history of the biological sciences contains a never-ending series of errors and detours that can be exclusively traced back to disruption of research by vestiges of idealistic philosophy. This understandably applies to a quite particular degree in *psychology,* although until quite recently it was surprisingly also true of *psychiatry.* But in *physiology* even the first great classic proponents (Johannes Müller and Claude Bernard) made the logically vain attempt to create the peculiar hybrid theory between idealism and materialism that we refer to as *vitalism.* The logical rupture within vitalism resides in the fact that, in a completely dogmatic fashion, it tears specific life processes out of the great framework of naturally explicable phenomena and declares them to be the immediate consequences of a supernatural principle. Viewed from the standpoint of the history of philosophy, this can be recognized as a direct descendant of the Platonic idea. In the section dealing with evolutionary theory and the methods of comparative ethology [section II], we shall see in some detail the disastrous inhibitory effects on research that were brought about, particularly in this field, by the logical schism within vitalism.

The closer scientific research approaches *human beings* as its focus of attention the more it becomes engaged with philosophy. For this reason, there is a decline in the success with which philosophical conviction and daily research activity can be assigned to two different worlds and kept neatly separated in the manner shown by the Kantian physiologist of my acquaintance. Therefore, the more natural science becomes the science of mankind, the greater must become the disruptive influence of any schism between the philosophical worldview and practical research activity. Accordingly, there is an even greater need for a categorical demand for complete identity between theory and practice. Credit is due to the pragmatic philosopher E[dvard] Baumgarten for recognizing with particular clarity and presenting in a convincing fashion the enormous and confusing barrier that is thrown up to any attempt to synthesize the natural sciences and the humanities by that widespread and longstanding habit of leading a philosophical double existence. During his time at Göttingen, Baumgarten brought together for evening discussions very bright scientists and men of letters in a small group with a highly unusual name. It was, in fact, called the "Club Against the Scandal of Double Logic"! Later, following the lead established by this promising germ, Baumgarten and I worked together in our double institute, which was to serve this very purpose—a synthesis between the natural sciences and the humanities. In this institute, which was established in Königsberg (in Kant's shadow, so to speak), philosophical anthropology was combined with comparative behavioral research as the natural science of mankind. Despite the short lifetime that was granted to this cooperative enterprise, it bore a certain number of fruits and will continue to do so.

As has already been said, any philosophy that deserves to be called the "philosophy of the natural sciences" is a burning necessity for inductive research, which is becoming progressively more specialized and whose individual branches therefore require with ever-increasing urgency an integrating, coordinating authority. But it is not only the *scientific* members of humanity who need a leading authority for their collective quest for knowledge. From the time of its earliest origins, philosophy has had the task and duty of providing guidance for the *moral behavior* of human beings! The ancient philosophers were quite plainly aware of this responsibility. All of the great philosophers of antiquity were *teachers* of people; they were *preachers* of morals, of *ethics*. But humanity today finds itself in a state of chaotic dissolution of all internal and external ethical laws. This has gone so far that, without any pessimistic exaggeration of the situation, one must say that it quite directly threatens mankind (or at least our present stage of civilization) with *destruction*. The analysis of these manifestations of ethical decay in the social behavior of human civilization and the presentation of its various causes will occupy us in the last part of this book, which deals specifically with human beings. As we shall demonstrate in great detail, the human species represents a highly adventurous, precarious "construction" of the evolution of species. Our species represents something *forbidden* to the extent that certain degrees of freedom in our behavior, which are straightforwardly constitutive for humanity, are due to a process that is dangerously close to being pathological and abnormal and which we refer to as *domestication*. Both the gifts of domestication, to which we owe our humanity, and the dangers of domestication, which have already begun to represent a direct threat to our existence, have arisen in us primarily through inherited *deficits in innate species-specific behavior patterns,* through "mutational loss" of "instincts." Domestication-induced mutational losses in "instinctive" behavior patterns on the one hand brought with them liberation from the constraints of automatic and reflexive processes, thus making a major contribution to the transition from animal to human. On the other hand, wherever they affect species-specific behavior patterns with a *social* function and *ethical* behavior, domestication-induced mutational losses can all too easily become a serious threat to the future survival of the populations concerned. They can become genuine *lethal factors* in the sense of modern genetic theory. The German historian Oswald Spengler was the first to notice the fateful regularity with which human societies and cultures are afflicted with decline and destruction at precisely the point at which the external circumstances for a further flowering seem to be *secure* and at which they enter the stage of so-called *civilization.* Spengler sees the internal cause for this regularity, which is historically undeniably real, in "senescence" of human populations, in a "logic of time," which is for our purposes entirely mystical. We shall be able to show that the real cause of the decline of human civilizations lies, with a probability approaching certainty, in the spread of domestication-induced deficits in innate species-specific *social* behavior patterns. In other words, the cause is to be sought in the disruption of innate *ethical* behavior. If comparative behavioral research has a claim to becoming the natural science of mankind, the justification for this resides not least in the fact that it is really in a position to trace a considerable part of these manifestations of ethical decay, which threaten the survival of modern humanity, back to physiological causes, which are therefore *in principle amenable to modification.* The forlorn fatalism exhibited in such an extreme form by Spengler is the last intellectual position that any inductive scientist worthy of the name would ever adopt. Thus it is that, in the service of our nation and of humanity, we see the greatest value and the most immediate task facing our own line of research as residing in the investigation of domestication-induced ethical deficiencies of human civilization and in its exposure to correction.

It is an historically well-established and quite undeniable fact that all previous cultural systems have gone to rack and ruin after reaching the stage of civilization. In our own cultural epoch, the situation is different *for the first time in history!* Inductive scientific research has appeared on the scene and, for the first time in the history of mankind, there are numerous people who know *why* cultural systems have so far always perished after attaining the stage of civilization. A natural science dealing with the laws governing human societies, namely sociology, has emerged, and has provided inductive evidence for the causes behind the decline of civilizations. Comparative behavioral research is now in the process of discovering further causes. A *recognized* cause is, ipso facto, no longer the same cause as it was prior to knowledge of its existence. It is but a small step from recognition of the cause of momentous bolts of lightning to the construction of Franklin's lightning conductor. It is an inherent feature of inductive scientific research that it can provide solutions to previously unavoidable blows of fate. It offers humanity the possibility, perhaps no more than a slim chance, of avoiding this time round the threats that destroyed all previous civilizations and to break out of the previous cycle of cultural systems that prevented further progress by humanity. On the other hand, however, inductive scientific research has itself added immeasurably to the previous dangers confronting mankind. Together with the heavenly gift of its collective treasure of knowledge, it has also presented humanity with the fatal gift of technology. Scientific research has changed the face of the globe and has remodeled and distorted human social structure in an unpredictable and clearly *pathological* fashion. The power over the natural environment that technology has now given to mankind and the new degrees of freedom of action that it has opened up have been abruptly thrust upon us. These developments have, so to speak, exacerbated the weaknesses in human ethical inhibitions already brought about by the effects of domestication. It is by no means a bitter joke but bitter seriousness when I claim that modern humanity finds itself in a similar situation, and confronted with the same dangers, as a horde of juvenile vandals or even chimpanzees which by some misfortune have come into possession of several loaded submachine guns. Modern humanity possesses unexpected new possibilities for everything "that we call sin, destruction—in short, evil." In our interaction with these possibilities, there has been complete failure not only of the instinctive, innate ethical inhibitions of human beings but unfortunately also of the traditional ethics of the great world religions, particularly that of Christianity. Least of all, however, has the ethics of idealistic philosophy been able to ward off the dangers born of the materialistic-scientific way of thinking.

Anybody who is a convinced materialist and a convinced scientist should feel a binding duty to fight against any threats to humanity that have arisen because of *our* research, which has in a literal and metaphorical sense provided us with the weapons of our own destruction. The constitutive threat to mankind lies precisely in the fact that any new freedom of thought and action must be *paid for* with the weakening of rigid but biologically proven structures. As has already been said, the domestication of mankind has only been able to open up new degrees of freedom in behavior by disrupting the rigid sequences of innate, species-specific behavior patterns and reflexes that undoubtedly possessed a particular value for the survival of the species. In fact, the scientific way of thinking has had very similar effects on the quite different, much higher plane of acquired behavior patterns. Natural science was able to provide mankind with new degrees of freedom in thought and activity only by destroying ancient and traditionally deeply rooted dogmas of religions and earlier schools of philosophy. But such destruction could not occur without affecting both innate response patterns and traditionally inculcated customs *that are of very great ethical and social significance.*

Any modern philosophy that would really deserve to be called the science of human nature—the science of a modern-day, technologized, corrupted, and threatened humanity that nonetheless bears the potential for undreamed-of progress—would be faced above all with the ambitious and responsible task of *creating a new system of human ethics.* This would have to replace that which was shattered to make way for the new freedoms of thought and activity and, which, indeed, *had to be* shattered. We would be the last to regret that this shattering occurred and we would certainly not wish to reverse the process! To regret the loss of rigid but reliable instincts amounts to regretting the fact that we humans diverged from animals. It is, in principle, very similar to regretting the shattering of traditional dogmas and taboos by scientific research. But the free act of shattering what was there before brings with it the responsibility, the unconditional obligation, to perform the equally free act of creating something new to replace it. But so far scientific thought has *not* fulfilled this obligation. Three fourths of the modern human population are held back from stealing and killing only because they believe that the eye of God is all-seeing and that eternal flames await offenders. Science has taken away the belief in this God and unfortunately also the belief in the flames of hell. Into the bargain, it has thrust marvellous technical developments into our hands for our misdeeds. Humanity as a whole finds itself at a critical and threatening stage of its mental development similar to that through which most of us must pass at the time of puberty. At this stage, people often *discard* the ancient, traditionally transmitted ethical ideals *before establishing their own ethical attitudes.* In the process, they are exposed to specific, quite severe dangers.

But the good old maxims such as "Thou shalt not steal," "Thou shalt not kill," and "Thou shalt not bear false witness" in fact have a justification *other* than the belief that anyone who disobeys them will end up in hell. This new justification is superior to the old belief in that it is true and not simply a fable dreamed up by adults in order to intimidate children who behave badly. The laws of ethics are founded on the *reality* of organic creation, on the eternal laws of development from lower to higher that govern all forms of life. This justification is *understandable* to any normal human being and can be *taught.* Nobody has to accept this as a mystic revelation. In fact, humanity should not *believe* in these ethical laws, but merely need to *know* them as facts: "For if I know a thing, I do not need to believe it!" Nowadays, scientific sociology has established a broad and solid foundation of knowledge about these laws, while our own discipline has discovered additional, more primitive laws of human social behavior at another, more basic level. In the final chapter of this book, we shall deal with those eternal and most general laws of creative higher development of organisms. There, it will be shown that on closer examination all human ethical laws, without exception, prove to be special cases or consequences of much broader and more general laws that govern the creative development of all living things. The investigation and clear abstraction of these laws *remains* the task of the natural science of the human species, remains as our task. But the task of establishing a *doctrine* on this basis is no longer that of the *researcher* but that of the *sage.* It requires extension from the objective observation of the existence of a great natural law into an obligatory "*Thou shalt.*" A new *system of human ethics* must be established as a free act in recognition of the real confused and threatened status of mankind. This is not a task for the natural sciences, for sociology, or for comparative ethology; it is a task for the science of human nature. It sounds almost arrogant to imagine that "something may be taught, which will improve and convert." Nevertheless, the attempt to tackle this titanic task is now the unconditional duty of the humanities. Prospects for a solution to this problem exist only if the humanities are open to the *reality* of the most internal workings of the human organism and if a *materialistic* philosophy develops from a genuine synthesis with the natural sciences.

II

Biological Prolegomena

5

General Attempts to Define Life

As we have already seen in the chapter dealing with the hierarchical system of the natural sciences [chapter 3], all life processes are in the final analysis chemical and physical processes of a quite special kind. Even today, we are still far from being able to state *what* it is that makes these processes special. It is, however, certain that the total phenomenon that we refer to as *life* involves a harmonious interaction between *a very large number* of physico-chemical processes *of many different kinds.* For this reason, all attempts to arrive at a *definition* of life through straightforward characterization based on a *single* principle are doomed to failure. In any case, every life process represents the end product of a unique *historical* development. For this reason alone, it is impossible to provide a simple, implicit definition of life. This is just as impossible as providing a definition of the Cheops pyramid or of mankind, both being phenomena that owe their particular properties to the combination of an incalculable number of converging causal chains. Nevertheless, attempts have been repeatedly made, and continue to be made, to take such complex phenomena, whose peculiarity resides in the *confluence and interaction of a multitude of constitutive individual processes* and to *define them on the basis of just one of them.* We shall be encountering many attempts to define mankind which in this way arbitrarily pick out individual phenomena to characterize human nature. In confronting these, we shall make use of these attempts to define mankind in the same way as we shall here make use of the better-known attempts to define life itself. In fact, all of these attempts to encapsulate a complex organic entity using a definition based on *one* of its constitutive aspects always fail to reflect the nature of the systemic whole and instead portray only *one facet* influencing the nature of that entity. Almost all attempts at a definition *in the humanities* have seized upon such individual aspects of life with intuitive, "prophetic" certainty. Even if the *definitions* that they have formulated are inadequate, we can nevertheless make use of them by simply stringing them together and thus compiling a list of the essential properties of life.

One such aspect of life that has been repeatedly incorporated into definitions is *metabolism.* A living organism is not an object in the true sense of the word, like a stone or a table, but a *process,* like a flame or a river. It maintains its individuality although it consists of different material components at any given moment of its existence. Goethe,* with the profound intuitive sight of a prophet, clearly recognized this constitutive aspect of life: "Indeed, I know whence I came!/Insatiable, like a flame/I devour myself and burn out./All I touch is set aglow,/Yet ash all that I bestow:/A flame I am beyond doubt."

Through all living substance there is a continuous flow of matter that is taken up and incorporated into that substance in a chemically modified form, while at the same time other matter is continuously expelled from the living substance. The first process of *annexation* of previously inanimate matter and its incorporation into the living substance of an organism always represents a synthesis of *more complex* and *species-specific* chemical compounds from unspecific, simpler components. In this way, the latter adopt the structures and properties of living material, being *made to conform* and thus themselves turning into living substance. This process of being made to conform in a specific way is labeled with the Latin term *assimilation.* The most significant processes of assimilative incorporation of organic substance are carried out by *plants.* By exploiting light energy with the aid of chlorophyll, they are able to generate all the compounds contained in living protoplasm from atmospheric carbon dioxide and simple salts contained in the soil. As is well known, animals cannot do this and for their own construction they always need compounds previously synthesized by plants, such as proteins, carbohydrates, fats, and so forth. Nevertheless, in this case too the actual process of assimilation represents a genuine synthesis of complex matter from simpler components, as the compounds supplied by plants must be extensively decomposed before they can be recombined for incorporation into the animal organism. In other words, all living substance must *die* and be reassimilated as inanimate matter when it is transferred from the framework of one organism to another.

In contrast to assimilation, the degradation and elimination of "spent" matter always takes the form of a *simplification* of chemical structure, which can proceed as far as the production of the simplest binary compounds such as CO_2 and H_2O. As a result, the substances concerned entirely lose

**The verses cited here are, in fact, from Nietzsche and not from Goethe.*

their species-specific character and they lose their resemblance to living substance, which is why we refer to this overall process as *dissimilation.* Assimilation and dissimilation usually proceed hand in hand. However, whereas dissimilation can also occur *in the absence* of assimilation processes, assimilation is necessarily associated with dissimilation processes. Assimilation processes, which—as has just been explained—always require construction of more complex chemical structures, are always *endothermic* processes of chemical combination. In other words, they require an input of energy in order to take place. Conversely, the chemical processes of dissimilation are *exothermic,* liberating energy when they occur. Therefore, it is a fundamental requirement that living substance must always be dissimilated in order to permit assimilation to take place, with the proviso that a gain in energy or living substance must always occur. Just as a merchant must always *sell* goods in order to generate free capital to buy new wares, an organism always obtains the free energy necessary for new assimilation through processes of dissimilation. Here, there is a further analogy to a flame, with which the heat generated by combustion is required to raise the temperature of as yet unburnt material to the ignition point so that it will burn in its turn. Accordingly, so to speak, the continual death of living substance is required to generate the energy required for the maintenance of living processes. For this reason, paradoxically, it is precisely the most conspicuous "life processes," particularly energetic manifestations such as body heat, movement, or muscular power, that are consistently based on processes of *dissimilation,* or molecular death. It was this fact that moved Claude Bernard to pronounce his well-known aphorism: "*La vie, c'est la mort.*" With all of these dissimilation processes that generate vital energy, the highly specific analogy to a flame resides in the fact that the great majority of them are *oxidation processes.* There are only a very few "*anaerobic*" microorganisms that can live in the absence of air that obtain the energy they need to live from other sources. They do this predominantly through nitration of nitrogenous substances, that is, using processes that are related not to combustion but to decomposition.

A further parallel between life and a burning flame resides in the following fact: Metabolism, which continuously generates from the interaction of endothermic and exothermic chemical processes the excess "operating capital" of freely available energy necessary for the continuation of the "business," *will tolerate no interruption!* If we cut off the supply of assimilable substances, dissimilation will continue to operate only as long as the reserves of the organism hold out. The organism burns itself up and then life is extinguished, just as a flame is extinguished when the combustible material runs out, and neither of them is able to rekindle itself. Just as combustible material can only be set alight through contact with a firebrand, the chemical substances of which living material is composed can only be converted into living substance if they are taken up and assimilated by a living organism.

At this point, however, the analogy begins to break down. Whereas combustion can be initiated by many other processes, particularly through the provision of heat energy from *whatever* source, *life processes can only be kindled by life processes.* So far, we know of no exception to this law! We do not know whether the flame of life on our planet was kindled by some as yet incomprehensible kind of self-ignition, through an entirely speculative process of "spontaneous generation," or whether faintly glowing sparks of this fire fell to earth. Both possibilities exist and at least one of them certainly must have occurred. But we can at least state, with a probability approaching certainty, that this process occurred a long time ago and *only once.** As we shall see in the chapter dealing with the evolutionary history of life, the vast numbers of higher plant and animal species are clearly derived from *one* common root.

A further essential property of all living organisms is the endeavor to *disperse* whenever possible, to *spread out* like a brush fire and—whenever permitted by environmental conditions—to *propagate* in an unlimited fashion. As has already been noted, under favorable living conditions metabolic processes always yield a positive balance of assimilation. Like a merchant who aims to make a profit out of buying and selling and to *expand* his business using this profit, all living matter has the inherent tendency to use the excess of assimilation over dissimilation to generate a surplus of species-specific living substance. In the life of the individual, this process results in *growth;* in the life of the species, however, it results in *propagation.*

"The tender spot, which gave life's sprout,/The gracious force, which then thrust out,/To take and give, itself to build,/At first at home but then afield. . . ." With these verses, Goethe has sketched the endeavor inherent in any living organism of generating a surplus of its *own* living substance. This essential endeavor of life to expand has been expressed by Nietzsche in a *broader* and much more abstract form in his concept of the *will to power.* He virtually used this as a definition of life or at least equated the two, when he coined the lapidary maxim: "Life is the will to power." There is, indeed, a close parallel here when an animal species, such as the rabbit in Australia,

**Note in the margin here:* effectively; later attempts perhaps eliminated by competition. ?bacteria polyphyletic?

encounters particularly favorable living conditions and then expands to cover the continent with billions of individuals. This resembles the expansion of a kingdom, such as that of Genghis Khan, overcoming the resistance of its competitors and extending out to reach its natural boundaries. Alternatively, it is similar to a business enterprise that achieves a monopoly through some stroke of fortune and increases its operating capital, and hence its power, to a massive extent. But, in a particular sense, these phenomena of growth and acquisition of power by an animal species, by a kingdom, or by a business enterprise are not quite normal; they are not "intended by the designer" (p. 226 et seq.).

Under normal living conditions, the tendency of any animal or plant species to expand in numbers is adjusted in such a way that some equilibrium exists with the *enemies* of that species—be they competitors for space, competitors for food, or predators. Massive population explosion of a given species, such as that of the rabbit in Australia, only occurs as a result of *disruption* of this biological equilibrium state. In this case, it occurred because the species concerned was artificially introduced to a region in which the usual competitors for space and predators were completely lacking. Such a population explosion of a given species is therefore something almost pathological. In the end, it turns out to be of no lasting benefit, just as the analogous sociological processes in world politics or in economic activity fail to produce any lasting, biologically healthy benefit to those who gain power. In these cases, too, the extravagant nature of growth leads to *disruption* of equilibrium in organic entities. In the final analysis, this always leads either to the downfall of the wildly expanding system or to the scaling down of its power to a physiological level that is compatible with the overall entity. Life as a whole can only achieve successes in further expansion as an *overall entity* consisting of animal and plant species in a biological equilibrium with one another. Similarly, in the specific human examples, a genuine and lasting gain in power is only possible within the framework of states of harmonious social equilibrium. Nevertheless, these almost pathological cases of wild expansion of a given organic system are generated by a basic tendency that, as such, is a property of life as a whole.

Life also exhibits analogies to a flame in the tendency toward unlimited expansion. Its expansion can be compared with the spread of a brushfire and life that chances upon a suitable territory that is not yet inhabited by other life forms behaves in a fashion very similar to fire. In fact, however, there are analogies in the inorganic realm that correspond to a far more important aspect of growth and reproduction, in that they are also based on processes of incorporation that show certain parallels to assimilation. A *crystal* growing in the mother liquor extracts from the disordered swarm of molecules dancing around it in the amorphous solution individual particles which are then incorporated *in an orderly fashion* into the strict framework of its own structure. In this case, the analogy to assimilation is far-reaching in that the "imposition of conformity" is equivalent to a *higher state of organization* with a greater and therefore generally less probable level of complexity, as is the case with all organic processes of assimilation.

It is precisely *this* feature that is *lacking* in another inorganic analogy to assimilative growth, which instead provides a better parallel to the process of expansion. If *amorphous* tin is brought into contact with a piece of tin in a crystalline state (of the kind that we see in the technically processed metal), the latter begins to decompose to the amorphous state at the point of contact and ends up as an amorphous white powder. The crystalline ordering of the molecules that originally held the piece of metal together disappears. This is in some ways similar to the way in which the organized framework of a knitted sock dissolves into "amorphous" wool as soon as it is disrupted in a single place, thus providing the *starting point* for rapidly spreading decomposition. Because conversion of metallic tin into the amorphous form thus operates in an "infectious" fashion and can hence have extremely vexatious consequences, it was early on given the name of "tin disease." This is, perhaps, the best illustration of the analogy between this process and organic phenomena. In recent years *research on viruses* has uncovered processes that do indeed show a certain affinity with this crystallographic analogy for the assimilative expansion of life. The virus responsible for tobacco mosaic disease consists of a *single* crystallizable chemical compound, probably an amino acid. Even the smallest quantity of this substance, probably down to a single molecule, has the ability to "assimilate" and to grow. When injected into a living tobacco plant, the virus removes the components from which it is made from the substance of the living cell and grows into long, branching chains of crystals. In this case, too, an essential character of genuine, organic assimilation is lacking in that there is no generation of a *higher* complexity of chemical structure, no production of higher-level states of organization. This character, in fact, already has an analogy in the growth of a crystal out of an amorphous solution.

In conceptual terms, the two great consequences of the surplus of assimilation over dissimilation—*growth* and *reproduction*—cannot be clearly separated from one another. With some bacteria lacking a cell nucleus which have the ability to grow in length in an unlimited fashion to form filaments under favorable nutritional conditions, growth and reproduction literally coincide. This is not very different at all from our inorganic analogue of "tin disease" or from the (as it were) semiorganic analogue of the tobacco mosaic virus. Higher unicellular organisms with a cell nucleus similarly grow continuously, but they *divide* as soon as they have reached a given size. In this case, increase in the number of individual organisms shows a closer parallel to the reproduction of multicellular animals and plants. A really sharp separation between growth and reproduction is first found with *multicellular* organisms in which the greater complexity of the systemic entity requires the *individuality* of each single organism. Literally speaking, it is in fact almost a contradiction in terms to refer to a unicellular organism that divides daily or even several times a day as an individual, as something that *cannot be divided!** It is only with the more complex structure of the higher multicellular organisms that the individual organism as a harmonious systemic entity becomes indivisible and it is only from this point on that a special mechanism for reproduction, sharply separated from that of growth, becomes necessary. The essential processes of life that have been discussed so far—metabolism and expansive, assimilative growth—have quite close parallels in the inorganic realm, as with the analogies of the flame and the crystal. By contrast, the manner in which higher organisms reproduce themselves is something that has no parallel in the inorganic world. Complex, harmonious systemic entities that generate simpler offshoots carrying the complete potential to develop into an equivalent complex entity—this is something that simply does not exist in the inorganic realm.

A further essential property of all life forms is the *holistic nature* of every organismal system. We can define a *holistic entity* as a closed system of structures and functions within which *each of these has a causal relationship to all others* and whose interaction is directed at the *maintenance* of these reciprocal relationships and thus at the preservation of the system as a structured entity. Such *regulatory systems of universal reciprocal causal connections* also exist in the inorganic world, as was noted by Wolfgang Köhler in his theory of "*physical Gestalts.*" A soap bubble, the electric charge on a spherical conductor, the solar system, and, indeed, the entire universe are all inorganic systemic entities of this kind. In each of them, everything is interrelated and, because of this universal interconnection, they maintain or restore their Gestalt as long as the harmony between the components is not greatly disrupted. If a small fraction of the charge on a spherical conductor is removed at any given site, this intervention affects the distribution of electricity at *all* other points. The equilibrium is initially disrupted, but it is at once regulated in

**Note in the margin here:* Mention: A corpse first occurs with *Volvox*.

the direction of the previously existing harmony such that the remaining charge shows a regular distribution over the entire conductor, as was previously the case. In a closely comparable fashion, the forces of surface tension and gas pressure regulate the "Gestalt" of a soap bubble. The liquid skin immediately regulates any inequality in its thickness that arises through removal or addition of fluid. A uniform distribution of the substance is restored with the remaining elements available and the same spherical shape is retained regardless of whether gas is blown into or extracted from the bubble. The spherical shape is also restored if external influences, such as air pressure from one side, have modified it. As a regulatory entity, the solar system can withstand major disruptions without losing its holistic harmony and it can "swallow" entire celestial bodies. Although a new comet that has wandered into the system alters the trajectories of *all* of its component stars to a small degree, it itself becomes a part of the system of universal causal connections. It is incorporated into the overall entity just as harmoniously as an additional droplet of soapy water is taken up into a soap bubble or a small additional charge is absorbed onto a spherical conductor.

However unquestionable it may be that such inorganic systems represent "total entities," in fundamental terms they provide relatively superficial analogies to the entity of a living organism. The analogy is by no means as substantial as that of a flame or of a growing crystal. Even with the very simplest of organisms, the regulative preservation of the individual and of the species is ensured by a "machinery" of cooperative organs and functions whose complexity greatly exceeds that of the most richly structured inorganic entity. The simplest bacterium and the lowest yeast cell are both immeasurably more complex systems than our entire solar system taken at once! There is not only a *quantitative* difference in that, even in the lowest organism, the *numbers* of components, the numbers of their interactions, and the numbers of *kinds of different* components are several hundred times greater than in the most complex "physical Gestalts" in Köhler's sense. Over and above this, there is a massive *qualitative* difference in that each of the "components" itself represents an entity, a harmonious system of universal reciprocal causal relationships. Indeed, this level in turn consists of further "components" to which the same applies and this principle extends through an immense hierarchy of graduated "subentities" extending down to the molecular level.

Within the exceedingly complex structure of this hierarchy of graduated subentities, every subentity is decidedly "constructed" to perform a quite specific *function* in the service of the governing entity. In other words, in the course of an historical developmental process a *division of labor* between the components of the entity has emerged. In the course of phylogenetic progress, these components have become increasingly *different* from one another, so that we refer to the overall process as one of *differentiation.* Goethe has aptly encapsulated this particular process in his definition of development as the *differentiation and subordination of parts.* In the course of the progressive division of labor between subentities, each one of them becomes increasingly and at the same time *more exclusively* differentiated to perform a specific function. For this very reason, every subentity becomes increasingly more *dependent* on all the other components and thus on the overall entity. The further the differentiation of an organ proceeds, the greater its subordination to the function of the whole must become.

As an extremely important consequence of this historical development of any organic system through the differentiation and subordination of its parts, it becomes possible (indeed, necessary) to raise the question of *purpose* in relation to all of its specific structures. We are confronted with the question of the so-called *finality* (from the Latin *finis:* purpose, goal, end) of an organ, a behavior pattern or a given structural feature. This question regarding finality *in the sense of preservation of the species* has of course nothing to do with the assumption that there is a supernatural "factor," inaccessible to causal research, that "directs" or "sets the goals" for organic processes. Instead, it amounts to a question regarding the species-preserving function served by the feature concerned, which—in the course of its phylogenetic development—led to its differentiation to acquire a particular form. When we ask *why* a cat has sharp claws, the question has the same significance and receives the same answer as when we inquire into the species-preserving function of these structures that resulted in their differentiation in that particular form in the course of phylogenetic development. The question *why* is therefore *only meaningful* if there has *really* been a process of development leading to the performance of a particular function that furthers the preservation of the organic systemic entity. If we attempt to apply this finalistic question to features or parts of inorganic entities, it immediately becomes clear that this is a pointless undertaking. Why is a soap bubble round? Why is the charge on a spherical conductor uniformly distributed? These are questions to which there is simply no answer.

The holistic nature, the finality, and the historical development of organisms—the three essential properties of life for which anyone is tempted to seek a definition—are so intimately linked together that none of them could emerge without the other two. Without the species-preserving, regulative maintenance of the system by the appropriate functioning of the subentities, organic systemic entities would be just as unthinkable as any finality of the components (or any kind of finality) without the existence of those entities. And both of these would be unimaginable in the absence of the developmental process that has led to the emergence of the harmonious entities represented by living organisms. This triad of historical development, holistic cohesion, and finality of all living organisms, which is an essential feature of life, fundamentally influences the kind of *analysis* of organic systems, the questions that are posed, and the methods that we apply in our field of research. This is true to such an extent that—in the interests of comprehensibility of later sections of this book—it seems advisable to devote three special chapters of Biological Prolegomena to them [chapters 6, 7, and 8] and to the conceptual and practical methods that they require.

One fundamental property of life which, as far as I am aware, has *so far never* been taken up in any of the definitions attempted in natural philosophy is the developmental sequence *from the simpler to the more complex,* which can be identified in virtually all branches of the phylogenetic tree of organisms. As has been particularly emphasized by [Max] Hartmann, a sharp distinction is necessary between the question regarding the evolution of species, referring to the causes of adaptation and the purely *species-preserving* finality of organic developmental processes, and the question as to why this developmental process, in the overwhelming majority of cases, leads to more richly structured, more complex, in sum "*higher*" forms of life. This form of directionality in phylogenetic development has nothing to do with the purely *species-preserving* purposiveness in the structure and function of organisms. It must be considered and investigated as a phenomenon in its own right.

The two terms *development* and *evolution* are, in fact, quite unsatisfactory in linguistic terms. Both are derived from a concept of unfolding in the *individual* lives of animals and plants, in which the original germ indeed gives rise only to an end product that was, so to speak, laid down in the basic plan established in the fertilized egg. In this respect, however, the phylogenetic origin of a higher form of life from a simpler ancestor has *nothing* in common with the development of a chicken from the egg. A bird was *not* present as an outline plan in the reptilian ancestors from which it originated, just as an outline plan for the human being was not present in the chimpanzee-like anthropoid ancestor from which we emerged. The most intrinsic feature of the process involved here is that it continually *generates something that is qualitatively entirely new.* The new product is in no way preformed or otherwise contained in the material from which it emerges, contrary to the false impression given by the words "development" or "evolution." In our language, there is no verb for this process other than *create* and no noun other than creation. In literal terms, the word *ectropy*, derived from Greek roots (ἐχ, out: τρέπω, to turn, revolve), means the same as development or evolution, but it has the great advantage in conceptual terms in that one will immediately think of the opposite term, *entropy*.* The entropy principle, the third main principle of physics, reads as follows: According to the laws of *probability*, all phenomena must progress from the improbable to the probable and therefore, because *order* is generally improbable, from the ordered to the disordered. With respect to the *transformation of energy*, in any process of force transformation there must be a gain in *disordered* molecular motion in the form of *heat* that can never be completely reconverted into another *ordered* form of energy of motion. Ultimately, therefore, over the course of time it is probable that *all* energy of motion will be irreversibly converted into heat. Eventually, all energy gradients must disappear, all movement must cease, and the universe must ultimately grind to a halt in "heat death," at a uniform, ungraded temperature.

Without offending against the principle of entropy in the physical sense, through its continuous process of development all organic creation achieves something that runs exactly counter to the purely probabilistic processes in the inorganic realm. The organic world is constantly involved in a process of conversion from the more probable to the generally more improbable by continuously giving rise to higher, more complex states of harmony from lower, simpler grades of organization. In using the terms *ectropy* and *ectropic* in the following text, we have in mind precisely this direction of phylogenetic development of virtually all living organisms running counter to the general probability of chaos.

As has already been indicated, the problem of the ectropic *direction* of phylogenetic development has nothing to do with the question concerning the species-preserving *finality* of organic structures. The process of so-called *adaptation* of an organism to perform a particular function in a given habitat is by no means directly identical to the development of *higher* forms of life. The view that higher organisms are *better* adapted, better suited for the struggle for survival is quite erroneous. The *species-preserving* purposivity of higher organisms is no greater than that of the lowest forms of life, and Jakob von Uexküll was entirely justified in stating that all living organisms are *equally well* adapted to their environments. In fact, one could more justifiably reverse the widespread view and state that the survival of higher forms of life is generally *more* threatened than that of lower organisms. To

**Note in the margin here:* Limitation of the entropy principle in modern physics only classical!

put it quite crudely: The human species is without a trace of doubt more directly threatened by extinction than the slipper animalcule *Paramecium* or the filamentous alga *Spirogyra!* Even in the realm of human life itself, the highest achievements of ectropic creation have nothing to do with advantages for the preservation of the species. In very many cases, they arise not to serve species-preserving purposivity but in spite of this and to its detriment. Even in the realm of prehuman organisms, ectropy leads to developments that are "only just" compatible with species-preserving purposivity, which sets their upper limits. Certain examples of this will be presented later on. The same principle applies in an even more extreme form to the sphere of human creativity. It would amount to a major misconstruction of the nature of all of the products of the highest ectropic creation, for example, to inquire into the survival value for the species *Homo sapiens* of Leonardo da Vinci's *Last Supper* or one of Beethoven's symphonies.

The ectropically directed development of all organic matter is the basic phenomenon underlying everything that we humans perceive as *values.* Most of us do not realize that a decisive *value judgment* is involved when the first volume of an animal encyclopedia refers to *lower* animals! The value judgment of human beings has simply expressed at a conscious level something that *all* living organisms have been doing since the very beginning in developing from lower to higher levels. The *tension* between lower and higher, between that which is and that which should come to be, is also the strongest driving force behind human creation of values. In the last section of this book, dealing with human striving after values, we shall return to consider the ectropy of the organic world in more detail, so no more will be said at this point. [*See footnote on p.* xliii.]

We have now arrived at the last, most puzzling and—from the point of view of comparative behavioral research—most important of all properties of life. This property, which has been involved more than any other in attempts to define life, is the subjective experience of living organisms. Among all of the properties of life, it is without doubt the existence of the *experiencing* subject, the presence of an "ego," that represents the sharpest distinction between the living and the nonliving. At the same time, it is also the property that is least accessible to the forms of thought and intuition with which we are endowed. *Causal,* physiological investigation of life processes as physicochemical reactions does not bring us one whisker closer to an understanding of the nature of subjective *experience* of the *soul.* One side of the nature of all life processes that appear as experience in the realm of our egos is that they are simultaneously and unconditionally *physical* processes in the central nervous system. Without any doubt, the physical and the experiential processes represent *only one single* phenomenon. Yet there is no logical connection between these two incommensurable sides of that phenomenon that is accessible to our mental examination. As Max Hartmann has said, the relationship between the parallel physical and experiential processes is alogical. We can clearly express this relationship by means of the following thought experiment: Let us assume that we have succeeded in applying the methods of physiological causal analysis to arrive at a complete explanation of the physiological side of some given nervous process accompanied by particular experiential processes. Even if we assume that we have *really* succeeded in formulating a *complete* explanation extending down to physicochemical processes as far as the atomic level, this would still have brought us *no closer* to an understanding of why these *causally completely understandable* processes are associated with these particular experiential processes and with no other! Although we have cogent reasons for our conviction that only *one* real process is in fact involved, the experiential side of this is accessible only to the inward-directed, subjective view. By contrast, the physiological aspect of *the same* process can only be approached through causal analysis involving the methods of inductive natural science.

The *parallel occurrence* of all mental experience with physical life processes has led many great thinkers to the assumption that this principle is *reversible*, that every single life process is accompanied by experiential processes. As has already been demonstrated in some detail in the Philosophical Prolegomena, in the chapter dealing with the consistent hierarchical system of the natural sciences, this conclusion is erroneous. Just as it is true that not all physicochemical process are life processes, it is also true that not all life processes involve *experiential* phenomena! It lies in the nature of the ectropic development of living organisms that there is continual generation of qualitatively new features that simply did not exist previously, not even in a primordial form. Understandably, it is precisely the highest levels of ectropic creation that distinguish living organisms from the inorganic realm. Nevertheless, it would be a quite indefensible mistake to conclude from this that the distinguishing feature concerned must therefore apply to *all* living organisms as an essential character of life. Developmental transitions *within* the series of living organisms also represent sharply delimited steps. The continuous *quantitative* increase in the complexity of organisms generates a quite *discontinuous* augmentation of functions that are *qualitatively* completely new. For any unbiased thinker, it must be a self-evident fact that the life processes of the lowest animals that lack a nervous system, like those of the plants, proceed *without* accompanying experiential processes. Conversely, no researcher concerned with the real world who has personally observed the behavior of living organisms such as dogs and chimpanzees could doubt for a moment that they possess subjective experience closely similar to our own. Without a shadow of a doubt, the transition from nonexperiential to experiential processes took place as a qualitative shift like many others *within* the series of living organisms. Even if, from a human standpoint, this transition seems to be just as massive and as significant as that from the inorganic to the organic, we must not allow ourselves to be misled into equating the two shifts.

Subjective experience and its relationship to parallel physical processes is of decisive significance for the theoretical and practical methods of comparative behavioral research. This is true to such an extent that, for an understanding of later sections of this book, it is necessary to devote a special chapter to it here in the Biological Prolegomena, as for the historical development, the holistic nature, and the finality of life processes. The principles of our research strategy include recognition of the fact that we should by no means reject the experiential side of psychophysiological processes as an important source of knowledge. Instead, such processes should be investigated, *simultaneously* wherever possible, from both objective and subjective standpoints.

We have now encountered no less than seven features that are fundamental to life. Almost every one of them has been used in the attempt to provide a definition of life. Metabolism, assimilative expansion, a holistic character, species-preserving finality, an historically unique evolutionary history, and subjective experience have all been cited in such definitions. As far as I am aware, only the phenomenon of ectropy has never been involved in this way. At the end of this chapter, it only remains for us to demonstrate briefly why all of these attempted definitions are doomed to failure as isolated undertakings. Definition of life exclusively according to the phenomenon of *metabolism* is not feasible because there are inorganic analogues that show the same phenomenon. The same applies to the tendency of all living organisms to show *assimilative expansion.* In this case, too, we have recognized the existence of inorganic analogues which would erroneously also be included in such a definition of life. However, if we were to take as a basis for a definition the special form of reproduction that is found with higher muticellular organisms, this would exclude the inorganic realm. Not even the most complex and most highly regulated inorganic systems are capable of replicating themselves by means of simplified offshoots. Even planetary systems do not lay eggs that give rise to new systems. But bacteria and yeast cells are also incapable of doing this, and such a definition would therefore also exclude from the realm of living organisms certain systems that undeniably belong in this category. It is similarly true of the regulative holistic character of living systems and of their historically unique origins that inorganic analogues can be found. Finally, this also applies in a certain sense to system-preserving purposivity. As a result, none of these phenomena taken in isolation suffices for a definition of life. The situation is somewhat different with respect to the phenomenon of ectropy. There are quite a number of living organisms that have arisen through phylogenetic modifications that undeniably represent *processes of reduction* that have proceeded in a pronouncedly nonectropic direction, from the complex to the simpler. With such living organisms, however, special circumstances are involved.* As a general rule, only *parasites* have undergone development in the direction of a pronounced simplification of their plan of organization. In parasitic worms, the digestive tract and the nervous system have been reduced to the point of complete disappearance. In the parasitic crustacean *Sacculina carcini,* the entire complex organization of the crustacean body, which is still completely represented in the larval stage, has completely disappeared in the adult. The latter consists of no more than a bundle of funguslike threads, which penetrate the body of the host (the shore crab) in all directions, even reaching the tips of the extremities.

While parasitism leads to various kinds of reduction, conversely under certain conditions pathological reduction in individual cell groups in the bodies of higher organisms can lead to a kind of parasitism. *Tumors,* or cancers, which play such a major and pernicious role in human pathology, arise because certain cells *reverse* their differentiation and subordination to the systemic entity. Like undifferentiated *embryonic* tissue, they begin to undergo new waves of cell division. The further this process of reduction proceeds, the greater becomes the degree of *immaturity* of the cancerous cells (in the pathologist's terminology), and the greater is the *malignancy* of the tumor. The latter behaves exactly like a parasite within the host's body, growing at its cost and eventually destroying it. Genuine processes of reduction of preexisting complex structures are also found in the form of so-called *domestication effects* in domestic animals and in human civilizations. All of these nonectropic features of living organisms have one thing in common: *they are incapable of existing in their own right.* In other words, they are only conceivable as long as they can parasitize an organic system that has not been affected by such reduction effects, or can at least—like a human being in relation to a domestic animal—*replace* the lost functions as a "symbiont."

**Note in the margin here:* As long as they have survived—"reduction" proceeding to extinction should not be excluded!

It is true that we cannot *define* life according to the phenomenon of ectropy, because such a definition would unjustifiably exclude from the ranks of the living not only the tapeworm, *Sacculina,* and the domestic pig but also asocial criminals affected by reduction of social behavior patterns. Nevertheless, this approach to a definition does concern something that is of more fundamental significance for the nature of life than the other definitions derived from a single constitutive aspect of life processes. In the course of its progress and proliferation, any development running counter to ectropy would gradually lead from the organic back to the inorganic level and hence to death. It is possible that *viruses* are engaged in precisely this trend. They are regarded by some research workers as *reduced* components of living cells that have become detached from and independent of the cell body in a manner analogous to the separation of tumors from the holistic system of a multicellular organism. In the last part of this book, we shall have to deal with the effects of domestication-induced reduction in the species-specific *social* behavior of civilized mankind. The survival of modern human cultural systems is just as immediately threatened by such effects as is that of a multicellular organism by a malignant tumor consisting of reduced cells developing within its body.

As can be seen, life cannot be encompassed by an implicit definition; that is, by one which sets out from a single principle. Life is characterized by an infinitely complex interaction of many different processes, not one of which can account for life in isolation. Even this combination of various phenomena that are essential features of life cannot be taken as basis for its definition for the simple reason that we can by no means claim that we have recognized *all* such phenomena. There could still be many more that are just as important as those that have been listed here but which have so far escaped our attention. In any case, in our present state of knowledge we are far from being able to review all of the interactions and interrelationships even between the best-known and best analyzed features of life. Of course, this fact certainly does not permit us to claim that life is *fundamentally* inaccessible to explanation and definition, that is, simply a miracle. Quite to the contrary, we have every reason to assume that the great laws of the two exact sciences are subject to no exceptions in the realm of life processes. In other words, life as such is fundamentally accessible to explanation.

6

The Unique Historical Origin of Organisms and the Phylogenetic Approach

The laws governing the inorganic world have no historical component. Individual *things* in the inorganic realm, such as a stone, a planet, or a solar system, do have a historical past that accounts for their existence in a particular condition. However, in the inanimate world even a chain of events extending over eons would never give rise to new, more complex, and more restricted laws. Universally, the kinetic energy of a moving body is and always has been M $v^2/2$, its mass [M] multiplied by half the square of its velocity [v]. The combination of hydrogen and oxygen has always yielded water and will always do so; it does so everywhere, in Altenberg just as in the Orion nebula. It can be *inferred* from the nature of the elements that this *must* be universally the case and why this is so.* By contrast, it is *not* an inherent feature of the elements constituting living protoplasm that they should have given rise to the particular organisms inhabiting our planet. It cannot be inferred from the properties of carbon, nitrogen, sulfur, oxygen, and so forth, that they would necessarily give rise to human beings or oak trees. The fact that human beings or oak trees evolved is simply the result of a unique, specific historical process that followed a particular course and no other. The results were not at all predestined to follow a particular course by the properties of the stages that provided the basis for the evolutionary process. Under somewhat different conditions, quite different results might have emerged. The fact that this is, indeed, the case is illustrated (for example) by the evolution of koala bears and eucalyptus trees in Australia. The rich confusion of causal chains that have contributed to a successful outcome, which have led to particular results and no others, contain such a large element of *chance* that the so-called "historical" or "nonrationalizable residue" is very large. As a result, we shall never fully know *why* human

**Note in the margin here:* But: it is not an inherent feature of protons and electrons that these particular elements: In other [. . .] rare but nevertheless: The org. itself

beings and oak trees evolved in the Palearctic region, whereas koala bears and eucalyptus trees evolved in the Australasian region.* Insofar as we are able to rationalize the particular form of individual organisms *at all,* this can only be accomplished by reconstructing the historical course of their evolution. Take, for example, the question of *why* human beings possess a hyoid bone with one large pair and one small pair of "horns" attached to the base of the skull by ligaments. The legitimate answer, really addressing the main cause, lies in the observation that this is so because human beings are descended from gill-breathing ancestors and because the first two gill arches underwent a particular process of transformation of form and function that eventually led to the human hyoid. *Every* question about the causes underlying the particular condition of *any* structure or function of an organism will receive a similar, fundamentally historical answer. All causal analysis of the structures and functions of any living organism is unconditionally obligated to engage in investigation of the phylogenetic origins of the features concerned. This is most particularly true of the investigation of the innate, species-specific behavior patterns of animals and humans, which is the primary task of comparative ethology. For precisely this reason, we must deal with the nature of the phylogenetic approach in considerable detail.

How do we in fact *know* that life on earth has evolved through an historical process and does not owe its existence to a unique act of creation that laid down a particular set of forms for all time? How do we arrive at the assumption that all plant and animal species, in their multitude of different forms, represent the living, continually developing and further subdividing tips of the branches of an ancient phylogenetic tree and that they are therefore *related* to one another? As a rule, the educated layman will tend to assume that the source of our knowledge of this fact resides in geology and paleontology, that is to say, in the investigation of the earth and its layers, along with the fossilized remains of prehistoric plants and animals contained in those layers. Yet, however important the results from these branches of research may be for the theory of evolution, the concept of evolution did not in fact arise from them. Instead, in a quite remarkable fashion, the initial impetus came from *systematic zoology.* The description and ordering of concrete individual things in this case led to recognition of one of the really great laws of nature in a virtually compelling fashion. The way in which this happened provides one of the classic examples of the manner in which the *nomothetic stage* develops in an organic fashion from the *idiographic* and *systematic* stages in the natural sciences.

**Note in the margin here:* The same applies to minerals. Formation is just quantitative.

To the educated ear, attachment of the label "*comparative*" to anatomy, morphology, histology, physiology, and even psychology has a very familiar ring. Nevertheless, very few nonbiologists have any clear idea of the meaning of the verb "compare" in this context. Even the editors of the American *Journal of Comparative Psychology* were unversed in this respect. In the sense understood by the evolutionary biologist, comparison means *the ordering of living organisms according to their degrees of similarity, which have arisen as a result of their phylogenetic relationships.* As the author of a genuine precursor to the concept of evolution, Johann Gottfried [von] Herder provided an illustrative analogy to the nature of this methodological approach. Let us assume that we have arrived in a previously unfamiliar farming village and that we are contemplating the similarities and differences between the inhabitants. We then find among these inhabitants a whole series of groups in each of which the members are far more similar to one another than they are to those of other groups. Smaller groups of children show an even higher degree of similarity to one another and two of them are virtually identical. Under these circumstances, we will not go far wrong in assuming that the latter two children are twins, that the groups of similar children generally contain siblings and that the larger groups of people resembling one another represent genetically related members of individual clans. Even if we should make a mistake in drawing these conclusions, the error will usually only concern the *degree* of relationship. We shall sometimes mistake siblings for cousins and vice versa, but only in relatively rare cases will a *chance* correspondence in characteristics create the impression of a close relationship where none is, in fact, present. But now we notice among the blond-haired inhabitants of a village—let us say in northern Friesland—a clan of conspicuously dark-haired, dark-skinned people. On inquiry, we discover that most of them have Italian names. Without a doubt, therefore, they are descended from an Italian who turned up in the village at some point and then took up residence there. If we now follow this train of thought further, we find that *all* northern Frisians and *all* southern Italians resemble one another more closely than they do members of the other group! These similarities and differences apparently indicate that the *degree of genetic relatedness* among all Italians and all Frisians is greater than that between Italians and Frisians. Taking it even further: It is almost impossible to confuse an Italian with a Frisian, but the former could more easily be taken for an inhabitant of southern France and the latter for an Englishman. Herder therefore arrives at the conclusion that all human races are genetically related to one another according to the degree of their similarity and have arisen from a common ancestral stock in an ever-branching family tree. In this way,

Herder came extremely close to a formulation of the theory of evolution in its modern form. However, in his conclusions he stopped at the level of the species, which for him was something unchangeable that had been created by God at one point in time and had remained static ever since. It remained for Charles Darwin and, independently, [Alfred] Wallace, to follow Herder's train of thought through to the end and to recognize the fact of the common origin of living animals and mankind. But the *method* that led to this result, which has remained the method underlying all phylogenetic research right up to the present time, is in no way different from the procedure followed by Herder. The procedure that we use to arrive at our interpretations of phylogenetic relationships is always the same. In our comparisons of living organisms, we seek corresponding characters that owe their correspondence to the fact that they have been inherited from a common ancestor. In other words, they arose in a branch of a phylogenetic tree that existed at a time when the two or more modern forms were not yet separated from one another, thus representing a single "twig" of the tree. Such characters are referred to as being *homologous* and we shall later consider in detail the methods that must be used to distinguish them from correspondences of other kinds. First of all, however, we shall take a concrete example and, in a didactically simplified fashion, show how phylogenetic analysis leads from the comparison of homologous characters to the reconstruction of an evolutionary tree for a given group of plants or animals. The example selected is that of the vertebrates, because these are organisms for which the nonbiologist has a relatively clear idea of their structure.

We can begin by following Carl von Linné [Linnaeus] in his general systematic division of the vertebrates into five classes. It is seen that one great group of vertebrates breathes through gills and possesses a single-chambered, venous heart, two eyes constructed according to the principle of the camera, a static organ equipped with three semicircular canals, a labyrinth, a propelling fin at the end of the tail serving as the main organ of locomotion, and so on. We refer to this class as fish, or Pisces. Members of a second class only breathe through gills in the juvenile stage and their gills differ in construction from those of fish. They possess two lungs, the static organ is supplemented by a hearing organ with an eardrum, and instead of paired fins these creatures have four legs with four clawless digits on the front pair and five on the hind pair. The ribs do not enclose the lungs in the form of a rib cage and the air that is breathed is pumped in using the musculature on the floor of the mouth. The skin is naked, only slightly cornified, and kept slimy through the secretion of numerous glands.

For this reason, the amphibians (Amphibia) are dependent on the immediate availability of water. In reptiles (Reptilia), four extremities are similarly present and the limb skeleton consists of bony elements homologous to those of amphibians. In reptiles, however, both the fore- and hind limbs bear five digits and these—in contrast to those of amphibians—are armed with claws. Reproduction takes place by means of thin-shelled, yolk-rich eggs and the offspring that hatch from them are completely developed, showing no difference from the adults. The eyes possess lids, which are lacking in fish and amphibians, and the eardrum is recessed. The skin is covered with a thick cornified layer and reptiles are therefore to a large extent independent of water. The ribs surround the lungs in the form of a more-or-less complete rib cage which is expanded through muscular contraction to *suck* air in, just as is the case with birds and mammals. The next two classes of vertebrates contain warm-blooded species. In the birds (Aves), the skin is covered with feathers, the forelimbs have been converted into wings, the hind limbs possess four toes, there is a four-chambered heart, and all of the features listed for reptiles are also present. The mammals (Mammalia) possess hair, nourish their offspring with the secretion of milk glands, and so on.

Even in this brief outline of the five classes of vertebrates, which has been deliberately simplified, it was unavoidably necessary to use characters that clearly reveal the *inequality* of those characters. For example, *one* class is characterized by fins while all of the others are united by the possession of four limbs, which is doubtless a highly significant feature. For this reason, they are combined in the Tetrapoda, which defines them *collectively* as a group equivalent to the fish. Birds, reptiles, and mammals resemble one another more closely than any one of these groups resembles the amphibians, and so forth. But now, following the lead established by Herder, we shall engage in a critique of Linné's systematic arrangement of the vertebrates. Let us assume that the broader the *distribution* of a homologous character is among the members of a cohesive group, the *older* it is. The vertebrates share many characters that are doubtless homologous. They all have in common a basic plan of the axial skeleton and of the nervous system, combined with the origin of both of these from the endoderm and the ectoderm, along with the development of a mesoderm and the body cavity or coelom (figure 1). This form of development of the germinal layers is found in various patterns in all animals that we refer to as Chordata because of their common possession of an axial skeleton, the *dorsal chord.* The lowest chordates are "still" far from being something that we would generally regard as "vertebrates." The entire group known as Tunicata—which includes the sea

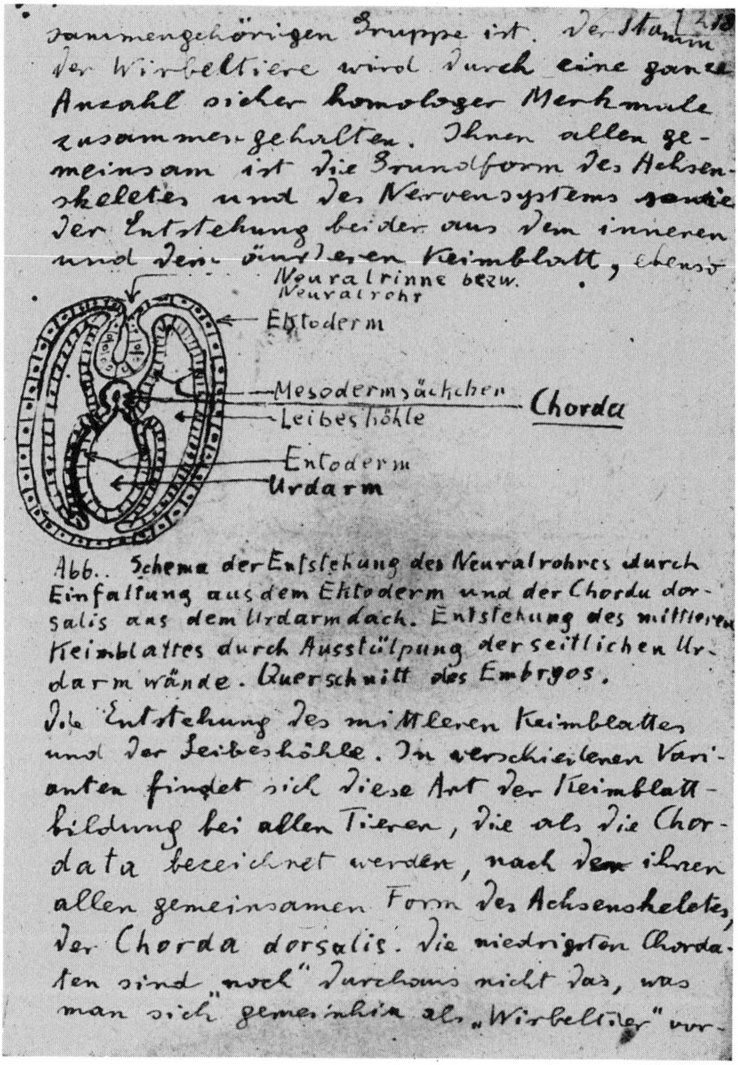

sammengehörigen Gruppe ist. Der Stamm [218]
der Wirbeltiere wird durch eine ganze
Anzahl sicher homologer Merkmale
zusammengehalten. Ihnen allen ge-
meinsam ist die Grundform des Achsenskeletes und des Nervensystems sowie
der Entstehung beider aus dem inneren
und dem äußeren Keimblatt, ebenso

Abb.. Schema der Entstehung des Neuralrohres durch
Einfaltung aus dem Ektoderm und der Chorda dorsalis aus dem Urdarmdach. Entstehung des mittleren
Keimblattes durch Ausstülpung der seitlichen Ur-
darmwände. Querschnitt des Embryos.

Die Entstehung des mittleren Keimblattes
und der Leibeshöhle. In verschiedenen Vari-
anten findet sich diese Art der Keimblatt-
bildung bei allen Tieren, die als die Chordata bezeichnet werden, nach der ihnen
allen gemeinsamen Form des Achsenskeletes,
der Chorda dorsalis. Die niedrigsten Chordaten sind „noch" durchaus nicht das, was
man sich gemeinhin als „Wirbeltier" vor-

Figure 1
Diagram of the origin of the neural tube by infolding from the ectoderm and the spinal cord from the gastrocoele. Formation of the central blastema by extrusion of the side walls of the gastrocoele. Cross-section of the embryo.

squirts, salps, and Larvacea—are shapeless, headless marine organisms. To the layman, their phylogenetic relationship to the vertebrates is far from obvious. Only the Larvacea have a persistent chord in their propulsive tail, whereas the sea squirts have become sessile "filter feeders" that lead a life very similar to that of the sponges (for example). But all of the tunicates share with the lower vertebrates a further, extremely important characteristic. In the most anterior part of the digestive tube, there are gill slits which allow respiratory water that enters through the mouth to escape through the sides of the body. However, these gill slits do not open directly to the outside. Instead, they lead into a common antechamber, the so-called peribranchial cavity, which communicates with the surrounding water through a single expiratory aperture. Tunicates *lack* a head, a segmented structure, bilateral symmetry, and a whole series of additional features that are characteristic of vertebrates. Although the tunicates generally resemble invertebrates in their general appearance, the so-called lancelet (*Amphioxus*) represents a genuine connecting link between invertebrates and vertebrates. This remarkable creature, which is of great "fascination" for those concerned with phylogenetic reconstruction, possesses an utterly fishlike body with close similarities in its segmentally organized musculature, dorsal chord, neural tube, body cavity, mouth, and anus. The gill apparatus resembles that of the tunicates inasmuch as there are *rows* of small gill openings instead of the serial gill slits of fish. Further, the gill openings do not open directly to the outside, as in fish, but into a peribranchial cavity, as in tunicates. Because of their laterally undulating movements, they closely resemble fish in their general habitus, as is indicated by their common name. But they lack the most important of the organs found in true vertebrates: a *head!* The dorsal chord extends right up to the very front of the body and the anterior part of the neural tubes terminates with a barely perceptible thickening. That part of the neural tube that is located *in front of* the anterior end of the dorsal chord to form the actual *brain* of vertebrates is completely lacking. Accordingly, the group consisting of two or three lancelet species is referred to as Acraniata (lacking a skull). It doubtless represents a branch of the evolutionary tree of chordates that is *systematically equivalent* to the Craniata, the group of vertebrates possessing a skull. All craniates possess a head with two eyes and two labyrinths composed of semicircular canals. The neural tube always projects in front of the anterior end of the dorsal chord, where the so-called *hypophysis* or pituitary gland is located. The gill apparatus opens directly to the outside world through gill slits, and at least the first pair of these is always retained in the form of ear openings. The general structure of the body—with segmentally arranged musculature, a body cavity (coelom) and other features—shows a basic pattern that is entirely consistent with that of *Amphioxus.* All of these characters and many others are certainly homologous. If we now follow Herder's initiative and (so to speak) *link up* living organisms according to their shared features, as shown in the outline given in figure 2, the Craniata emerge unequivocally as a common stem. But, as we shall see, this stem does not in fact divide up into five equivalent branches. Instead, within the "class" of the "fish," the craniate stem divides to yield a group of unusual creatures that stand in contrast to all other craniates just as the lancelet is opposed to all craniates. This is the group of round-mouthed fish (Cyclostomata), which contains only two genera and about four species, including the river lamprey, the sea lamprey, and the hagfish. These animals lack a large number of important organs that are common to *all* other vertebrates. Above all, they lack the *jaw apparatus,* formed from the palatoquadrate and the mandible on each side, which is found in all other members of the Craniata. They possess only a round mouth opening, resembling that of many worms, equipped with a ring of keratinized teeth. In addition, they lack the two pairs of extremities that are found in all vertebrates, from fish to humans, inasmuch as they have not undergone secondary reduction. Such reduction has occurred in the conger eels among fish, in caecilians among amphibians, and in slowworms and snakes among reptiles. Instead of the paired nasal openings found in other craniates, the lampreys have an unpaired olfactory pit. Further, their labyrinth contains only two instead of three semicircular canals and the brain lacks the hindmost part, such that this region, giving rise to the last three pairs of spinal nerves, is (so to speak) "still" spinal cord. The structure of the gill apparatus is extremely interesting. The adult fish possesses gill slits that open freely to the outside. Although there are considerably more of them than in any of the true fish, the basic structure is the same. In the larval stage of lampreys, by contrast, there is a peribranchial chamber, as in *Amphioxus* and other tunicates, which clearly supports the possibility that the vertebrates are derived from similar forms. Without a doubt, the lampreys represent a branch of the evolutionary tree that is systematically equivalent to all other vertebrates, which are known as jawed vertebrates or Gnathostomata. An attempt has been made to interpret the lampreys as true fish in which the gnathostome features listed above have all been lost through processes of reduction. This can be seen to be greatly contrived, as paleontology makes a decisive contribution here: The oldest fossil fish from the Silurian are derived from round-mouthed forms with numerous gill slits that lack paired extremities. They are surely very closely related to modern cyclostomes.

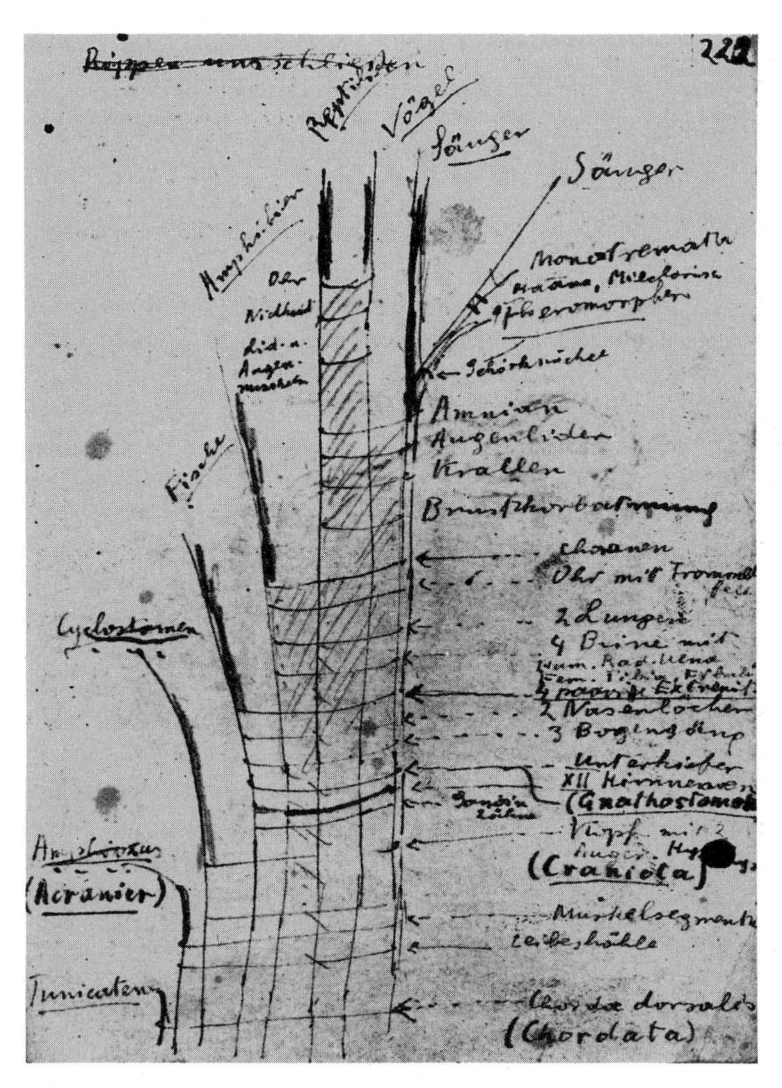

Figure 2
Diagram of the evolutionary tree of the vertebrates.

It seems somewhat incongruous, not only to the nonbiological layman but also to a scientist concerned with phylogenetic analysis, that a division into Cyclostomata and Gnathostomata gives the impression that the lampreys are less closely related to other fish than the latter are to humans. For this reason, it is advisable to discuss a particular difficulty of phylogenetic systematics at this early stage. The lamprey is, indeed, obviously a "fish" with its eel-like body, its big fish eyes, its actively breathing gill slits, its silvery skin, its two-chambered heart, its median fin and its sinuous pattern of movement. In examining a phylogenetic tree, however, we must always remember that it is *at least* a *three-dimensional* structure.* In terms of its evolutionary level, that is to say, when seen in what is in effect a transverse section of the phylogenetic tree, the lamprey is not very far removed from the ancestors that it shares with the gnathostomes, and these ancestors were doubtless "fish."†

Further common characters of the amniote stem, which anyone can recall from his own observations, are the possession of five digits on the anterior and posterior extremities, the presence of eyelids, inspiration of air with the aid of the rib cage, and the cornified layer of the epidermis, which renders these animals largely independent of water. Further, in all three amniote classes the skin bears appendages that are developed in connection with the cornified layer, such as the digital claws, scales, feathers, and hair.

Among the amniotes themselves, the birds and reptiles—often collectively referred to under the name Sauropsida—are more closely related to one another than either is to the mammals. In their bony structure, particularly that of the pelvis, the birds exhibit a great number of definite homologies to the dinosaurs of the Jurassic period, and they are without a trace of doubt descended from them. By contrast, the mammals originated from the reptilian stem very "deep down." The advanced mammal-like reptiles, or Theromorpha, from which they evolved show a quite remarkable mixture of primitive features of cotylosaurs or stegocephalians with mammalian features. They share with the latter a great many definitely homologous characters in the skull and dentition. Above all, they have already undergone the peculiar functional modification that converted bones of the *jaw*

**Note in the margin here:* It should be noted here that living species are always "twigs" and never ancestral forms.

†*At this point, pp. 226–229 (roughly two typewritten pages) of the original manuscript were missing. Fortunately, the same theme is covered in: K. Lorenz, Vergleichende Verhaltensforschung. Vienna, Springer-Verlag, 1978, p. 66, et seq. The illustration of an outline evolutionary tree in that publication (Fig. 6, p. 65) is also the same as figure 2 of this book.*

hinge, the articular and quadrate, into two ear ossicles, the malleus and incus. The theromorphs and the most ancient mammals were already present in the lower Triassic, that is, in the period immediately following the Permian, the formation containing the cotylosaurs. This was long before the upper Jurassic, which was the period when the birds originated. The extremely close relationship between members of the Sauropsida can easily be illustrated for the layman simply by demonstrating the structure and function of the eye and eyelid system in a lizard and a bird. In both, the nictitating membrane—which has usually been reduced in mammals—keeps the transparent cornea moist and clean. Further, both possess the same complex muscular system that moves the opening between the eyelids jointly with the eyeball whenever the eye is moved. By contrast, in all mammals (including humans) the opening between the eyelids is firmly attached to the bony cavity of the eye, such that the eyeball moves independently of that opening. Such complexes of characters that are completely homologous in birds and reptiles are found with equal frequency in the anatomy and in the behavior of these animals, as we shall see in due course in the special part of this book. The concept of the Sauropsida as a group that is, at most, of equal systematic significance to that of the mammals is entirely justified.

The mammals can thus be seen as a relatively ancient branch of the amniotes that diverged at an early stage. The *warm-bloodedness* of the mammals and birds is surely not a homologous feature derived from a common ancestor, but was independently "invented" in the two lineages. It is, so to speak, a didactically unfortunate coincidence that members of the mammalian subclass *Monotremata*—which, like all primitive amniotes, have a single common opening for feces, urine, and reproductive products and lay eggs—also happen to have a *bill.* The combination of egg-laying with the presence of a beak has, in many circles, led to the opinion that these most remarkable mammals represent an intermediate between mammals and birds. In reality, the peculiar primitive characters of the monotremes are, of course, *reptilian* features, while the bill is analogous, not homologous, with that of the birds. Embryos of the platypus have dental laminae that later degenerate. These are quite similar to those of true mammals, rather than resembling those of reptiles, and there is a particularly close resemblance to those of the extinct mammalian order Placodontia. In systematic schemes, the remaining "true" mammals are not usually contrasted as a unit with the monotremes, although the remaining two subclasses of *marsupials* (Metatheria) and *placentals* (Eutheria) share a large number of features that are lacking in monotremes and are familiar to anyone from the overall image of a mammal.

As we have seen, categorization of the various kinds of vertebrates, in the complete absence of any preconceptions, according to general and special homologous characters does not in fact yield five equivalent "classes." Instead, it yields a diverse, treelike structure with a highly irregular branching pattern which differs from the regular architecture of real trees in one essential point: The *thickness* of the branches, which in our diagram represents the number of animal types belonging to any given lineage, is not at all proportional to the *length* of the branch, which in our diagram indicates its geological *age.* This is not the case with a real tree, in which an old branch that diverges low down on the trunk must necessarily increase in thickness as its length increases. Instead, quite "thin" branches consisting of only half-dozen species, such as that of the cyclostomes, extend from the Silurian right up to the present time. It is necessary to emphasize this for the following reason: There is a widely distributed conscious or unconscious assumption, even among those who are professionally concerned with systematics, that the number of living species belonging to a given group exerts an influence on its so-called *systematic value.* But it is, in fact, a matter of taste for the individual to decide whether a group should be given the value of a phylum or class, or indeed that of a subclass or order. The cyclostomes have actually been accorded all four of these labels in different classificatory systems. It is possible to have entirely different opinions about the value of the *characters* of a group. The only objectively determinable, *quantitative* parameter that can be used to assess its systematic value is its *age.* However, as can be seen from the diagram in figure 2, this age is not even approximately equivalent for any two branches in the tree. In other words, there are no two groups that really have the *same* systematic value.* For this reason, a systematic classification into equivalent groups is a—necessary—*artifice* that gives rise to certain theoretical difficulties, to which we shall devote a special discussion later on. Only one point should be noted here: there *is* no "natural system" other than graphic representations such as that presented here.

In fact, the categorization of animal species without any preconceptions according to general and special homologous characters leads on to something else, something of greater importance: *without* the need for any *knowl-*

**Note in the margin here:* This means only at the time of division. It may be said that Cyclo = Gnatho. *Editors' note: At the time of divergence in the phylogenetic tree, it cannot be said whether the animals are still cyclostomes or already gnathostomes.*

edge of the fact of phylogenetic descent, we obtain the treelike branching structure that evoked the concept of evolution. It is almost inconceivably improbable that this particular and extremely regular distribution of ancient and more recent characters could have arisen *other* than through common evolutionary origin. There is no other explanation that could comfortably accomodate the extraordinary regularity of the facts. Even with the laughably small number of characters presented in our outline of the evolutionary tree of vertebrates, which has been greatly simplified for the benefit of the layman, it is highly improbable that there could be any explanation for the observed regularities other than that evoking commonality of descent. When confronted with the massive factual evidence from the fields of comparative morphology and anatomy, which fills entire libraries, this probability shrinks to zero. Even if comparative systematics were the *only* discipline providing evidence from which the facts of evolutionary descent could be derived, the probability that living organisms are descended from a common ancestor would amount to certainty. But it is not the only discipline of this kind. The evidence from comparative morphology is joined with that from *paleontology*, the discipline dealing with extinct organisms that are preserved as fossilized remains in various geological strata. This additional testimony agrees point for point in thousands and thousands of individual details with the other evidence without any contradiction of any kind. Let us, for example, compare a short summary of the conclusions derived from paleontology regarding the origins of the individual groups of vertebrates with the phylogenetic tree presented above on the basis of comparative morphological evidence. In the Silurian, we find only fish that lack paired extremities and have no jaw apparatus, while showing a further genuine feature of cyclostomes in possessing a large number of gill slits. The first gnathostomes appear in the Devonian, in the form of sharks. In the Carboniferous, we see the origin of the crossopterygians or coelacanths, which doubtless led on to the tetrapods; the peculiar dipnoans, which were equipped with lungs; and finally the oldest true tetrapods, the stegocephalians (amphibians with a carapace). The first reptiles, cotylosaurs related to the stegocephalians, appear in the Permian. In the Triassic there are advanced mammal-like reptiles (Theromorpha) that exhibit unmistakable homologies with corresponding features of mammals in the structure of the skull and, as already mentioned, in the jaw hinge and in the ear. Also present in the Triassic, however, are the first true mammals. In fact, they occur in the same geological formations as those in which the theromorphs reach their peak, namely in the so-called Karroo Formation of the Cape region of southern Africa. The peak of the reptilian stem is found in the Jurassic. At that time, gigantic dinosaurs existed and one major group of them, the Ornithischia, shares a whole series of definitely homologous characters with birds. In the Jurassic shale of Solnhofen, we have also found an impression of the primitive bird *Archaeopteryx* that shows all of the details of the feathers. This creature still possessed teeth in its jaws and also had a long, lizardlike tail containing many vertebrae but fringed on either side with a row of feathers. Finally, from the Cretaceous onward we find entirely typical representatives of all five great "classes" of vertebrates.

As can be seen, if we take our outline evolutionary tree based exclusively on morphological facts, we only need to add the six successive geological epochs (Silurian, Devonian, Carboniferous, Permian, Triassic, Jurassic) in the form of horizontal lines in order to be able to read off the times of origin of the various lineages at the margin. Paleontology thus provides incontrovertible evidence for the correctness of our assumption that the *more widely distributed* homologous characters are *older* and should be allocated a *basal* position in our outline tree. The complete correspondence between two conclusions derived from two utterly different factual sources increases their probability to the level of an historical certainty.

But over and above this we also have evidence derived from plant and animal geography. In combination with the history of the continents, this evidence reveals to us how geographically isolated lineages have followed their own evolutionary pathways. Reliable historical information from these sources agrees *completely* with the conclusions drawn from the two main lines of evidence presented above. Then we have the evidence provided by embryological development. Here, we see the remarkable tendency, common to so many living organisms, to recapitulate certain characters and stages of their evolutionary history in the course of their individual development or *ontogeny.* This is the tendency which gives rise to the rudimentary appearance of gill slits in human embryology as in the development of the chicken, which repeats the formation of reptilian dental laminae in the bird embryo and which recapitulates in some form the *gastrula,* the primordial form of the coelenterates, in the earliest developmental stages of all multicellular organisms. We also have the evidence of serological cross-reactions which demonstrate the *chemical* similarity between the species-specific proteins of related animal species—including chimpanzees and humans. This is supplemented by modern genetic theory, which not only explains the origin of new species but virtually permits experimental demonstration of the process. Finally, as an additional modest contribution, we have the evidence from comparative behavioral research. As we shall see repeatedly in the course of this book, this approach permits the demonstration or confirmation of certain fine details of phylogenetic relationship in a particularly precise fashion.

This all leads us to the following conclusion: the fact of the evolutionary origin of living organisms is historically not just as convincing but considerably more convincing than any other historical knowledge, such as particular, well-documented events in human history. This point can be crudely expressed in the following way: It is *highly improbable,* for example, that Julius Caesar never existed and that all of the documents that we possess dealing with his life and works owe their origin to a plot to misinform us hatched by historians. But it is *impossible* that a counterfeiting evolutionary biologist smuggled *Archaeopteryx* into Jurassic shale and advanced mammal-like reptiles into the Karroo Formation, that is to say, into precisely those geological strata in which these transitional forms between two classes would be expected according to the comparative morphological facts. The "theory" of evolution is *not, in fact, a theory;* it is reliable human knowledge insofar as any human knowledge can be reliable!

Because comparative behavioral research tackles its problems with precisely the same theoretical and practical methods as comparative morphology, it is indispensable for an understanding of the following text to provide some additional comments about special details and particularly about individual difficulties associated with these methods. Naturally, the phenomena that these two disciplines have to explain are far more complex than is indicated in our simplified outline of the comparative evaluation of homologous characters. We are aware of several additional, more specialized, and complicating principles that must be considered in arriving at our interpretations. It is unfortunately also necessary to consider these because they are evidently unknown to many psychologists and philosophers who accept the theory of evolution and attempt to tackle the problems of the evolutionary origin of humans. Understandably, such ignorance can lead to simplistic misinterpretations.

One major complication inherent in the methodology of character evaluation resides in the following: It is by no means the case that all characters that have arisen in the course of evolution do us the favor of persisting in that particular form in all descendants of the relevant ancestral form. In our outline evolutionary tree of the chordates, we have *as far as possible* included only those characters that have persisted. Nevertheless, it has doubtless not escaped the attention of the critical reader that, for some characters in our diagram, we indicated a general distribution *to which exceptions exist.* As we have indicated, the main feature that characterizes the tetrapods—the possession of four limbs—is by no means present in all members of that group. Caecilians, slowworms, and snakes have no legs at all. One bird, the kiwi (*Apteryx*), has no forelimbs, while two groups of mammals—the whales (Cetacea) and the sea-cows (Sirenia)—have no hind limbs. What gives us the right simply to *neglect* this ancient and doubtless important feature in these forms and, despite its occurrence, to allocate them not only to the Tetrapoda but to quite specific subgroups thereof, with the slowworms even placed in a suborder (Lacertilia)? This right comes from the well-founded opinion that in all of these forms the fully developed limbs that were present in their ancestors were *reduced* in the course of their evolution. But how do we know that such reduction has taken place? In the first place, both embryology and paleontology can inform us about the secondary disappearance of a character. Through their possession of obvious limb buds, the embryos of limbless lizards indicate their evolution from four-limbed ancestors. Further, to take the example of the whales, fossil remains convincingly document the progressive reduction of the hind limbs and pelvic girdle. But even without these aids, comparative morphology in isolation can reliably reveal that the absence of a particular character is to be explained through secondary disappearance and that it should therefore not be granted any great value in *systematic* terms. In fact, the *modification* of one character can be inferred by the *unchanged form* of all other characters. Or, to put it more precisely: the *relative* modifiability or conservatism of any single character, and hence its "taxonomic respectability," must be determined in any individual case by surveying those aspects for *all other* characters of the animal group concerned. The *same* character can be very conservative in one group and yet extremely variable in another, such that its systematic value is utterly different in the two cases. The slowworm shares a large number of special and highly specialized characters only with the lizards, while it shares only one feature—the lack of paired extremities—with the cyclostomes. It is therefore almost infinitely improbable that the absence of legs in this case is an ancient, primitive "pretetrapod" feature. It simply cannot be assumed that the host of other features corresponding to those of lizards evolved in that particular form in a lineage that has had nothing to do with the ancestral sequence leading to the lizards since the time of divergence of the cyclostomes. Seen in the context of the many other ancient features—such as the absence of jaws, the presence of only nine cephalic nerves, the possession of a peribranchial cavity, and so on—the lack of extremities in cyclostomes has a taxonomic significance that is quite different from that seen in the context of the many special and modern lizardlike features of the blindworms. But the same applies to the evaluation of *any* character of *any* species and the great art of comparative phylogenetic analysis lies in the ability to determine correctly for *all* of the characters of a group the relative significance of each one of them. It is therefore *fundamentally* impossible to portray the phylogenetic relationship of a large group of organisms, or indeed of the entire animal or plant kingdom, in a table of characters, as has been attempted by many systematic biologists. Any such table, constructed in the form of an identification key, is misleading in that it attributes to every character a *constant* significance in systematic and phylogenetic terms. In reality, each such character displays a different rate of change in every different plant or animal form. Accordingly, any attempt to pack the living, growing diversity of organisms in rigid boxes must *necessarily* lead to some degree of distortion of the real factual state of affairs.

Determination of the relative value, and hence the relative degree of *conservatism,* of every single character out of the continuum of relationships among all characters of the group concerned is therefore the *most important precondition* for correct interpretation of the phylogenetic relationships within that group. In the truest sense of the word, it demands of the researcher a *holistic* appreciation of a maximum number of characters in a maximum number of representatives of the group under investigation. There are two requirements which must be met for this precondition to be present and which must be possessed by the researcher if he is to be up to this most difficult and most stimulating of all the tasks involved in phylogenetic analysis. Firstly, a well-developed innate capacity for *Gestalt perception* of such holistic relationships is necessary. In the Philosophical Prolegomena we have already discussed in detail, in the chapter dealing with the achievements of so-called *intuition* (pp. 56; 62 et seq.), the part that is played by Gestalt perception in this particular field of scientific research. Secondly, a scientist engaged in phylogenetic reconstruction must of course possess knowledge of a "wealth of characters." He should ideally be *familiar* with *all* representatives of the group investigated and with *all* the homologous characters that can be evaluated *on the basis of personal experience!* The optimal background for the reconstruction of phylogenetic relationships is possessed not by somebody who has traced a single organ or a single character in all of its manifestations through a very large systematic assemblage but by an investigator who is *thoroughly familiar* with a relatively *small* taxonomic group. Ideally, as has been said, this knowledge should encompass all representatives and all of their characters. As will be discussed in more detail in the section on evolutionary history and the methods of comparative behavioral research, it was precisely this insight that motivated two brilliant phylogeneticists, C[harles] O[tis] Whitman and O[skar] Heinroth. These two scientists, both equipped with a particularly well-developed capacity for Gestalt perception of holistic relationships, endeavored continuously to increase the range of their comparative set of characters. Ultimately, this led to the inclusion of *behavioral* characters for phylogenetic inference. In this way, comparative behavioral research not only originated as a daughter discipline of comparative morphology but also arose, so to speak, to serve the latter. As a result, comparative behavioral research not only inherited the concepts and methods from that source but went on to refine them and make them more precise. In this particular area, faced with the complexity that applies to all comparative research and with the necessity of determining correctly the taxonomic value of individual features on the basis of a holistic appreciation of a wealth of characters, certain comparative behavioral investigators have been enormously successful. They have penetrated more deeply into fine aspects of phylogenetic relationships than any purely morphological phylogeneticist before them. This is precisely because they were able to exploit, *in addition* to the range of features of bodily construction, behavioral characters of living organisms on an inductive basis.

A *second* major complication in the reconstruction of phylogenetic relationships through comparison of characters arises because of the phenomenon of so-called *convergence* that affects certain features. This is of special significance in comparative behavioral research, not least because this approach is often in a position to exclude the possibility of convergence in certain fine systematic details more reliably than is ever the case with a purely morphological comparison of such "microdetails." The term "convergence" can be defined as follows: In the course of adaptation for a particular function, different groups of animals that are not closely related can independently arrive, so to speak, at "the same invention." In this way, they develop similarities in specific features of morphology and patterns of movement that *cannot* be traced back to a common phylogenetic stage of development. The wings of hummingbirds and of certain lepidopterans, the hawk moths, show a close correspondence in their construction. Both are undoubtedly "wings," yet these organs are not at all *homologous* but merely *analogous*. Mere analogies between organs can lead to such extensive, convergent correspondences that they seem to the uninitiated to be really "the same." Let us take, for example, the *heads* of a mouse, a frog, a squid, and a grasshopper. With their mouths surmounted by eyes and a brain capsule, all of these structures seem to match so directly that even an expert can have difficulty in remembering how their development, which was crucial to evolutionary advance, took place *twice* within the phylogenetic tree of animals. As is unequivocally demonstrated by the study of embryology, in worms, molluscs, and arthropods the head quite literally develops *at the opposite end of the body* in comparison to that of the chordates! It has been established beyond doubt that there was a major division between two branches early on in the phylogenetic tree of the animal kingdom, closely following the emergence of the Coelenterata. The coelenterates have a simple, saclike body which arises through "gastrulation" (see figure 3), that is, through invagination of the "blastula" to form a "gastrula." This process is repeated in some form in the embryonic development of all higher multicellular organisms. The blind alimentary cavity has only a single opening, the so-called *blastopore* (primordial mouth), which corresponds to the site of invagination of the blastula. The development of a *second* opening at the opposite end to the blastopore converted this blind cavity into a digestive passage. At this point, the ancestral stem of the metazoans was, so to speak, confronted with the need to make a choice between anterior and posterior. To put it more precisely, the second body opening was "invented" by *one* animal lineage as a mouth and by the *second* as an anus. The blastopore, which originally served the functions of both mouth *and* anus, hence came to adopt opposite functions in the two lineages. The lineage of the Protostomata, in which the blastopore (or protostoma) became the true mouth, gave rise to the worms, the molluscs, and the anthropods. By contrast, the Deuterostomata—creatures with a second mouth—gave rise to the arrow worms (chaetognaths), the echinoderms, the acorn worms (Enteropneusta), and the chordates. In cases where the sense organs and the nervous system subsequently became more advanced, for understandable reasons they became concentrated at the end of the body that led during movement, close to the "mouth," thus giving rise to the structure that we refer to as a head in the two different lineages.

Another classic example at a higher systematic level can be mentioned here. In adaptation to life in the open sea, an order of reptiles that lived during the Jurassic and an extant order of mammals—the ichthyosaurs and the whales, respectively—developed a very close similarity to fish in their body form and pattern of movement. The similarity between these two groups in fact went even further, to the extent that such a respected paleontologist as Steinmann was led astray by this convergent similarity and interpreted the whales as descendants of the ichthyosaurs! If a researcher is really acquainted with a wealth of characters, however, such "major" convergence will very rarely lead him astray. This is because the number of group-specific characters will necessarily always be a thousand times greater than the number of characters shared with a member of another group as a result of convergence. An ichthyosaur is marked out as a reptile

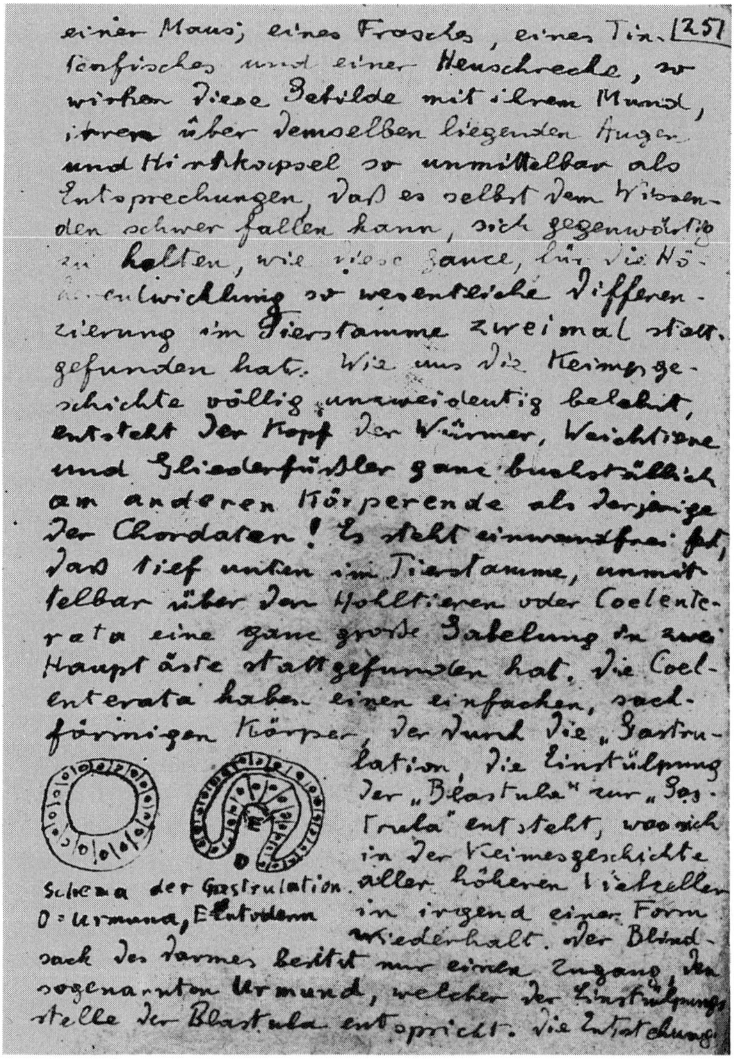

[25]

einer Maus; eines Frosches, eines Tin-
tenfisches und einer Heuschrecke, so
wirken diese Gebilde mit ihrem Mund,
ihren über demselben liegenden Augen
und Hirnkapsel so unmittelbar als
Entsprechungen, daß es selbst dem Wissen-
den schwer fallen kann, sich gegenwärtig
zu halten, wie diese Ganze, für die Hö-
herentwicklung so wesentliche Differen-
zierung im Tierstamme zweimal stattgefunden hat. Wie uns die Keimesgeschichte völlig unzweideutig belehrt,
entsteht der Kopf der Würmer, Weichtiere
und Gliederfüßler ganz buchstäblich
am anderen Körperende als derjenige
der Chordaten! Es steht einwandfrei fest,
daß tief unten im Tierstamme, unmittelbar über den Hohltieren oder Coelenterata eine ganz große Gabelung in zwei
Hauptäste stattgefunden hat. Die Coelenterata haben einen einfachen, sack-
förmigen Körper, der durch die „Gastrulation, die Einstülpung
der „Blastula" zur „Gas-
trula" entsteht, was sich
in der Keimesgeschichte
aller höheren Vielzeller
in irgend einer Form
wiederholt. Der Blind-
sack des Darmes besitzt nur einen Zugang, den
sogenannten Urmund, welcher der Einstülpungs-
stelle der Blastula entspricht. Die Entstehung

Schema der Gastrulation.
O = Urmund, E = Entoderm

Figure 3
Gastrulation diagram. O, blastopore; E, endoderm.

by thousands of characters and a whale is characterized to an equal extent by mammalian characters, whereas only a few externally conspicuous features are common to both. In just the same way, the head of a grasshopper resembles that of a mouse only to the extent that a brain and two eyes along with a few additional paired sense organs are present above the mouth. By contrast, thousands and thousands of characters indicate that one is the front end of a protostome while the other is the anterior end of a chordate.

Such distinctions are much more difficult in cases of "minor" convergence between closely related animal forms. For instance, two groups of birds—the heron group and the roller group (Aquintocubitales and Coracorpitres, respectively, in Fürbringer's classification)—have both given rise to a group of predatory birds that grasp their prey with the claws on their feet. The first group contains storks, herons, ducks, and geese together with the birds of prey (Falconiformes) with which we are concerned here. The second contains rollers, kingfishers, hummingbirds, goatsuckers, swallows, and other birds, including the owls. The last show convergent similarity in so many details to the "diurnal raptors" that Linnaeus referred to them as "nocturnal raptors" and placed both groups in the same order. The correspondence between them extends into fine details that apparently have nothing at all to do with the similarity in their feeding behavior. This applies to the structure of the cere on the beak, the arrangement of the feathers around the eyes and ears, certain locomotor patterns, and various other features. For this reason, it is indeed the case that "even the cleverest man may be fooled." It requires extremely subtle deliberation based on a great number of characters to demonstrate that birds of prey and owls are no more closely related to one another than are, for example, storks and hummingbirds! Understandably, the danger of confusing convergent features with homologous characters increases as the degree of relationship between the original ancestral forms increases. A superb example of triple convergence is provided by the eagles, which have been interpreted by most systematists as just one group erroneously designated as a family (Aquilinae) or even attributed to the single genus *Aquila.* The "eagles" in fact belong to no less than *three* separate families that have independently given rise to large birds of prey. Largely for aerodynamic reasons, they have developed a high degree of similarity to one another. The kite group (Milvinae) gave rise to the sea eagles, which are in fact linked to their family by obvious intermediates such as *Milvago.* The buzzard group (Buteoninae) gave rise to the golden eagle, the imperial eagle, the booted eagle, and other forms. Finally, the hawk eagle, the crested eagle, and the somewhat less specialized crowned eagle and harpy eagle are, without a shadow of doubt, derived from the hawk group (Accipitrinae). It is utterly characteristic of the power of comparative behavioral research that it was precisely with respect to such questions in fine systematics that those best acquainted with the *living* animals arrived at this attribution, or—to put it more aptly—at a subdivision of the "Aquilinae." Quite independent of one another, Heinroth, Heck, and Antonius arrived at exactly the same conclusions with respect to the relationships between these convergent forms. Among professional systematists in the field of ornithology, E[rwin] Stresemann—who has close connections with comparative behavioral research—took these interpretations into account in his reworking of bird classification in *Bronn's Klassen und Ordnungen des Tierreiches.*

A further good example for such multiple convergence is provided by ducks of the family Anatidae which used to be classified in the goose subfamily (Anserinae). The most primitive form of the bill, which is most widely distributed within the order, is the well-known duck bill with its sieving apparatus consisting of fine horny lamellae on the upper and lower beak and the tongue. Its highest degree of differentiation is found in the shoveler (*Spatula*). Now in all groups of the Anatidae that have secondarily become terrestrial *grass eaters,* an analogous change in function has led to conversion of the lamellae of the bill into tough horny teeth that are suited to the cropping of grass stems. Within the family, this modification has taken place independently and convergently in no less than three and perhaps even in four separate lineages. It has occurred in sheldgeese (*Chloephaga*), to which the blue-winged goose (*Cyanochen*) is allied, and all of these forms are derived from the Casarcinae. Secondly, it has occurred in all "true" geese of the subfamily Anserinae, such as *Anser, Chen, Eulabea, Branta* and so on. Thirdly, it has taken place in the very primitive spur-winged goose (*Plectropterus*), which is related to the Cairininae. Fourthly, it has occurred in the very isolated Australian magpie goose (*Neochen*), although this form is possibly related to the Anserinae.

With all such cases of convergence it is always adaptation to a quite specific similar *function* that serves as a determining factor in the evolution of characters to render them similar in specific respects. All experienced phylogeneticists are therefore extremely careful in evaluating any similarity in characters between two species that is connected with adaptation for a *function* that is common to both. In fine systematic analyses, it is frequently necessary to rule out the evaluation of certain characters, despite the fact that they are quite probably genuinely homologous, simply because the possibility of covergent adaptation cannot be reliably excluded. Precisely with such issues in fine systematics, which are of great significance for phylogenetic reconstruction, comparative research into *behavior* can yield highly reliable and exact results in individual cases. This is because this approach can reliably *exclude convergence* in certain contexts. This applies to behavior patterns whose survival value resides in the elicitation of social responses from conspecifics, as is the case with all so-called *display patterns.* These include threatening and submissive postures, the begging patterns of young birds, and the many reciprocally releasing instinctive behavior patterns of this kind that are involved in mating, functioning like a tooth-and-cog mechanism. Analogous systems of the latter type are known from cuttlefish, spiders, insects, fish, reptiles, birds, and mammals. All of these behavior patterns, which elicit a response from a conspecific in the form of an unconditioned reflex, operate like *signals.* In this respect, they show an extensive functional analogy to the symbols of human *language.* As in language, the *significance* of an individual animal signal is a pure *convention* between the transmitter and the receiver! The tail-wagging of a dog expresses friendly arousal whereas that of a cat indicates animosity, and both are correspondingly "understood" and answered by the innate schemata of all conspecifics. This fact is exclusively attributable to the particular evolutionary history of the stimulus-transmitting instinctive motor patterns and the stimulus-receiving schemata. As far as the form of the given motor patterns is concerned, the significance could just as easily be reversed or entirely changed. In other words, the specific releasing *function* of such a display pattern bears no functional relationship to its *form,* in contrast to the situation with all mechanically operating motor patterns or organs. After all, one cannot *discern* the significance of a display pattern merely by looking at it, just as one cannot understand a word in a foreign language if one has no knowledge of its historically determined conventional significance. On the other hand, it is precisely with such motor display patterns that one can confidently exclude *convergence,* because there is no operational relationship between form and function. As with a word in two different human languages, it is almost infinitely improbable that identity or even approximate similarity between two functionally analogous motor display patterns of two animals species will arise purely by *chance.*

When fish belonging to the families Cichlidae, Cyprinodontidae, Gobiidae, and Percidae fight, they all exhibit the same threat display of orienting themselves side to side and beating with their tails. On observing this, we no more doubt the existence of a common phylogenetic origin than a philologist would doubt the existence of a common "ancestral form" as the source of the words *Mutter* (German), *mater* (Latin), μήτηρ (Greek), and мать (Russian). Phylogeneticists and philologists draw their respective conclusions for the same reasons and with similar complete justification. Hence, with very few exceptions, similarities between motor display patterns with the same significance are *always based on homology.* This important capacity for reliable exclusion of convergent adaptation in many cases permits comparative behavioral investigators to make certain statements about fine details of phylogenetic relationship between closely related animal species that are seldom permissible in the field of comparative morphology.*

There is another phenomenon that has nothing to do with *adaptation* to a particular *function* that generates "convergence" in a somewhat wider sense. Precisely for this reason, it constitutes a complicating obstacle to phylogenetic investigation, although this only applies to the study of the finest relationships between closely related species. Such closely related plants and animals, which show extensive structural similarity in the *genome* (the totality of the hereditary material in the chromosomal apparatus of the cell nucleus), tend to undergo mutations in the same "gene," at the same hereditary locus. Because such modification of the hereditary material thus affects genes that are, indeed, *homologous* in phylogenetic terms, they are commonly referred to as *homologous mutations* in modern genetics. But this label can easily lead to misunderstandings inasmuch as a parallel change can occur *independently* in two *different* lineages. For example, in golden pheasants (*Chrysolophus pictus*) the so-called *obscurus* mutant has demonstrably occurred several times independently in breeding lines in captivity. The same applies to the black-winged mutant of the peacock (*Pavo cristatus*), which similarly shows dominant inheritance. But, as with different breeding lines of *one* species, "homologous" mutations can also produce the same features in *closely related* species without the carriers of the character sharing a common ancestor. This means that these characters, *in a phylogenetic sense, are not in fact homologous.* For example, the feature of so-called "*mottling*"—evenly distributed darker feathers among the wing coverts—occurs in various wild dove species and in several breeds of the domestic pigeon as a consequence of independent "homologous" mutations. Such "homologous" mutations have doubtless been repeatedly confused with phylogenetically homologous characters by very able systematists and phylogeneticists. For instance, in standard genealogical reconstructions for *domestic dogs,* all breeds with pug heads have been derived from the Molossian dogs of the Romans, whereas in reality the pug-headed character arose as an independent mutation in widely different dog lineages. Accordingly, we prefer to use the term *genetic*

**Note in the margin here:* Important! Because there is a bridge between genetics and phylogenetic reconstruction at the finest level of systematics!

convergence instead of homologous mutation, which is misleading in a phylogenetic context. Genetic convergence is an extremely vexatious obstacle for fine systematics because it can generate a high degree of similarity and create an illusion of phylogenetic relationship. A number of examples of this will emerge in the discussion of the fine systematics of the dabbling ducks (Anatinae).

In addition to the two above-discussed complications inherent in any phylogenetic assessment—relative fluctuation in the consistency of an individual character and the necessity to exclude the effects of convergence—it is necessary to mention a third source of complexity and difficulty in phylogenetic analysis. This is important enough that any comparative behavioral investigator should become familiar with it. The difficulty lies in the fact that a "phylogenetic tree" has a *multidimensional* structure. Its branches extend not only in all dimensions of *space* but also through the dimension of *time.* The resulting structure can be represented only poorly in human conceptual thought and it is even more difficult to portray it in the linear sequence of spoken and written words. Indeed, it remains difficult to reflect it graphically in two dimensions on paper. Even in a three-dimensional model, phylogenetic processes cannot be depicted completely satisfactorily, although this is the best available approach. Many zoologists are not fully aware of this great complexity of the real situation. There are continuing attempts to express the phylogenetic relationships between living animal species by means of a *linear* sequence. Because this matter is also of great significance for comparative behavioral research, it needs closer examination here.

Above all, anybody who attempts to organize animal species using the well-known metaphor of a tree should never forget that modern species can never be derived *from one another.* All living species should, of course, be visualized as the terminal twigs of the tree. Ipso facto, they should never be arranged in a linear sequence as chronological sections of the tree, as if they were *temporally successive* stages of evolution or *segments* of the tree. This is not permissible even when the branch whose tip is represented by a living species originated from a common trunk low down in the tree and when that species still shares a relatively large number of characters with that past segment of the tree from which it diverged. As *forms alive today,* all modern species lie in *one* plane, representing endpoints to which various diverging and branching lineages ultimately lead. This plane represents the present condition. In a graphically symbolic form, this plane can be represented as a layer which gives the tree the form of a stone-pine or umbrella acacia in which all of the terminal buds lie at approximately the same height. The advantage of this arrangement is that time can be represented in an illustrative fashion by the vertical axis. This is why the diagram on p. 107 was also structured in that way. Alternatively, the phylogenetic tree can be represented as a bush trimmed into a spherical form, with living species depicted as terminal twigs in the outer surface. Successive time intervals are then represented in the form of concentric spherical layers, although this is not so readily understandable. In addition, one loses the undeniable advantage that the *degree* of evolutionary development—that is, the relative distance separating any species from the segment of the tree from which it originated—can be represented on the vertical axis.

A linear arrangement of living animals constitutes a gross distortion of the real state of affairs because it places the terminal twigs *within the tree.* Yet the amoebae and lancelets alive today are modern organisms from the year 1948. Even if they possess a *larger number* of ancient ancestral features and have developed relatively few new ones, no single species has really remained the same as the ancestral form. Not one of them completely *lacks* new, highly specialized and modern characters and it would be utterly misleading to label such an organism *as a whole* with the adjective "primitive." *There are no primitive species alive today but merely primitive characters.* Some living species have retained a greater number of such characters than others.

It is essential to recognize how easily a *series* of similarities between related plant or animal species can create the illusion of an ascending scale of evolutionary stages. Only then is the investigator protected against the error of interpreting such a series as an actual evolutionary sequence. Let us assume that a biologically successful ancestral form has given rise to a large number of species which are all equally related to one another, to the extent that they have all diverged in different directions from one common ancestor (figure 4). When expressed according to the metaphor of a three-dimensional phylogenetic tree, such a group of species forms a kind of brush composed of unbranching lineages. The terminal points are in some cases very similar but in others distinctly different. The points lying farthest apart on the brush are the most different from one another, but all of them are "linked" by a large number of so-called transitional forms. But let us now assume that most of the divergent species in such a "brush" have become extinct, so that only one *series* has remained in which each species shares more characters with its two immediate neighbors than with any other species (figure 5). When confronted with such a series, the investigator will always be tempted to interpret it as a genetic sequence of stages. This will particularly be the case if one end of the series is "closer to the root" than the other, that is to say, if fewer characters have changed from the

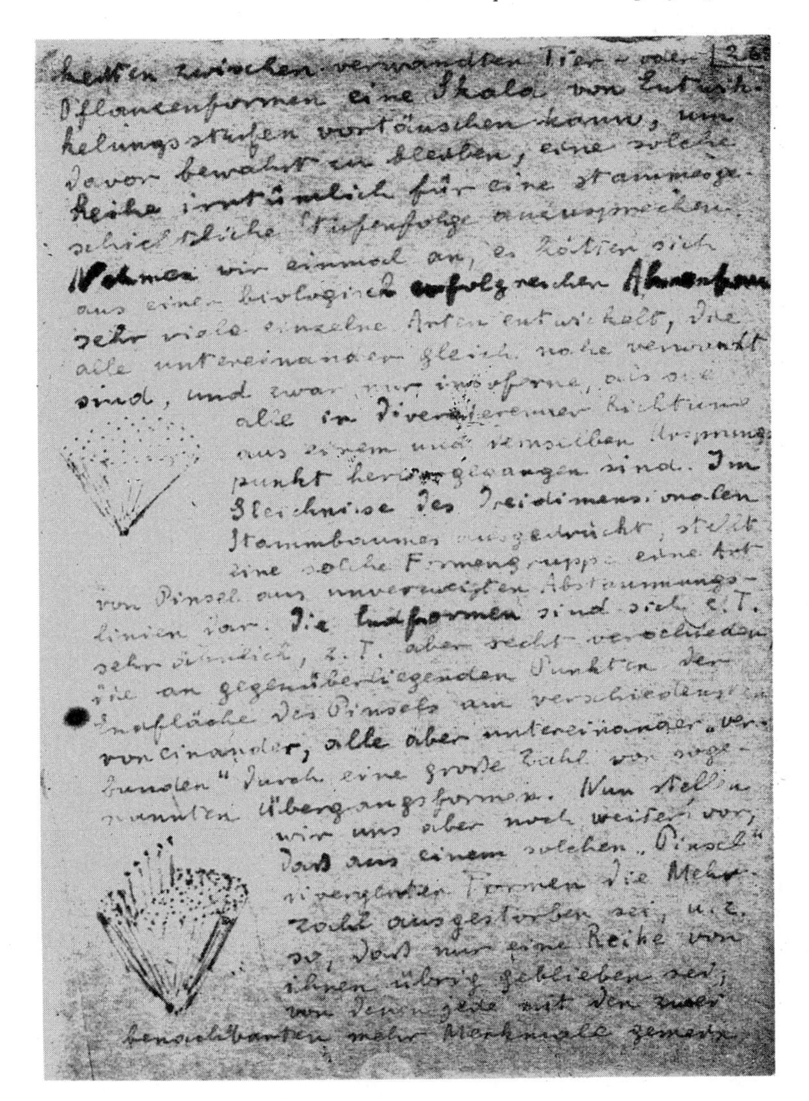

heiten zwischen verwandten Tier- oder [26]
Pflanzenformen eine Skala von Entwick-
lungsstufen vortäuschen kann, um
davor bewahrt zu bleiben, eine solche
Reihe irrtümlich für eine stammesge-
schichtliche Stufenfolge anzusprechen.
Nehmen wir einmal an, es hätten sich
aus einer biologisch erfolgreichen Ahnenform
sehr viele einzelne Arten entwickelt, die
alle untereinander gleich nahe verwandt
sind, und zwar nur insoferne, als sie
alle in divergierender Richtung
aus einem und demselben Ursprungs-
punkt hervorgegangen sind. Im
Gleichnisse des dreidimensionalen
Stammbaumes ausgedrückt, stellt
eine solche Formengruppe eine Art
von Pinsel aus unverzweigten Abstammungs-
linien dar. Die Endformen sind sich z.T.
sehr ähnlich, z.T. aber recht verschieden,
die an gegenüberliegenden Punkten der
Endfläche des Pinsels am verschiedensten
voneinander, alle aber untereinander „ver-
bunden" durch eine große Zahl von sogenannten Übergangsformen. Nun stellen
wir uns aber noch weiter vor,
daß aus einem solchen „Pinsel"
divergenter Formen die Mehr-
zahl ausgestorben sei, u.z.
so, daß nur eine Reihe von
ihnen übrig geblieben sei;
von denen jede mit den zwei
benachbarten mehr Merkmale gemein-

Figures 4 and 5

ancestral condition in the former. In the figure, this has been symbolized by the curvature of the terminal surface of the brush, such that the upper end of the series lies somewhat *higher.* Such "brushes" and "series" are an extremely common occurrence in the fine systematics of higher animals. In his well-known theory of the *Formenkreis* (circle of forms), [Otto] Kleinschmidt for the first time provided a pertinent analysis of the phenomenon discussed here. He showed that in a particularly large number of cases it happens to be the middle of the brush that has disappeared, leaving a circle of forms in which there is "proximity" between any two neighboring species as in the hypothetical case discussed above.

Hence, living species never form a genuine phylogenetic sequence. Such a sequence can only be seen in a concrete form with fossil evidence in which successive strata really contain a sequence of forms that are phylogenetically derived one from the other. The classic example of this is provided by the paleontological sequence documenting the evolution of the horse. With a purely comparative morphological approach, however—and the same applies to comparative behavioral research—a phylogenetic sequence can never consist of entire, living animal species but only of characters or of reconstructed ancestral stages based upon them in order to trace the course of phylogenetic development. These abstract concepts, such as the reconstructed ancestral stages for cyclostomes, fish, stegocepahlians, reptiles, and mammals, can indeed yield a phylogenetic sequence, whereas this is not true of living representatives of these groups. In fact, it is generally improbable, even with fossil forms, that the preserved remains really correspond to the main stem of the phylogenetic tree rather than to side branches.

But, if we are restricted to pure comparative morphology or to the characters derived from comparative ethology, is it at all possible to distinguish a genuine phylogenetic series from the illusory sequence of the "brush phenomenon"? With the aid of the following indicators, this is indeed possible. If one attempts to group the species in a brush using the method of the phylogenetic diagram shown on p. 107, it at once becomes clear that this is not possible in the sense discussed there. In the brush, it is not possible to distinguish older from younger characters, only widely distributed from uncommon features (figure 6). All of the characters *overlap* in a quite diffuse manner, as is shown in the upper left diagram of figure 6. The basic phenomenon whose discovery led to the concept of the phylogenetic tree was that not just one but several features *exclusively* characterize any given group. But this *regularity* in the distribution of characters, which is essential for the reconstruction of genuine phylogenetic sequences, is completely *lacking* not only with the "brush" but also with the illusory sequences and *Formenkreise* discussed above. This enables us to distinguish them from real phylogenetic sequences on a purely comparative basis without recourse to paleontology or ontogeny.

We have now become acquainted with three major complications of all comparative research which must also be mastered by the comparative behavioral investigator: firstly, the inconstancy of the value of any individual

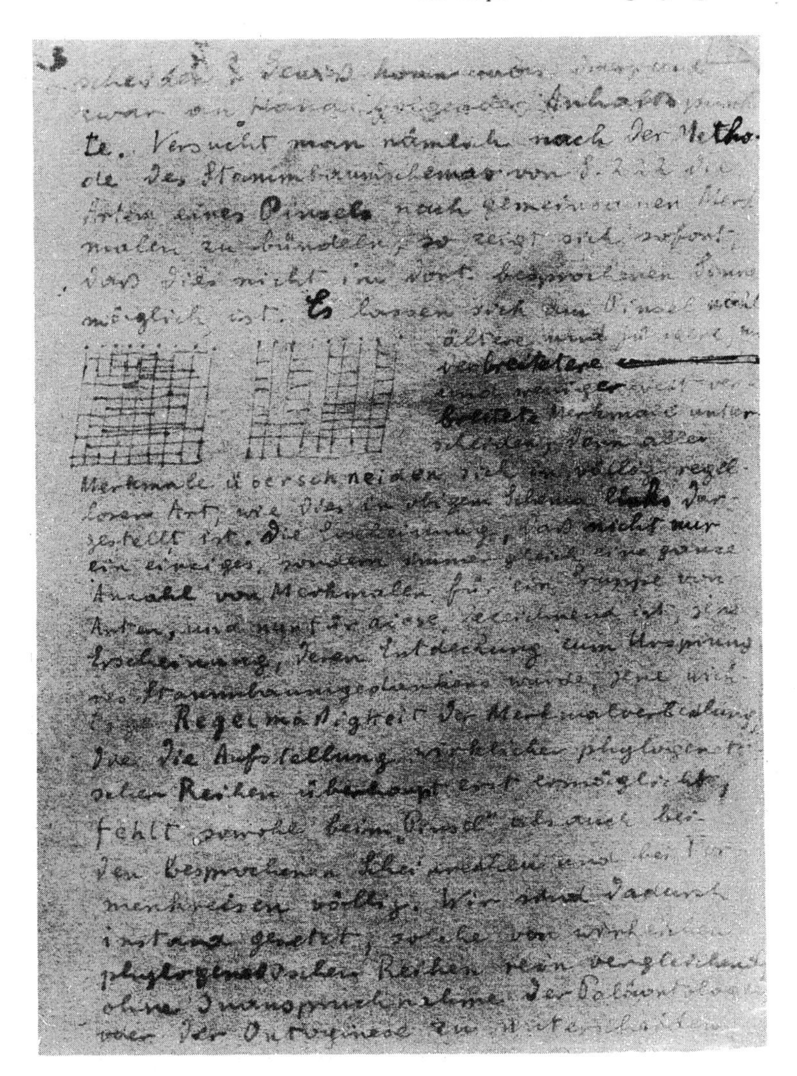

Figure 6

character, which requires determination of its taxonomic relevance within the overall character set of the group concerned (pp. 113–116); secondly, the phenomenon of convergence (pp. 116–123); and thirdly, the multidimensional nature of the phylogenetic tree (pp. 123–126). Because of the need for simultaneous consideration of these and other complicating factors, comparative research is an undertaking that demands great discretion and a gift for synthesis. It requires a great deal of self-criticism and, above all, a relatively well-developed talent for the perception of an entire Gestalt.

Such abilities, taken all together, are often referred to as "systematic finesse." This expression itself indicates the extent to which intuition and Gestalt perception play a part in the work of the phylogeneticist. In the face of the magmalike flow of the process of evolutionary development, it is an unconditional necessity to have a procedure of thought and operation that is holistic in the truest sense of the word. The almost lawless plasticity of organic origin certainly does not facilitate the task of conducting a historical reconstruction of the course of evolution.

For this reason, one persistent principle governing the process of evolution—one could almost say the only principle to which there is no exception—is all the more important. This is Dollo's law of the *irreversibility of adaptation.* Although it has no theoretical basis at present, it is an empirically binding rule that phylogenetic processes leading to the differentiation of characters and organs *are never reversed to lead back to the original condition.* This law is expressed particularly clearly in cases where the external appearance and function of an organ undergoes an *apparent* reversal to regain a phylogenetically earlier form. This is, for example, the case with the forelimbs of the whale, whose function and external appearance have indeed regained the form of the most primitive paired extremities of the vertebrates, that of fish fins. Even if there are still certain differences of opinion about the exact course of its differentiation, the tetrapod limb is undoubtedly derived from a fish fin. Now in whales—along with sea-cows (Sirenidae) and seals and sea lions (Pinnipedia) among mammals; and icthyosaurs, plesiosaurs, and morosaurs among reptiles—the extremities of originally terrestrial quadrupeds have been converted to structures entirely resembling fins because of a secondary readaptation to aquatic life. But the route that they followed in the process was entirely different from that along which the tetrapod extremity was originally derived from the fish fin. However finlike they may seem on the outside, all of these secondary forms of the extremities—in some cases showing extensive convergence—prove to be true tetrapod limbs in their internal structure. The skeleton of the forelimb consists of the shoulder girdle, humerus, ulna, radius, carpals, metacarpals, and phalanges, like that of all other tetrapods, and the same applies to the hind limb (hip girdle, femur, etc.).

This peculiarity of phylogenetic change, that it never returns along the route it originally followed and hence never *reverses* anything in the strict sense of the term, but always builds upon that which is already present, has a fortunate consequence for research. Historically developed characters are often present as successive layers, virtually like geological strata or the growth rings of a tree. As long as they are not detrimental to the survival of the species, which would set in motion selection against them, all historically *ancient* organs and structural characters are carried along almost eternally. It should be noted that organic evolution includes not only "constructions" that serve a direct adaptive function but also some that have not become so poorly adapted that they have led to extinction of the individuals carrying them. Although unused structures can disappear rapidly if they present a significant obstacle to the survival of the species, adaptively useless characters that are *neutral* can be maintained over vast periods of time. So-called *rudimentary* organs represent a phenomenon that owes its existence entirely to the workings of Dollo's law of the irreversibility of adaptation. Although it is conceptually distinct, the process by which a rudiment is generated, the slow "atrophy" of an organ cannot in practice be distinguished from the process of a *change in function.* Just as an unused room can be employed as storage space or an old suit can be used to make a scarecrow or dusters, organs that are no longer used can almost always find "applications" of some kind. The redundant primary feathers of the ostrich (*Struthio camelus*) have taken on a role in courtship; the rudimentary, unpaired third eye of the vertebrates has been converted into the pineal gland; the tail of large mammals has lost its original function in locomotion and has become a fly whisk; vestiges of the gill arches have been converted to the hyoid and larynx in all mammals; and so on. Rudimentary organs which have "retired" from their original function and have developed some minor "secondary occupation" in this way are much more *common* than those that are carried along without serving any function at all. This led Jakob von Uexküll to make the doubtless exaggerated claim that there are in fact no rudimentary organs. When faced with such tenets, which are always guided by the vitalistic belief that all living organisms are occupied by a directing "factor," one must always remember the principle stated above, that even *nonfunctional features* may be preserved as long as they present no threat to the survival of the species. Genuinely rudimentary organs do, in fact, exist and they include not only cases that are neutral with respect to selection but even some that are demonstrably *detrimental* to the survival of the species (even if they do not lead to the effective involvement of selective processes that "eliminate" them by suppressing either the character or the species). The appendix (*appendix vermicularis* [sic]) on the human caecum is a *demonstrably* completely nonfunctional rudiment. Even if it does not threaten the existence of our species, its presence has a cost to mankind amounting to the loss of thousands of otherwise healthy people every year. It therefore displays an undeniably *negative* selective value. In the last part of this book, we shall see how much more serious a threat is posed to the survival of our species by certain "instincts" that have lost their original function.

The peculiarity of evolutionary change expressed in Dollo's law, that it never returns along the path that it originally followed,* is doubtless also responsible for the fact that the structure of organisms carries so much historically determined baggage. As we have seen, this is often not to the advantage of the species concerned, but it is very fortunate for anyone who wishes to investigate their historical origins. The following point must be clearly recognized: Without such carriage of historical features, determined by the irreversibility of the evolutionary process, it would in fact be *utterly impossible* to draw any conclusions about the phylogenetic *homology* of characters and organs. If the forelimb of the whale could find its way *back* to the fish fin, we would have no possibility of ever finding out that it originated from the arm and hand of a tetrapod running around on dry land. On p. 102 et seq., we have discussed the lawfulness seen in the distribution of older *more general* and younger *more special* characters, the primary principle which permitted comparative morphologists to recognize the fact of evolutionary descent. This principle is, in the final analysis, *identical* with the law of irreversibility in the differentiation of characters. Without it, any inferences regarding the *temporal* sequence of development of individual characters would be utterly ruled out.

The great and magnificent law of the irreversibility of all organic evolution suffers only one exception, which is no more than apparent. This is due to a process that we must examine in more detail here because it is of major significance for the evolution of human beings from anthropoid apes. This extremely peculiar process of modification of species occurs in an analogous fashion in a whole range of animal species that in many cases show no close phylogenetic relationship to one another. Briefly stated, this process leads in an abrupt evolutionary transition to a condition in which sexual maturity is achieved earlier than was previously the case with the species concerned. In other words, maturity is reached at a developmental stage that is markedly different from the previous definitively adult form. Further, the remaining ontogenetic stage leading to the original final form is *suppressed*. In this way, a form that was previously just a transitional stage in individual development becomes the definitive form for the species. Understandably, this phenomenon is particularly conspicuous and clearly definable with animal species in which the successive developmental stages

**Note in the margin here:* Back mutations do occur!

are sharply distinguished from one another by a particularly large number of characters. In other words, it is clearest in species that undergo a distinct change in form (*metamorphosis*). In such species, a previous *larval form* abruptly becomes the final stage of the species. This entire phenomenon of persistence of the juvenile or larval form is referred to as *neoteny,* an unattractive term because it is a linguistic hybrid (from the Greek νέοζ, for young, and the Latin *teneo,* to keep).

The best information available for the phylogenetic development of neoteny is provided by the behavior of the newts and salamanders (Urodela), with which the phenomenon was first discovered and investigated. *Occasional* achievement of sexual maturity by gill-bearing, aquatic larvae is known for a whole range of newt species and even occurs in such a clearly terrestrial species as the fire salamander (*Salamandra maculosa*). Initially, however, this does not exclude preservation of the potential for continued development to the terrestrial end form. By contrast, with the Central American axolotl (*Amblystoma tigrinum*), neoteny is the rule. The fact that the rare terrestrial form, with its salamander-like external appearance, belongs to the same species as the broad-headed, gill-bearing larval form was discovered only at a relatively late stage. Even in *Amblystoma,* the potential for development into the terrestrial form is maintained. The transition can be imposed upon even old, sexually mature individuals of the aquatic form, either by gradual dessication of their aquatic habitat or (interestingly) by administration of thyroid hormone. But there are a whole series of newts and salamanders that no longer develop into the gill-less terrestrial form at all. There are convincing arguments to support the interpretation that this is by no means a primitive condition in these species, but represents genuine neoteny that has become obligatory. Species of the American–East Asian genus of "giant salamanders" (*Megalobatrachus*), the olm (*Proteus*), the American "mud puppy" (*Necturus*), the siren (*Siren*), and others are all forms in which persistent larval features have become definitive characters of the species. Partial neonteny, with a persistent potential for complete development, is also known for insects. A particularly good example is provided by the midge genus *Chironomus,* whose aquatic larvae regularly become sexually mature in inland lakes and breed from generation to generation without pupation or metamorphosis to the imago. But there are also a large number of large groups of animals, such as the radiolarians and the crustacean order Cladocera, which are interpreted by modern phylogeneticists as obligatory neonate forms. In fact, there is no lack of informed support for the notion that the entire class of insects is derived from neotenous larvae of the millepedes. These larvae, with their six legs and a whole series of other features, indeed show quite astounding similarity to certain primitive insects (Apterygota).

Now neoteny may happen to affect an animal form whose larva exhibits numerous recapitulations of ancestral characters, as discussed on p. 112, perhaps even resembling the ancestral form far more closely than the adult does. Such a case provides the *apparent exception* to the law of irreversibility of adaptation which required us to discuss the phenomenon of neoteny here. For example, the neotenous form of the axolotl—with its broad head bearing the eyes on its upper surface, its well-developed gill arches, the form of its mouth, and the broad, laterally flattened propulsive tail—possesses a whole series of ancestral features from the era of the stegocephalians. These features *are no longer present* in the *non-neotenous* form of the axolotl. In the absence of any knowledge of the phenomenon of neoteny, anybody acquainted with Dollo's law of irreversibility *would have to reject* the derivation of the gill-bearing neotenous form from the terrestrial axolotl. The origin of the former from the latter would require the *reappearance* of ancestral characters that have apparently already disappeared, thus providing the illusion of a genuine reversal of phylogenetic change along its original path. In just the same way, *convergence* arising through the process of neoteny can generate the illusion of a phylogenetic relationship that does not, in fact, exist. All neotenous newts and salamanders, which are certainly derived from quite different orders, were originally combined together in the group "Perennibranchiata."

On the other hand, wherever a larval form exhibits more phylogenetically new characters that have arisen as special adaptations to the conditions of larval existence—so-called cenogenetic characters (from the Greek χαινόζ, new)—neoteny gives rise to something "entirely novel." For example, the property of increasing the number of body segments in the course of ontogeny is a very ancient feature of the segmented worms which has been retained by crustaceans and other descendants of the Annelidae. Similarly, the large number of segments in primitive crustaceans and in millepedes is doubtless an ancient heritage from the segmented worms and should hence be interpreted as a primitive character. As has happened with the water fleas in the group Cladocera already mentioned above, the larva of a branchiopod that possesses several hundred legs in the adult condition may become neotenous at the so-called metanauplius stage with seven to nine pairs of legs. When this happens, within a very short space of time (in geological terms) an extraordinary, entirely novel dwarf form is produced. This form is markedly different from all of its close relatives and even differs from the ancestral form in many ancient and otherwise very conservative characters.

The apparent exception to Dollo's law thus occurs *only* in cases where neoteny affects forms whose juvenile stages are rich in ancestral characters. In other words, it only involves "palingenetic" characters that are recapitulated from ancestral forms. It was necessary to discuss all of this in some detail because *human beings correspond to the juvenile stages of anthropoid apes* in a large number of features, both in morphology and in the realm of innate behavior. Because a number of these persistent juvenile features of humans are ancestral characters, authors who failed to take into account the principles of neoteny set out above have repeatedly cited them as arguments against our derivation from anthropoid apes. In reality, they do not provide such negative evidence. In the last part of this book, we shall need to return to consider these characters and their phylogenetic significance in great detail.

A second major law of all phylogenetic change, which is at least as significant as Dollo's law of irreversibility, is the *ectropy* of all organic development. We have already discussed this in chapter 5 and little need be added here. The general directional tendency of all organic evolution from the simpler and "lower" to the more complex and "higher" in itself has nothing to do with the directional constancy of the law of irreversibility. Although it is not justifiable to speak of a *law* of ectropic directionality, because of the many exceptions already mentioned on p. 96, the regularity that is seen is nevertheless so general that the phylogeneticist generally tends to rely upon it confidently. Within a group of related forms, the most complex is interpreted as the younger, more derived species. This has proved to be particularly justified in the stratigraphic evaluation of many indicator fossils from the mollusc phylum, such as with ammonites and various snails.

For behavioral research, the function of ectropy—that of generating novel, original, and higher states from simpler forerunners—is of great methodological significance. The relationship between the higher forms of life and the less differentiated forerunners from which they developed shows a remarkable analogy to the relationship between the organic and inorganic realms, as discussed on p. 87 and p. 90. Just as all living processes, so to speak, represent a special case of more general and ancient physicochemical processes, every living organism is in a certain sense a special case of the more primitive group from which it arose. All the properties and laws that prevail in the simpler precursor are, in a manner of speaking, still present at the higher level. But it would be misleading to claim, for example, that birds are "really no more than" reptiles. I would remind the reader of what was said on p. 42 about our use of the words "actually" and "only." Precisely with respect to what it *really* is, that is to say, with respect to its real distinctions, a higher form is always *more* than the ancestor from which it originated. It is something completely new, something that has never existed previously. The special features of the "higher" form are subject to laws and harmonies that are not only narrower and more specialized but simultaneously more complex and "higher," governing the newly developed entity of the higher organic system. Like the newly developed system itself, these laws and harmonies did not exist previously. Because of their more general validity, the lower, simpler laws are also included in the more specialized, complex laws and in fact provide the foundation for them.

This undeniable fact leads to an inevitable methodological conclusion. This is that an understanding of the narrower and more specialized laws of a higher form of life is just as dependent on knowledge of its more general, simpler precursors as is quite generally the case for successively inclusive natural laws that are investigated within the consistent hierarchical framework of the natural sciences. It is precisely with respect to assessment of the relationship between humans and their animal ancestors—the most central problem of comparative behavioral research—that these laws of ectropy are of decisive importance. Experience has shown that the strong emotional aversion shown by many otherwise quite reasonable people to the undoubted fact of our origin from ape ancestors is rooted in a misunderstanding of the process of "evolution." Or, to put it more exactly, there is a lack of understanding of the constitutive *creative capacity* of ectropy. As has already been shown with sufficient clarity, the fact that humans are derived from ancestors resembling the chimpanzee certainly does not mean that we are "really only" chimpanzees. It is precisely because of the most specialized and highest laws that govern us that many people would feel insulted if they were labeled as apes. And it is precisely with respect to the laws of human morality and ethics that we are *really* not apes but something entirely novel! Nevertheless, *alongside* these highest laws—or rather *within* them—the old, more general original laws that prevailed prior to the origin of humans still operate. *While it is true that not everything that is human is found in animals, everything that is animal is found in humans!* Knowledge of this animal within the human being is an unconditional prerequisite for understanding—indeed even for the *discovery* of—the novel, higher laws that first appeared with the emergence of human beings.

We have had to devote a very long chapter to the properties and laws inherent in the unique phylogenetic history of living organisms. This is because the methods of comparative ethology, as a completely phylogenetic science, are fundamentally influenced by these properties and laws. We now come to a second constitutive feature of living matter to which very similar considerations apply. The properties and laws governing all organic *entities* are also quite basic for the conceptual and practical methods of our discipline. For this reason, as has already been indicated on p. 91, we must devote an extensive chapter to them.

7

The Organism as an Entity and Analysis on a Broad Front

Anyone who looks openly at organic creation is struck by the wonderful and awe-inspiring *harmony* of organic systems. "How everything is woven tight, all acts and life as one unite." The immeasurably complex interaction of the parts or "subentities," the self-regulating equilibrium of complementary processes, the ability to restore the integrity of the harmonic system after the loss of parts or other disruptions: all of these are phenomena for which a causal understanding is still a long way off. But it is precisely this awe-inspiring nature, this *beauty* of organic harmonies that has misled many thinkers into interpreting the harmonic integrity of organic beings as something that is above or beyond nature. We shall return to this in our discussion of *vitalism*. For many authors, the word "entity" has become a slogan that is introduced as a pseudo-explanation of a complex organic process in the same unacceptable manner as the "extranatural factors" of life force, entelechy, instinct, and so forth. In chapter 5, we defined an entity from our causal-analytical standpoint as a system in which all the component parts are involved in a network of reciprocal causal influences. Entities are *regulatory systems of universal, reciprocal causal connections.* As was stated in that chapter, an organic entity *contrasts* significantly with all analogous systems of the inorganic world in that the component parts and the reciprocal relationships between them are fundamentally *different* from one another. As we have also already seen, this diversity among subentities is the outcome of an *evolutionary process* in the course of which a progressive division of labor has led to differentiation and subordination of the component parts. It is therefore quite misleading to say that the entity "is constructed from its parts." It is at least equally justifiable, and in fact a better reflection of the true course of events, to state that the parts are constructed by the entity. After all, they *arise* from the entity! One of the high points in the history of psychology is that it was *psychologists* who first correctly recognized this relationship between the entity and its parts. In the Philosophical

Prolegomena (p. 60), we discussed the criteria established for the *Gestalt* by [Christian von] Ehrenfels: the primacy of the entity over its parts, supersummation, and transposability. In fact, these criteria are characteristic not only of the holistic function of Gestalt perception as a special case of organic integrity but of all organic entities.

The primacy of the entity over its parts is perhaps even more evident in the phylogenetic and ontogenetic development of organisms than in the establishment of Gestalt perception. An egg is both in a literal and in a metaphorical sense a rounded entity which clearly possesses far fewer qualitatively distinguishable parts than the organism that emerges from it. This entity is thus just as clearly present *before* all of the many parts that are constructed *from* it in the course of embryonic development! Without losing for a single moment the character of a systemic entity, the egg turns into a caterpillar, the caterpillar into a pupa, and the pupa into a butterfly as the individual organs "develop" from the entity. In other words, division of labor increases and the parts become increasingly different from one another, at the same time becoming more directly dependent upon the entity. Exactly the same considerations as those applying to embryonic development are also valid for the relationship between the entity and its parts in the evolutionary history of organisms. We know virtually nothing about the first forms of life on our planet. Nevertheless, we can confidently state that they were self-maintaining, regulatory systems and that they possessed fewer differentiated parts than modern higher organisms. This is just as certain as the fact that in the creative progressive evolution of organisms these simple entities existed earlier than the differentiated and subordinated parts that arose from them at a later stage.

Ehrenfels' second principle of the *supersummation* of an entity is equally widely applicable to the organic world in general. A systemic entity with its complete, mighty hierarchical sequence of subentities is far, far more than the mere sum of its parts. Or, rather, it is only the sum of its parts for anyone who is blind to Gestalts. For anybody who is able to see a structured Gestalt, it is far *more.* Like *all* organic phenomena, it is a unique, historically developed achievement whose nature cannot be derived from the properties of the individual elements that it contains. Karl Marx's saying that quantity gives rise to *quality* is nowhere more appropriate than when applied to the phylogenetic developmental processes of organic creation.

In essence, we have already become familiar with *transposability* as a characteristic that is an intrinsic, constitutive property of every organic systemic entity: all of the elements involved in the construction of any organism are, of course, continually *changing*. Even individual organs change in the course of ontogeny and even the individuals that constitute the systemic entity of a species are continually changing. In species with a generational change, even the forms of successive generations differ distinctly from one another and yet the harmony of the self-maintaining, regulatory entity remains the same. The *melody* of the overall process represented by the life cycle of a plant or animal species remains constant in the entire flow of matter or itself develops in a creative manner to new melodies and harmonies of a higher order.

As is commonly the case with major discoveries, the discovery of the Gestalt in the field of perceptual psychology led to certain exaggerations in the generalization of the newly recognized principles. The dominant role played by constitutive properties of the *entity* as such in Gestalt perception led to the erroneous conclusion that the concepts of Gestalt and entity are identical. Wolfgang Köhler himself used the two words almost as synonyms. For example, he used the term *physical Gestalts* to refer to all of the regulatory systems that can be identified in the inorganic world as analogues of organic entities (p. 87, p. 90)! This can be seen as an unjustifiable, or at least inappropriate, extension of the concept of the Gestalt. As a *perceptual* phenomenon, the Gestalt is no more than a clearly circumscribed *special case* of organic holistic processes. Although it is undeniably true that every Gestalt is an entity, it is far from true that the reverse applies in the sense of Gestalt psychology. After all, there are an infinite number of extremely complex entities that we are quite unable to perceive as Gestalts. For any analytical natural scientist, who needs more narrowly and sharply defined concepts for his work, the more restricted concept of the Gestalt as a phenomenon that exists exclusively in the realm of perceptual function is quite indispensable. Clear definition of concepts is all the more necessary here because in idealistic-vitalistic circles the same misuse of the Gestalt concept occurs as with that of the entity. Many authors also present the Gestalt as something that is simply inexplicable, but above all as something that is not accessible to causal, "atomistic" explanation. When Gestalt psychology demonstrated the fallacy of the methodologically flawed monistic explanations of the so-called "scientific" psychology of the turn of the century, there was an understandable tendency to throw overboard not only the fallacious atomistic concepts but also quite justifiable attempts to find causal explanations. This was an unfortunate result of the conflict between vitalism and mechanism, which we shall encounter in a similar fashion in other related areas.

Even somebody with a purely atomistic-mechanistic orientation who is engaged in research into the world of organisms and has no interest whatsoever in the purely *subjective* phenomenon of the *Gestalt* must have acquired a thorough understanding of the *objective* nature of the *entity* as a "regulatory system of universal reciprocal causal connections." This is because this nature imposes quite specific, tightly circumscribed *methods* of analytical procedure on the research conducted. That is true even if—as is always the case in comparative behavioral research—we adhere to the mechanistic working hypothesis that all processes of an organism are fundamentally traceable to material processes, to "elementary" processes in molecules and atoms. For methodological reasons, we must not begin our observation and research with these elements or even with a hierarchically higher "part" of the entity with which we are concerned. But these methodological reasons by no means reside in some mystical special law governing the organic entity, as vitalists and proponents of related idealistic schools like to maintain, but in undeniably objective facts that are accessible to any causal researcher following an atomistic approach. In the inorganic realm, for research into individual causes and their effects it is always possible to find or contrive concrete test cases in which the causal chain shows little or no *ramification.* If we wish to study the influence of gravity on the free fall of a solid body, we can *exclude* the resistance effect of air by letting the object fall in a vacuum. We can experimentally *isolate* a single "causal thread" and investigate the effect of a single causal factor. In the inorganic world, causal chains usually operate independently of one another without any significant interaction *between* causal factors. In most cases, too, there is no feedback between the effect and the process that caused it. When a stone is thrown, its inertia operates in one direction, the resistance of the air operates in an opposite direction and gravity operates in a third. The solid body obeys *each* of these forces in just the same way as if the other two were not present at all. In other words, the resultant of the effects of all of the causal factors involved provides a fully correct synthesis of their common operation. Because of this, purely atomistic causal research in the inorganic realm is generally entirely justified and successful. There are only a few inorganic systems whose analysis requires a more holistic conceptual and operational approach. By contrast, in all organic systems it is fundamentally *not* possible to separate the individual causal chains from one another. In a system of universal reciprocal causal connections, the exclusion or modification of *one* causal factor always entails simultaneous modification of *all other* causal factors. If we experimentally remove the thyroid gland from a rabbit, we do *not* immediately distinguish the effect of the thyroid gland from the effects exerted by other endocrine glands. If we remove the effect of air pressure, inertia and gravity have just the same effects on a solid body as they did before this intervention. Following removal of the thyroid gland, however, literally all of the remaining endocrine glands function in a completely altered fashion. Through our intervention, we have by no means removed just one causal factor. We have in fact modified *everything* in the entire system of generalized, reciprocal causal interaction represented by the organism, most particularly the endocrine system. With the exception of a few special cases that present particularly favorable conditions for analysis, it is not possible to eliminate a part from the entity and to understand it in isolation. This is because the part that we have extracted has ipso facto been modified in a barely predictable fashion and represents something that is totally different from that which was incorporated into the framework of the organic entity. Exactly the same considerations apply to the remainder of the systemic entity from which the part has been removed.

A network of reciprocal causal relationships unites virtually all components within an organism; every causal thread that produces an individual effect feeds back to a multitude of other causal factors. Because of this fact, we are obliged to use a quite specific method of causal research that we will refer to as *analysis on a broad front.* Even for a purely mechanistic research worker, there is no other method that will permit analytical penetration into systems with generalized, reciprocal causal relationships. This can be illustrated by the example of a very peculiar organic entity that has been chosen because it is the only case among organic systemic entities that we can really analyze down to the physicochemical level. We can do this because we have ourselves synthesized it from physicochemical precursors. I am referring to a *machine* constructed by human beings. Vitalists have repeatedly rejected our machine analogies as inappropriate because, in their opinion, one cannot compare the inorganic with the organic. But in so doing they forget the fact that all machines are nothing other than human *organs.* Any zoologist, doubtless with full justification, will as a matter of course interpret the shell of a snail—which is clearly constructed of nonliving, inorganic matter—as an organ of the snail. But there is no valid reason at all for seeing anything fundamentally different in a locomotive or a microscope! As I often used to say in lectures for the sake of effect, microscopes, locomotives, and cars are never found in the *wild state* like the aurochs or the Przewalski horse. It is, of course, completely clear to us that even the most complicated machines produced by humans do not even come close to the simplest known organisms in terms of the *complexity* of the universal network of reciprocal causal relationships that exists within them.

Even the automatically connecting long-distance communication system of a modern city or the calculating machine are immeasurably simpler than a yeast cell or an amoeba. On the other hand, such machines are far more complex than the most complicated inorganic systems known to us. Further, they possess two properties which are found exclusively in organic systems and which are lacking from the most complex, most regulatory "physical Gestalts." In fact, they exhibit *finality* in the sense of purposive survival value and they possess the hierarchical sequence of interdependent, fundamentally different subentities that we discussed on p. 137 et seq. For precisely this reason, they can be used to illustrate the difficulties that face any analysis of organic entities. At the same time, they illustrate the only methodological procedure that can cope with them.

Let us assume that, as inhabitants of Mars, we have just landed on earth with a spaceship, and we are faced with the task of "analyzing" a machine built by human hand. For the sake of general comprehensibility, we can take the example of a car. First of all, this thought experiment makes clear to us the purely *analytical* significance of the finalistic question. For sure, we will never learn to understand the relationship and interactions between the parts of our object until we have identified the *purpose* of the machine. We must first find out that it is a vehicle that is used as an organ of locomotion by the inhabitants. But we will have little success in further analysis if we remove a single component of the vehicle—such as a piston ring, the fuel jet, or the valve of a tire—and attempt to understand it *in isolation*. We will have even less success if we attempt to resynthesize the systemic entity from the few parts that we might, by mere chance, have managed to isolate. This is not an absurd jibe. It quite literally corresponds to the approach taken by certain lines of psychological research, such as the association psychology of the turn of the century. We shall be returning later to their monistic explanations.

It is fundamentally impossible to understand the function of a component of a systemic entity unless we can also *simultaneously* survey and interpret the functions of all other parts. The fuel jet and its function can only be understood if we have already understood how the downwardly moving piston sucks the mixture from it. But this, in turn, can only be understood if we have already worked out the function of the connecting rod, the crankshaft, and the flywheel. This is necessary in order to understand how the piston obtains the energy needed for its downward movement. Yet we will not even grasp the fact that the flywheel turns and thus delivers the energy necessary for the three passive phases of the cycle until, in addition to understanding other parts such as the valves, the camshaft, and the magneto, we have finally been able to interpret the function of the carburetor itself. Without simultaneous insight into the functioning of all other parts, no single part or individual function can be understood. No part can be analyzed *before* the others. This fact is also very clearly expressed by the observation that an instructor is continually obliged to *jump ahead* to refer to other parts that have not yet been explained and, in a quite unjustified manner, to take them as preestablished fact. This applies to any *didactic* presentation of such an organic entity, which must of course *start* at some arbitrarily selected point. Somebody lecturing about the four-stroke engine first describes the moving parts—the crankshaft, the connecting rod, the pistons, and the valves—and then begins to explain the four phases of the cycle. He might then say: "At this point, the exhaust valve opens," or, "This is when the gas mixture flows into the cylinder," or, "This is when ignition takes place," although the apprentice in fact cannot yet know why the valve opens, what the gas mixture is, or what is meant by ignition. The instructor hopes, so to speak, that the apprentice will *provisionally* have an approximately correct idea of what is meant by these preemptive expressions, which do not yet represent actual concepts, and that he will *reserve* a place in his picture of the system concerned for explanations that will not be presented until a later stage. Understandably, the didactic difficulty involved in the presentation of far more complex real organic entities is infinitely greater than in our grossly simplified example.

Of course, initial research into systems with generalized reciprocal causal relationships faces the same fundamental problems as we have just witnessed for instruction about them. The impossibility of isolating one part and understanding it *before* other parts makes it absolutely necessary *to begin the investigation with simultaneous observation of all parts at once!* Because one can never isolate a single causal thread from a system of universal reciprocal causal relationships, it is unavoidably necessary in the analysis of such a system to examine, as far as possible, *all* of the functions *simultaneously*. Tracing of the many causal threads that link them to all of the causal factors involved must also be tackled simultaneously. A physicist is in the fortunate position of being able to isolate a single causal thread. Without looking to the left or right, he can justifiably expose a long, linear causal chain in a single "deep drilling exercise." But the biologist is faced with something that O[tto] Koehler has aptly described as a *causal blanket*. He is confronted with a multitude of causal processes, each of which represents the end of a causal chain. Yet each of these causal chains leads back not just to *one* cause but ramifies to link up with a large number of causal factors. And these many causes are *the same* for *all* of the observed effects in the systemic entity of the organism!

An infinitely complex network of ramifying causal chains links every causal factor to every effect, and every effect feeds back to all causes. We find ourselves in the same situation as a man who wishes to unravel a tissue composed of a multitude of threads and can only succeed in doing so by watching all of the threads simultaneously and unraveling all of the threads at the same pace. As soon as the unraveling of a single thread is just slightly ahead of that of its neighbor, the side chains extending from this thread into the surrounding, as yet unraveled tissue inhibit further progress. Only when the neighboring tissue has been unraveled to some extent is it possible to proceed a little further with the pursuit of the first thread. But if the attempt is made to pull a thread further out *by force*, a thick clump is formed by side chains and by other, overlapping threads, preventing any further progress. If even more force is applied, it might be possible to pull the single thread out completely and to isolate it down to the end of its strongest terminal branch. But this can only be achieved *by tearing all of the side chains.* This last detail of our metaphorical example tells us that the forceful isolation of a single causal chain through atomistic research that is blind to the systemic entity always leads to *corruption* of the result obtained. This is because lateral inputs from causes contributing to the phenomenon investigated are neglected. In the process, one can identify *one* out of x different causal factors. The best that can happen is that this is the most important causal factor involved, but $x - 1$ further causes are omitted in an extremely misleading fashion.

Accordingly there is *only one methodologically legitimate approach* that will lead us to a genuine *causal* understanding of the systems with generalized reciprocal causal relationships that we refer to as entities. In analogy to the metaphor that has just been presented, we refer to this approach as *analysis on a broad front.* If our considerations are translated into the general practice of all biological research, the following applies: any organism that is taken as a subject for research must *first of all be grasped as an entity* before we attempt to understand individual components or contributory functions. This holistic examination, which is merely a *preparation* for further analysis, must also include observation of the encompassing entity constituted by the organism and its habitat. This represents a major prerequisite for understanding the *finality* of all its organs and their functions, which is itself the precondition for any successful causal analysis of those functions on a broad front. While pressing ahead with partial analyses across the spectrum of individual functions, as far as possible simultaneously from every conceivable direction, we should never lose sight of the way in which a given part fits into the overall entity. This is true even in occasional cases where favorable conditions permit particularly penetrating analysis of an individual component. In his book dealing with the Gestalt, [Ruprecht] Matthaei hit upon a very apt analogy for this kind of procedure. He states that the analysis of an entity must employ an approach resembling that taken by a painter. The first step is to make a rough sketch of the overall entity, after which we must progress at the same pace with the development of all parts of the picture until the finished entity is revealed to our gaze in all of its convincing vividness.

These general conclusions, which are imposed upon the process of analysis by the nature of all systems with generalized reciprocal causal relationships, must be borne in mind with special care in all *experimental* investigations. When observing the *effect* of a given intervention on an organic system, we must never assume that this intervention *alone* is the cause of the observed effect. Instead, any claim of this kind must always be preceded by the reservation "provided that all other (in some case unknown) causes *within* the organism operate in the same way as at the time of our experiment." This, too, can be illustrated with a metaphor! Let us assume that, at a given point in time, we have switched a set of points in a large railway complex and that this intervention has "caused" a collision between two trains. It is undoubtedly correct to state that our intervention was the cause of the collision. But in our experimental intervention this cause will only produce the observed effect if the settings of *all other* points in the paths of the two trains (still unknown to us at the time of the experiment) remain the same. Thus, to speak of *one* cause of the observed effect requires the unspoken assumption *ceteris paribus,* as discussed above. The truth of the matter is that the instantaneous setting of every one of the other unknown points is just as much a real cause of the collision as our deliberate modification of a single set of points! A genuine understanding of the causal origins of the effect of our initial, experimental causal intervention requires considerably more work. We must first of all conduct a vast number of additional experiments to discover the number, identity, and settings of all other points involved in the overall process in order to determine how a particular end result was caused.

A particularly good example of analysis on a broad front that is really appropriate to the overall entity is provided by research on *endocrine glands.* These glands, which release their secretions into the bloodstream, constitute an extremely complex system in which opposing, "antagonistic" functions of the individual glands operate in a regulatory fashion to maintain a finely balanced equilibrium in metabolic processes. This system of endocrine glands constitutes an entity in the truest sense of the term in that every part is literally linked to every other part in a network of reciprocal causal relationships. The function served by *each individual* endocrine gland first became clear, and *could* only become clear, when the function of *all* of them were known, at least in broad outline. It was first necessary to achieve a synopsis of the harmonious system of interactions between the glands as *a whole,* permitting a *simultaneous* view from all sides. We first needed to find out which glands belong to the system and then to conduct experiments to determine the influence exerted by increase or reduction in the function of any one gland on the functioning of every other one. Only then were we in a position to draw clear conclusions regarding the significance and function of the individual parts and of the overall entity.

Thus far, we have presented and discussed the general rule that in any organic system the individual components are all involved in reciprocal relationships. A point has now been reached where we must consider the exceptions and restrictions to this rule. It is very fortunate for analytical research that there are in fact exceptional cases in which it is both possible and methodologically legitimate to isolate a *part* from an entity and to investigate it in isolation. As it happens, the definition of an entity as a system of *universal* reciprocal causal relationships does not apply *without restriction.* For reasons of didactic simplicity, mention of this has so far been omitted. There *are* within organic entities *certain parts* whose form and function are not significantly influenced by the overall system but which *themselves* exert a major influence on the form and function of the entity to which they belong. Such parts are usually "entities" in their own right, but in the atomistic sense they behave as "parts" or "elements" of the encompassing entity rather than as "subentities" in the sense of Gestalt psychology. For such parts, we shall now introduce the term *relatively entity-independent components.* In their completed state, all fixed *structures* of organisms represent such relatively entity-independent components. In its completed condition in the adult organism—but of course not during its development!—the articulated skeletal apparatus of any vertebrate or arthropod can be regarded as a component, as a genuine *part* of the system. It has a *unilateral,* nonreciprocal relationship to the overall entity. It exerts an undeniably strong influence on the form and function of the entity without itself being subject to any substantial influence.

Understandably, the discovery of such relatively fixed and entity-independent structural elements greatly *facilitates* the analysis of all of the phenomena that are associated with and causally influenced by them. As a result, entity-independent components always constitute the points of attack, the levers for initial penetration of analytical research into the entity of the organism. In research, as in teaching, morphology begins with the *skeleton.* In a similar way, all branches of biological research first direct their attention to the *least variable* phenomenon within the entity that is being investigated. The presence of an entity-independent component always represents an exception to the rule of generalized reciprocal causal relationships and permits analytical penetration. This applies even to conceptual and practical approaches that do not acknowledge the nature of the entity and completely ignore the operational principles governing analysis on a broad front. Wherever entity-independent components are present, even purely atomistic research methods are completely successful and methodologically legitimate.

In the realm of research into animal and human behavior, there are three great schools (two of them still in full swing) characterized by a purely atomistic conceptual and practical approach which have achieved fundamental and everlasting successes through the discovery of an important entity-independent component. These are the association psychology of the turn of the century, behaviorism, and (by far the most important) the reflexology school founded by I[van] P[etrovich] Pavlov. The entity-independent components whose discovery led to the major advances in causal analysis achieved by these three incrementally related mechanistic schools were basically the same in all three cases: the *reflex* and the *conditioned reflex.* The reflex and the conditioned reflex are veritably classic examples of relatively entity-independent components. In section III (Historical Origins and Methods of Comparative Behavioral Research), we shall examine in some detail the unsuspected new possibilities for progress in genuine causal analysis of animal and human behavior that were opened up by the discovery of the conditioned reflex by Pavlov. On the other hand, we shall also need to show how the overly atomistic approach of the mechanists, which methodologically ignores the holistic character of the organism and offends against the laws of analysis on a broad front, lapses into certain fundamental errors. This always occurs at the point where causal analysis *passes beyond* the limits of the relatively entity-independent component, exceeding the exclusive confines within which the purely atomistic conceptual and practical approach is methodologically legitimate. As soon as attempts at atomistic explanations drift beyond the unidirectional causal chains of the entity-independent component into the confusing "causal blanket" of universal reciprocal relationships, they *necessarily* fail because only analysis on a broad front that is appropriate to the systemic entity can assist us here. Pavlov's monumental achievements in the investigation of unconditioned and conditioned reflexes fully deserve our admiration. At the same time, methodological and objective criticism is both necessary and justifiable wherever his statements are not confined to those entity-independent components but concern the *entity* represented by the organism itself.

As we have seen, a "relatively entity-independent component" has a *unidirectional* causal relationship to the system of which it is a part. For this reason, it can be investigated in its own right, that is, without simultaneous consideration of the causal blanket of the rest of the entity concerned. As has already been explained, it is customary to begin research or teaching with such "skeletal elements" or "fixed inclusions" of organic systems. This is justified because, when the complete organism is observed, an entity-independent component almost always serves only as a *cause* and virtually never as an *effect* of other phenomena. It is precisely for this reason that it so greatly simplifies our research effort and presents such extremely welcome starting points. In exploiting these possibilities for a method of analysis that ignores the nature of the entity, we must consistently remember that this method is, so to speak, *provisional*. It is *restricted* to a quite specific component and, in a certain sense, it is even *unbiological*. Entity-independent components behave more or less like inorganic *inclusions* in the organic entity. The skeletal element is in a quite literal sense such an inorganic inclusion in an organic system and this applies almost as literally to all fixed structures along with all of the functions associated with them. We must never forget that the *core problems* of life certainly do *not* reside in these relatively easily analyzable structures and structural functions. In effect, we merely attempt to "analyze away" such features in order to demarcate and hence come closer to those core problems. Of course, anyone concerned with research into an organic entity will happily seize upon any relatively entity-independent component that will, at least for a brief interlude, remove the need for the demanding and painstaking method of analysis on a broad front. But, while preoccupied with this interlude, the research worker must never lose the readiness to return thereafter to the only method that is appropriate for dealing with entities. In addition to the single causal thread that is being examined, all of the other threads within reach must eventually be picked up as well.

In our own field of research, the *instinctive motor pattern* along with *automatic rhythmic, centrally coordinated stimulus production* constituted a primary entity-independent component, a function mechanically attached to fixed structures. This presented the possibility for "deep drilling" with atomistic causal analysis and hence determined the nature and method of the entire line of research. We shall also begin our didactic presentation of the entity constituted by animal and human behavior with this particular "skeletal element." But we shall never forget, and the reader will never be allowed to forget, that this component is not the *only* one that is involved in the construction of the entity. We shall not lapse into the methodological error for which we must reproach the great mechanistic schools, namely, that of regarding the newly discovered element and explanatory principle as the sole essential feature and of attempting to synthesize the organic entity of animal and human behavior on this basis alone. Just as the discovery of an important entity-independent component can exert a major promoting effect on research, such monistic explanations can exert a major inhibitory effect.

8

Finality

As has been explained in chapter 5, it is the regulatory, self-maintaining entity of the organism itself which renders the question of *purpose* both possible and necessary. It introduces the concept of the *significance* of an organ, a structural feature or a function for the maintenance of the system. If we ask, "Why does a cat have sharp, curved claws?" or, "Why does a duck have a sieving apparatus composed of horny lamellae in its beak?," these questions make sense because a cat and a duck are self-regulatory organic systems. In the course of their phylogenetic development, the organs concerned actually arose *to serve* a particular function. Their differentiation took place in intimate association with the emergence of the function they now serve. It is only in the organic realm that we find such parts of an entity that were developed in just this way to promote the regulatory self-maintenance of the overall system. In the sense of Goethe's definition of development, they have become "differentiated" in association with an increasing division of labor and they have become progressively more "subordinate" to the entity. Even in the most complex inorganic systems, there is *nothing* that is really comparable to this! For this reason, the question of finality is fundamentally never applicable to phenomena in the inorganic realm. "For what purpose are planetary trajectories elliptical?" "For what purpose is oxygen bivalent?" "For what purpose do masses exert reciprocal attraction?" These and similar questions have no sense unless one wishes to give the naive anthropocentric answer that all of these things are as they are so that human beings can live on earth as the summit of creation.

Finality in the sense of survival of a system or a species exists therefore only in the realm of living organisms. Because there are no organic systems that lack such species-preserving finality, many observers of nature—in fact, precisely those who are most responsive to the miracle of organic creation—are inclined to indulge in an unjustifiable generalization of this statement. They maintain that every phenomenon that can be observed in an organism, every structural feature, and every behavior pattern must have a purpose and must further the survival of the species. Vitalists, who see the effects of a directional "life force" in every aspect of life, are particularly prone to such extravagant exaggeration of the principle of purposivity. They fail to recognize the fact, well known to any physician or any good zookeeper, that organic systems can easily become imbalanced. The maintenance of the system can be threatened by ludicrously small disturbances which can be easily corrected, even with very limited human insight into the relationships among life processes. When confronted with such disturbances, the "entity-preserving" and "directional" factors of the vitalists collapse completely. In what follows, we shall see repeatedly how often and how easily "infallible instinct" can be thrown off course. It is advisable to bear in mind Heinroth's aphorism, mentioned above, to the effect that the organic realm is characterized not only by features that serve a purpose but also by features that are not so unsuitable that they set in motion the mechanism of natural selection that would eliminate all carriers thereof. Those who worship finality fail to see the stubborn conservatism with which evolution can drag along, as worthless ballast, features that are neutral with respect to purpose.

Accordingly, there are some life phenomena for which one must pose the finalistic question and others for which we are *unable* to do so! It is far from easy to distinguish between these two possibilities in an individual, concrete case. But in all biological research the entire success of the analysis depends upon the correctness of this distinction. This is particularly true of the subdiscipline of *medicine,* in which the question "normal or pathological?" arises at the beginning of any investigation. For reasons that will be explained later, the same question is just as important in the daily work of the comparative behavioral investigator and this leads to a number of interesting covergences between medicine and behavioral research. Hence, at the beginning of our investigation into any individual case, we must decide whether a given phenomenon displays a system-maintaining effect or whether it is neutral or even detrimental with respect to the survival of the species. In obviously *pathological* cases, the absence of species-preserving finality is directly apparent. Nobody would inquire into the purpose of the flat feet or corns that some people have. For what purpose are squirrels sometimes rufous and sometimes black in coloration and why is the color not the same for all individuals as with almost all other wild animal species? This is similarly a question that obviously has no sense. The fact that rufous and black individuals survive equally well demonstrates the fact that the difference between the two colors is of no importance. As far as the survival of the species is concerned, it would make no difference if all squirrels were either rufous or black. Let is consider another question which is apparently equally pointless: Why does a pug have a curly tail? At first sight, it would seem that this question is also completely valueless. But if we remember that dog breeders will only breed from individuals with exactly standardized features, including the curly tail, it can be seen that this feature is of direct significance for the survival of the pug. Only pugs with a flawless curly tail have any prospect of producing descendants. Any vitalist, with his finalistic rather than causal approach, would doubtless object that a pug does not possess a curly tail in order to be selected by a commission of dog breeders. This is quite correct. But the pug has a curly tail *because* selection eliminates individuals that lack this feature. This is precisely the same reason as that which explains the possession by any organism of characters that promote survival of the species. The domestic cat, the species *Felis domesticus*, does not "know" why it has sharp, curved claws. In this case too, it is not *internal* factors with some kind of final "directionality" but purely *external* factors, namely those associated with selection, which have favored the development of the character from the raw material offered by mutation.

Wherever the relationship between form and species-preserving function is obvious, one is inclined to forget the underlying set of causal relationships. The short, robust, and bowed forelegs of the dachshund are so obviously well-suited for digging out foxes that it is all too easy to forget that this character was by no means developed as the result of a natural interaction with foxes. Instead, this feature owes its origin to exactly the same processes of mutation and "artificial" selection on the part of the breeder as the silly curly tail of the pug! As far as we know at this point, every adaptation results from the same process of mutation and selection. It is completely immaterial to the nature of this process whether selection leading to specialized species-preserving purposivity in the structure and function of the organisms concerned is brought about by a deliberate breeding strategy on the part of a human or by elimination through natural predators. It is in this sense that we should understand the statement made in chapter 5 (p. 91), that the question concerning the finality of a structural feature is identical with the question as to the *function* for which the organ concerned was developed.

For this reason, the other question as to whether a given structural feature serves a purpose with respect to survival of the species becomes easier to answer as its differentiation becomes more advanced and specialized. Understandably, the greater the specialization and the greater the complexity, the more improbable it becomes that this particular complicated and regular combination of characters could have arisen purely *by chance.* If we are confronted with a highly differentiated organ, such as the sieve apparatus in the beak of a shoveller or the tongue of an anteater, we will not doubt for an instant that the structure was developed to serve a function of great importance for the survival of the species. In such cases, therefore, we would not even bother to ask *whether* species-preserving finality is present. Instead, we would immediately turn to the problem of determining *how* this finality is expressed. By contrast, if we are faced with a character for which the survival value is not immediately obvious, quite complicated analysis is required to track down any purposivity that may exist. Take, for example, the question of why female dwarf cichlid fish of the genus *Apistogramma* have a yellow and black pattern of coloration. Only an exact analysis of the unconditioned reflex responses of the offspring, which respond to this particular color pattern by following, can answer the question as to whether this color pattern of "brooding" females has a survival value. This question can only be answered in this way for a large number of innate behavior patterns and the answer must be obtained right at the *beginning* of our investigation. Otherwise, our causal analysis will not be successful.

The question concerning finality is of extraordinarily *great* significance, particularly for progress with causal analysis in research into innate species-specific behavior patterns. If a tame jackdaw is flying over my head out in the open, its wings are shifted slightly backward to reduce the burden on the lifting surface of the tail. The tightly compressed tail is thus allowed to perform peculiar, lateral wagging movements. To anyone familiar with these birds, it seems at least probable that this motor pattern, which is regularly performed in a quite specific context, has some kind of "significance." But once again we can only find out what this is by studing the unconditional response that is given by conspecifics to this display pattern.

Finding a correct answer to the question about finality is also important for anyone conducting research into underlying causes. For the investigation of organic entities, it is just as *obligatory* as the method of analysis on a broad front. Nevertheless, the great importance that should therefore be attributed to the question about finality must not mislead us into believing that this is the only problem involved in research into all organic entities. Clear recognition of finality always has an "illuminating" effect in such research. In the truest sense of the phrase, things are really seen in a new light! But it is precisely the vivid experience of recognizing purposivity with respect to survival of the species that leads many research workers to forget that this step is far removed from an *explanation* for the phenomenon concerned. Many natural philosophers examine organic phenomena in the natural world from a *teleological* viewpoint (from the Greek τέλοσ, goal), that is to say in terms of a vitalistic concept of finality. Incomprehensibly, they regard the demonstration of the finality of a phenomenon as an explanation of it, conclusively satisfying their quest for understanding. Particularly within our own field of research into "instincts" there are many highly respected authors (McDougall, Bierens de Haan, and others) who treat an investigation into a given behavior pattern as concluded once its purpose with respect to survival of the species has been demonstrated.

It is not only in the formal logical sense that it is a gross error to answer the question "why" with the formula "in order to." This conceptual error in fact has almost catastrophic consequences for the *practical* research work of anyone who confuses and confounds finality and causality in this way. I shall illustrate the great difference that actually exists between finalistic and causal questions, and the great need for a sharp distinction between them in practical life and research, by means of a metaphorical example. Let us assume that I am driving a car through the countryside with some easily understandable purpose in mind, for example, to give a lecture at a congress. My thoughts drift into contemplation of the wonderful construction of my car and of the purposive design of its mechanisms. I revel in the holistic nature and careful planning of its construction and marvel at the functional service that my small car as an entity provides for the higher goals of my lecture tour. This is a whole series of finalistic interpretations which are in themselves utterly correct and scientifically legitimate. But does such recognition of finality render a *causal* understanding of the phenomenon concerned *superfluous?* Does the one *replace* the other in some way? The fact that this is *not* the case is made forcefully clear to me just as soon as the smallest defect, be it only a droplet of water in the fuel jet, *disrupts* the entire hierarchical edifice of finalities. The motor "coughs" a few times and then comes to a halt! Given this state of affairs, it does not help me in the slightest to know that I *should* arrive at a particular time at the venue of my lecture. The urgency of this finality has no influence whatsoever on the water droplet in the carburetor. The finality is, so to speak, *not available on demand.* It does not render the network of concrete causes superfluous; it does not intervene as a "factor" in the causal sequence of events. If I wish to attain the goal and purpose of my voyage at all, I am well advised to free my mind of any of my previous thoughts about finalities. Instead, I should direct all of my energies toward answering the question as to *why* the motor has come to a halt. Only *insight into the causality* underlying the origin of the purposive phenomena and—in this particular case—into the *disruption* of these causal chains will permit me to correct the defect and restore the impaired entity of the purposive system. Such insight by reflective human beings is the only "holistically constructive" factor that exists. Medicine, the queen of the applied natural sciences, is concerned almost exclusively with investigation of the *causes* of pathological disruption in order to be able to restore the workings of an entity that has gone astray. Even in the realm of living matter, the existence of finality does not free us from the rigid sequence of causal chains involved in life processes. Finality is unable to replace even the smallest component that has ceased to function!

Accordingly, for any research worker the demonstration of finality simply means that a holistic *system* of life processes is present. In no way does it provide insight into the universal reciprocal causal connections of the system concerned. *Demonstration of finality always indicates the existence of a problem for causal-analytical research.* Thus, contrary to the common assumption made both by vitalists and by mechanists, finalistic and causal considerations of any given phenomenon do not represent two mutually exclusive viewpoints. They do indeed represent two entirely different aspects and whenever they are confused and confounded the effect is quite detrimental—as, for example, in answering the question "why" with an "in order to." On the other hand, however, if the concepts and methods associated with finality and causality are clearly separated, they are just as compatible with one another as are the concepts of entities and analysis. For a holistic analysis of organic systems, the question of finality is quite *indispensable,* since the answer is, of course, required before we can appreciate the role played by a component within the entity concerned. In the investigation of organic entities, the finalistic and causal aspects must be studied simultaneously and in connection with one another.

9

The Mind-Body Problem

The concept of the *mind* is an age-old inheritance of ancient and oriental philosophy. From a tender age, everything to do with the problem of body and mind or soul has been influenced by every word of our parents and teachers, by the entire authority of the Christian religion and idealistic philosophy, by every book written by our great poets, and even by the expressions of idiomatic speech. As a result, the conviction has been hammered into us that the mind is something that exists in its own right and is independent of the body. In addition to this, any contemplation of the mind must necessarily be an introspective contemplation of one's *own* mind. And this very mind, the presence of an individual ego, is the *most certain* of all things. Indeed, it is the only thing that is beyond any doubt. The concept of the mind is one half of a pair of opposing concepts which could not exist at all without the counterpart, the concept of a mindless body. Now a mindless body is something with which we are familiar in real life. Not just every corpse but also any thoroughly unconscious body directly confronts us with the possibility that a human body can exist *without* a human mind. Yet this fact, together with the preconceived notion that mind and body represent equal opposites, leads to the quite unjustified conclusion that the mind in turn can also exist independently of its counterpart. The complete disappearance of the mind or soul from the dying body is interpreted not as its extinction but as its separation from the body. However, although there is abundant empirical evidence for the possible existence of a body *without* a functioning mind, the corresponding assumption that a mind or soul can exist without a body is purely dogmatic. It is based on no real experience, insofar as one refuses to accept the results of spiritual séances and similar absurdities as factual scientific evidence. But the existence of a "mind" independent of any bodily processes (particularly nervous activity) can be seen to be particularly improbable from observations of cases in which the mind suffers persistent and incurable modification because of purely bodily effects, following damage to the central nervous system. The "mind" of somebody who is paralyzed or has a major brain injury lacks its highest and most valuable properties and functions. Under such circumstances, someone who was previously morally and intellectually outstanding can become a morally uninhibited being, often exhibiting virtually diabolical wickedness. For anybody who is familiar from direct experience with such shocking personality changes, the assumption made by those who believe in the soul that the former noble, good soul of the invalid will be *restored* at the time of death is completely nonsensical! We must ask why the *destruction* of a brain by death should restore something that was so massively, irreversibly affected by partial disablement of the central nervous system in life? As inductively operating natural scientists, we must base our conclusions regarding the relationship between the mind and the body on the undoubted, universally evident empirical fact that spiritual processes simply *do not exist without* parallel processes operating in a living organism, indeed without processes operating in a brain. Without doubt, there is a completely inviolable connection between all spiritual phenomena and bodily processes. The ancient Chinese philosopher K'ung Fu-tzu—the Confucius of Western medieval philosophers—used a wonderful metaphor for this connection. He said that the relationship between the mind and the living body is like that between a cutting edge and a knife. A cutting edge without a knife is inconceivable, while a knife without a cutting edge lacks the most essential feature inherent in the concept of a knife.

Given such considerations, it is extremely tempting to equate the concept of possession of a mind directly with life itself. At first sight, it is indeed very enticing to equate or confuse the gigantic developmental leap from mindless life to the possession of a mind with that other, even greater leap from inorganic to organic matter. Countless great philosophers have done just that and continue to do so even today. *No* reasonable observer who is close to nature can doubt that all spiritual phenomena are *life processes.* But, as we have already seen in the chapter dealing with the consistent hierarchical system of the natural sciences [chapter 3], this statement can no more be reversed than the other, equally correct, statement that all life processes are in turn physicochemical processes. By no means are all physicochemical phenomena life processes. This applies only to a very few, special cases. Similarly, by no means are all physiological process of the body accompanied by spiritual processes. This, again, is true only of very few, special cases. It is part of the nature of organic evolution that novel features are added to that which already exists, completely modifying the quality and the nature of what existed previously. At this point, I would like to translate

Confucius' metaphor into genetics: The production of a knife—which includes the presence of a cutting edge as one of its essential features—is only conceivable if we take a long, flat metallic object that in its initial form is *not yet a knife* and hone a cutting edge. In exactly the same way, in organic creation, the qualitative leap to something completely new can only take place by the addition of something that did not previously exist to that which existed previously. Human beings arose from a quite chimpanzee-like ancestor by the addition of the human intellect to the chimpanzee mind that existed previously. This developmental leap from a being possessing a mind to one possessing a mind *and* an intellect is barely smaller than the leap from mindless life processes to those accompanied by a mind.

Experience tells us that many thinkers, especially those with vitalistic leanings, have a deep, emotional aversion to accepting the existence of life without a soul, particularly the existence of soul-less *animal* life. Indeed, the word *animal* itself is derived from the Latin *anima,* meaning soul! This attitude stems from certain forms of thought and experience which are innate to human beings and which are difficult even for a great thinker to shrug off. Human beings possess certain innate receptor correlates that respond specifically to the expression of spiritual processes by their fellows. These permit us to "understand" immediately certain expressions of life in our congeners and to experience them directly as *spiritual.* We can *share* the experience as something that is directly equivalent to our own spiritual processes. As has been said, this experience is *immediate,* that is to say, lacking any need to draw conclusions from rational analogies. The old idea that we can only experience the minds of fellow human beings by drawing analogies is *fundamentally wrong,* as was Helmholtz's belief that depth perception depends upon rational "conclusions!" Our shared experience of pain, joy, and other experiences of our fellow human beings is *not* based on conclusions drawn from our own spiritual life. Instead, our sharing of experience and emotion is quite definitely based on innate forms of experience which are a priori in Kant's sense in just the same way as our experience of space, time, and causality. The firm foundation for this claim will be examined in detail in the general part of this book,* in the section dealing with the innate releasing pattern. From an epistemological point of view, the immediate *obviousness* of the existence of experiencing fellow subjects is closely related to innate "a priori" forms of thought and interpretation. The closeness of this relationship is strikingly clear from the fact that the great idealists Kant

* *The author is referring here to a planned later section of the book that was never written. See the footnote on p. xliii.*

and Schopenhauer definitely do not treat the existence of comparable fellow beings as something of which we are aware only through the evidence provided by our sense organs. Both philosophers treat the souls of fellow human beings not as a mere property of the phenomenal world, whose relationships to that which exists a priori are fundamentally unknowable to us, but decidedly as something that belongs to the surely existing *intelligible* world. It belongs to that ideal-spiritual world to which the laws of reason, the a priori forms of thought and intuition also belong.

Now our innate forms of experience, which permit us to experience certain life expressions of other beings immediately as something spiritual, possess the character of so-called *innate releasing patterns.* As a result, at the physiological level they possess all the properties of the *unconditioned reflex,* while at the psychological level they represent highly species-specific forms of experience that are *anthropomorphic* in the truest sense of the term. For this reason, they exhibit the blind incorrigibility that is inherent to so many unconditioned reflex processes, whereas at the psychological level they are very prone to deception like all of the forms of experience that we label as "only subjective" in colloquial language. Hence, they respond extremely readily to environmental situations that in reality do not possess the features that would be required for "adequate correspondence" of these anthropomorphic forms of experience (to use Kant's expression). Or, to use the language of comparative behavioral research, they respond in situations other than those for which they were developed as receiving organs, as receptor correlates in the course of human evolution.

This kind of response by innate releasing mechanisms whose function is adapted to the expression of spiritual processes *in fellow human beings* is *erroneous* in the sense of species-preserving finality. One very common outcome of such responses is that spiritual properties are, so to speak, *projected* onto *nonhuman* organisms and even onto inorganic objects. For example, almost every superficial observer will experience the upright body posture of the crane as "proud," because there is an external correspondence to the expression of this sentiment by human beings. An eagle is perceived as "bold" because of the form of the slightly downturned, tightly closed angle of the mouth and because of the two bony struts above the eyes, which give the impression of furrowed brows. These features evoke a response from an innate schema adapted for the expression of fearless determination in human beings. Even features of the inorganic world are experienced anthropomorphically as possessing spiritual properties because of processes of this kind. For anybody who is fully familiar with the experiential processes concerned, a steep projecting cliff face or a dark, towering storm cloud have the same immediate and incorrigible experiential effect as a *threat* from a fellow human being. We speak of "happy meadows" and the poet sings: "The sea is smiling and invites us to swim."

Any natural scientist who is professionally concerned with *objectivizing* his own experience rapidly comes to understand that an experienced phenomenon will by no means always be accompanied by the real processes and properties that our subjective experience projects onto them. He will notice, for instance, that—from a purely physiognomic point of view—a crane seems to be particularly proud when it adopts a pronounced upright posture in the rain in order to make the water run off its feathers better. In reality, the crane's spiritual state, or (to put it objectively) its neurophysiological state, is that of a "soaked spaniel"! In the same way, anyone who lives for some time in the company of a tame eagle will unerringly notice that this beautiful bird continues to bear the expression of wild and fearless determination even when its mood reaches the heights of "warmth" and "tenderness." The expression is an unchanging property of its fixed head structure. Of course, it is even more obvious that a cliff is not really menacing and that the sea does not really smile. In all of these cases, insight into the function of innate releasing mechanisms can relatively easily correct the "false image" that is produced by the distorting effect of the "reading glasses" worn by our forms of experience.

As has already been mentioned on p. 43, correction of such false images is much more difficult in cases where we are confronted with behavior patterns of living organisms that *really* are functional analogues of human behavior with spiritual concomitants. When an amoeba responds to a prick from a dissecting needle with "painful" contraction at the point of contact, or when it "greedily" thrusts out a pseudopod toward its prey, these "negative" and "positive" responses or taxes are really analogous to equivalent human behavior patterns. They are hence analogous to behavior patterns that are associated with the experiential qualities cited in quotation marks. A comprehensive, thorough knowledge of the behavior of lower organisms is required for the researcher to realize that the responses of the lowest animals are far more closely related to responses of the human nervous system that are *not* accompanied by experience than they are to those that have such accompaniment. Even a purely physiological examination that is exclusively confined to the physical aspect of the process reveals that all human nervous processes accompanied by experience are quite unusual and particular. The lowest animals, such as protozoans and coelenterates, simply do not exhibit such processes and they do not even display functional analogues of them. While this book is being read, the reader's pyloric reflex is alternately opening and closing the outlet from the stomach and thus delivering to the duodenum finely gauged quantities of predigested food pulp. The reader has just as little experience of this process as does an amoeba of its highly appropriate responses to a prick from a needle or to prey. Even if we cannot completely exclude the possibility that these responses are accompanied by experiential processes, there is nothing that would justify such an assumption. It is, indeed, *possible* that every positive taxis contains an element of experience of pleasure and every negative taxis an element of aversion. It is impossible to attain a scientific resolution of this question and no sensible natural scientist would allow himself to be misled into making a definitive statement about the presence or absence of an experiential component in such very primitive responses of the lowest animals. However, if I were asked for a purely emotional, purely intuitive opinion about the potential existence of subjective experience in such organisms, my answer would be *negative!*

The assumption that experiential processes exist becomes far more realistic wherever *purposive* behavior is involved. Following E[dward] C[hace] Tolman, this concept is taken to include all behavior patterns with which an organism can reach the *same* objective by means of *variable* motor patterns. Take, for example, the case of a dog that begins by slinking around a rabbit hutch, trying to find a way in. After failing to find an entrance, the dog then attempts to rip open the wire mesh of the enclosure with its teeth and ends up by digging a hole under the fence in order to reach its prey. On witnessing this, any naive observer would conclude that the dog is *striving* to attain a subjective goal and, in particular, to obtain a pleasurable experience that is connected with the attainment of that goal. Instead, without any psychologization of the dog's behavior, we can draw the following conclusions: (1) The animal continues its activity until it attains a quite specific, previously determined goal (in this case, the killing of the prey). (2) The dog reaches this constantly maintained goal by means of quite different behavior patterns. It seeks a hole in the fence and would not attempt to rip a hole in it itself if one were already present, nor would it begin to dig if it were previously successful in ripping a hole in the fence. Tolman's definition of purposive behavior, cited above, is based on precisely these two facts, permitting a purely objective formulation of the concept of *purpose* and *means:* the constantly maintained final goal can be equated with the purpose and the previously observed, variable behavior patterns with the means. This equation of *objectively* observable behavior with *subjective* concepts that are originally derived from the experiential world of human beings is justified insofar as all human behavior that is directed at subjective goals is covered undeniably and without contradiction by the objectifying, purely "behaviorist" definition.

In addition, however, there is the following significant consideration: all of the behavioral variability that we can characterize as purposive is extremely closely connected with the phenomena of *memory* and *learning*. In an objective sense, a "mneme"—a memory of what has happened previously—is already present wherever the behavior of an organism is influenced *by what it has just done*. As will be shown in more detail in the special part, this is to some extent already the case with protozoans. For example, the stentor (*Stentor roeseli*) possesses an entire graded sequence of responses that it uses to "attempt" to escape from a disturbing stimulus. If the surrounding water is contaminated by a large number of grains of carmine, *Stentor* will first of all reverse the direction in which its cilia beat so as to change the flow from one that carries food particles inward to one that drives water away. If the contamination affects only a small part of the respiratory water, this will eliminate the disturbing stimulus. But if the stimulus persists after several repetitions of the first response, the animal shows another, more radical response in that it turns on its pedicle to face in a different direction. If it fails to find any water free of contamination after several such changes in position, it temporarily stops the beat of its cilia and contracts. The period of contraction increases with the number of repetitions of the response. If the disturbing stimulus is still present even after a large number of repetitions of this latter behavior pattern, the animal frees itself from the substrate and swims away. In this way, like the dog in our first example, *Stentor* only performs the next behavior pattern in the sequence if the previous pattern *does not achieve* the appropriate stimulus situation. Thus, the next response in the sequence will only occur if the previous response *brought about some modification* in the responsiveness of the organism. In the broadest sense, one can refer to this modification as a memory of the immediately preceding context, as an example of the most primitive form of negative learning. The organism, so to speak, collects negative experiences of *unsuccessful* behavior.

However, *learning* in the true sense is something fundamentally different and higher than this simplest form of influence by previous events. In fact, the dog in our example does not only learn what is "unsuccessful," as in the case of *Stentor*. Instead, it remembers over a long period of time which of the behavior patterns it used actually attained the goal. In a later section, we shall take a closer look at such learning through trial and error and at its physiological basis, the conditioned reflex. For the problems that are at issue here, however, the following context is of importance: Genuine learning based on the development of conditioned reflexes is virtually exclusively confined to situations where purposive behavior is involved. Only the attainment of a particular goal "conditions" an organism to perform a particular behavior pattern. In every practical case where an animal is trained to perform a given task, as in all human learning, there must be a *purpose* in the sense of a stimulus situation that is to be attained or avoided. Without the principle of the "carrot and the stick," a circus horse will learn no more than a human child. But this principle of the carrot and the stick is based on something that is quite fundamental to the nature of all experience and indeed for the mind-body problem as such. The principle of the attractive or repulsive stimulus that leads to the development of a conditioned response is, with a probability bordering on certainty, *identical with the principle of pleasure and aversion.* We have compelling reasons for equating the objective side of animal and human behavior with specific subjective phenomena in this way. First of all, in the case of human beings—that is to say, with the only species whose subjective experience is directly accessible to us—*all* learning by trial and error is inseparably linked to the subjective experience of pleasure and aversion. In fact, with humans the same applies to all purposive behavior across the board. But, as with higher animals, human purposive behavior is virtually nonexistent *without* the development of conditioned reflexes and *without* the participation of learning. After all, purposive behavior can only exhibit the plasticity, the adaptability that constitutes its most substantial survival value because of the fact that the blind principle of trial and error is supported by the purposive plasticity of learning.

For this reason, with all living organisms that possess some capacity for the development of conditioned reflexes, we can equate purposive behavior with behavior that can be modified by conditioning without thereby being guilty of any conceptual confusion or imprecision. Now, any behavior of humans and animals that can be modified by conditioning is doubtless based on *completely identical* physiological processes, namely, on the development of conditioned reflexes. But the parallels that exist between humans and animals in this respect go considerably further than this. It is not only the *general* process of development of conditioned reflexes that is identical in humans and animals; very many of the *special* goals that serve the conditioning process are simply the same. In humans, for example, the sensual pleasure that is the experiential accompaniment of the instinctive motor patterns of food intake and copulation represents the most important subjective motive for all behavior patterns that can be modified by conditioning. At the same time, the purely objectively definable goals of conditionable behavior of higher mammals are represented by instinctive motor patterns *that are not only functionally analogous but quite demonstrably phylogenetically homologous.* It is therefore far, far more than a simple analogy if we attribute to these higher animals experiences of pleasure and aversion that are similar to those accompanying the behavior patterns concerned in humans.

In addition to the *physiological* identity of the learning process in animals and humans and the *phylogenetic* identity of decisive instinctive motor patterns that serve as purposive factors, there is a third, equally compelling argument to support the assumption that animals experience pleasure and aversion. Contrary to the simple picture that has been presented so far, when it is examined quite precisely the experience of pleasure and aversion is *not directly* the subjective side of purpose-serving behavior patterns and receptor processes. Instead, experience is the subjective side of a *mediating agency* that *governs* these purpose-serving processes. The objective function of this mediating agency is to label these processes with a plus or a minus, so that they are acquire a positive or a negative value. This mediating agency has a very close relationship to the functions of *Gestalt perception!* We have already become acquainted (p. 28) with Wolfgang Köhler's statement that an organism is only able to acquire a particular character as a releaser of a conditioned behavior if it is able to perceive that character as a Gestalt. But because a Gestalt is an unmistakable complex quality consisting of a great number of individual sensory data that combine together to generate that quality, the mediating agency for experience has the possibility of labeling an entire large *complex* of processes with a big plus or minus sign according to its success or failure and according to its favorable or adverse contribution to survival. In fact, it is precisely this function that represents *the actual survival value of experience.* [Hans] Volkelt has encapsulated this in his statement that experience is "the most important mediating agency in nature, as it is the agency that permits the organism to connect a single conclusion with a multitude of conditions." We need only add to this that the conclusion is "toward" or "away from"! We can attribute experience to all organisms that are able to reach such a conclusion.

The unconditional linkage of all learning processes to Gestalt perception is an objective fact that can scarcely be doubted. But it is equally certain that the whip of aversion and the bait of pleasure can *only* serve their major adaptive function in cases where they operate as a *means of conditioning* to bring about an appropriate adaptation of behavior. Therefore, the laws of conceptual parsimony dictate that we can only assume that this most primitive form of experience is present in organisms that possess the capacities for the development of conditioned reflexes and for the perception of holistic Gestalts. For such capacities, however, a relatively *high* degree of development of the central nervous system and of the sense organs is an indispensable prerequisite.

It is sufficiently clear from what has been said that we can attribute subjective experience only to the higher animals and hence *only to a part* of the range of "animal" organisms. Experience tells us that even psychologists without vitalistic leanings will take objection to this and accuse us of erecting an arbitrary and unjustifiable boundary in a situation in which more far-reaching generalization is called for. The following reply can be made to this: from a biological point of view there is nothing more misleading than the old aphorism *Natura non facit saltum* ("Leaps never occur in nature"). To the contrary, nature *always* changes by leaps, which occur quite abruptly at any point where a quantitive modification leads to the sudden emergence of a new quality. The greatest leap of this kind leads from the inorganic to the organic realm and the second greatest leads from life processes that are unaccompanied by experience to those that are. Even the transition from higher animals, which doubtless possess subjective experience, to humans represents a considerable leap, although our own egocentricity perhaps leads us to overestimate the magnitude of this particular leap.

It is therefore an undeniable historical fact that many other abrupt qualitative leaps have taken place. Given this fact, what is to prevent us from assuming that experience in its simplest and most primitive form, as the "capacity for pleasure and suffering," emerged as an entirely novel, unprecedented phenomenon at some point in the course of animal evolution, perhaps quite abruptly? It could have been just like the abrupt qualitative transition that occurs with a progressively heated physical body which suddenly begins to glow and transmit light. There is nothing to *contradict* this assumption and there are very compelling arguments that *support* it. We have demonstrated exhaustively how experience, the whip of aversion, and the bait of pleasure can only serve an adaptive function in cases where they operate as a *means of conditioning* to bring about a plastic, appropriate adaptation of behavior through a learning process. A fixed, unmodifiable motor pattern can serve its adaptive function just as well without any accompanying experience. We also know that variability in behavior brought about by conditioning simply does not exist with the lowest animals, not to speak of plants. It is also known that even relatively advanced animals such as insects and molluscs lack *pain,* which is possibly the most primitive form of experience that exists. This is quite emphatically demonstrated by the grotesque sight of a grasshopper or a sea slug voraciously eating its own rear end.

Anybody who is really closely acquainted with a dog or a chimpanzee will be utterly convinced by sound human reason and the direct evidence of his own senses that these animals are equipped with genuine subjective experience. The idea that a dog or a chimpanzee could be a reflex machine lacking any conscious perception is, quite simply, *absurd.* This is true quite apart from the observation that such an idea is barely compatible with the historical fact that human beings, which are undeniably equipped with conscious perception, are equally undeniably descended from supposed machines of this kind. On the other hand, an observer who is really intimately acquainted with animal behavior will equally evidently arrive at the rationally justifiable conclusion that subjective experience is lacking in a paramecium or in any plant just as certainly as it is present in a chimpanzee. Subjective experience must have originated somewhere *between* the paramecium and the chimpanzee. In view of all of the considerations that have been set out here, it seems to be an eminently logical step to seek this origin at the point where the mediating agency of the pleasure/aversion principle quite abruptly acquires prominent adaptive significance through a combination of Gestalt perception with learning ability.

With all that has been said so far, it has been assumed as virtually self-evident that all processes of subjective experience are *simultaneously* genuine life processes and hence physical, material processes. We of course *know* very well that psychological and physiological phenomena are simply *two different phenomenal forms in which the same intrinsic reality is reflected in our world-image apparatus.* Nevertheless, despite this knowledge, the mental/physical duality of all neurophysiological processes that are accompanied by experience remains the greatest of all puzzles to confront the human mind. We can just grasp the real *identity* of subjective-experiential and objective-neurophysiological processes by bearing in mind the fact that we possess *two* forms of a priori intuition that are qualitatively completely different, two separate receptor systems for the *same* real datum. Each of these two forms of intuition can encompass only *one* aspect of an organic entity possessing subjective experience and our mental processes lack any possibility for constructing a bridge between them. This is true even of the thought experiment that was introduced at the beginning of this section as a utopian ideal case. In that case, we would have been able to analyze a physiological phenomenon down to its ultimate causal connections at the atomic level and hence achieve complete predictability of its subsequent course. Yet this full insight into the material side of the process would not have brought us one iota closer to an understanding of the question concerning the *relationship* between it and the psychological phenomena occurring in parallel. Nor would it have told us why a particular neurophysiological process is accompanied by one particular experiential process and no other. Conversely, insight into the chain of *psychological* causes, however complete, would yield just as little information as to why they are the experiential accompaniment of these particular neurophysiological processes and of no others. *The thing is in itself a unitary whole,* namely, the holistic Gestalt of a higher organism with subjective experience. But our *thought processes* can only approach this unity from two incommensurable sides. These are two sides for which there exist *receptor organs* or—to use Kant's terminology—a priori schemata in the form of innate forms of possible experience. It is just as if we were able to both see and smell a particular thing but were unable to link these two kinds of perception to that object in an informative fashion. There is, in fact, no conceivable bridge *between* the two a priori forms of intuition for the subjective contemplation of an organic entity equipped with subjective experience. The relationship between the subjective and objective sides of experiential life processes is fundamentally *alogical,* as M[ax] Hartmann has put it.

Although we *lack* any kind of *receptor organ* for the relationship between the neurophysiological and psychological sides of one given process, we are in no doubt about the real existence of such a relationship. There is no doubt about the fundamental extrasubjective identity of the thing that can only be grasped by the two unconnected receptors of our world-image apparatus as a phenomenon, or more precisely as two different, incommensurable phenomena. The colorful world of experienced *qualities* remains *fundamentally* inaccessible to quantitative causal analysis, despite the fact that the processes underlying this experience are, in their extrasubjective reality, identical with the fundamentally quantifiable, causally analyzable processes. The causal quantitative side of the phenomenon is equally inaccessible to the experiential aspect. It is as if we were to have before us a two-dimensional structure of some kind, for example, a contoured metal plate, but could see with only one eye, so that we could only examine one side at a time. Or, to take a more precise analogy, it is as if we lacked the bridge constituted by the preformed capacity for *cooperation* between our two eyes. This permits our Gestalt perception to register such a two-sided structure as a single object in three dimensions even when it is oriented in such a way that each eye can see only *one* side of it.

The *duality* of psychophysiological phenomena has led many great thinkers, with Descartes foremost among them, to assume that every single experiential process corresponds to a quite specific physical process, and vice versa. This gave rise to the theory that two causal chains, one physiological-physical and the other psychological, operate simultaneously without any logical or causal connection between them but running strictly in parallel, with an exact correspondence between their individual components.

In many respects, this portrayal does indeed exactly fit the relationship between physical and experiential processes. Even quite thorough investigation at first suggests that two causal chains are operating in parallel with a one-to-one correspondence between their components. It is just as justifiable to speak of causal chains for psychological aspects as for material aspects! A human being is *really* sad *because* he has suffered a disappointment and he becomes angry *because* somebody has aroused his jealousy, and so on. Such causal chains of psychological effects operate in parallel with causal chains of physiological processes on the physical side, and in fact it is often possible to analyze the one-to-one correspondence between their components, as we shall soon see. For very many psychologists, however, the following circumstance has generated great confusion in the conceptual approach used to deal with these chains operating in parallel: It often happens that one component in one of the chains can be readily grasped from either the psychological or the physiological side, whereas the *corresponding* component in the other chain can only be tackled with great difficulty from the aspects relevant *to that chain.* The opposite then applies to the next pair of components in the two chains. For example, with the two causal relationships A (psych.) → B (psych.) and A (phys.) → B (phys.) only A (psych.) and B (phys.) are readily amenable to observation, while the existence of a physiological counterpart to A (psych.) and of a psychological counterpart to B (phys.) remains concealed from the observer. For present purposes, we shall ignore the additional possibility that an experiential side to B (phys.) may be completely lacking. Let us now consider the case of a person who is subject to a very complicated experiential process and suffers a "psychological trauma," as a result of which he develops a major cardiac neurosis, without showing any noticeable further psychological change. In such a case, there is a great temptation to speak of a *psychological* cause of a *physiological* effect and many psychologists fail to recognize just how inadmissible this is, both logically and methodologically. In reality, it is of course the highly inaccessible *physical* side of the nervous processes accompanying the psychological trauma that serves as the cause for the objectively observable disruption of cardiac activity. Any mental *switch* from one causal chain to another, from one side of the metal plate to the other, is logically inadmissible and can lead to methodological confusion that has a highly inhibitory effect on research. Descartes, the founder of the theory of psychophysical parallelism, quite correctly recognized that this is so.

Let us now consider a further example that illustrates a methodologically correct approach to a complex psychophysical phenomenon. Quite unexpectedly, somebody receives a hefty slap on the ear, which initially startles him. Immediately afterward, he becomes angry and returns the blow with great satisfaction. We can quite justifiably regard this sequence of events as a causal chain of relationships that is physiological and at the same time psychological. From the physiological side, the causal chain has approximately the following form: The reverberation of the head and the marked excitation of receptors serving the sense of pain generate a "shock." This is essentially represented by excitation of the parasympathetic nervous system and a corresponding inhibition of the sympathetic system, accompanied by extensive inhibition of all higher functions of the central nervous system. Because of this disturbance of the balance between the parasympathetic and sympathetic systems, there is a loss of tonus of the body musculature and of the smooth muscles of the blood vessel walls. We can recognize the former from the sagging arms and jaw of the receiver of the slap, while the latter is revealed by his pale coloration. The inhibition of the brain centers is shown by the fact that he is immobilized "as if struck by lightning" and is momentarily incapable of any voluntary movement. He is not simply "as if paralyzed"; his higher brain functions are at that point in time, both literally and physiologically, paralyzed. However, just fractions of a second later, the described states of excitation and inhibition swing back in the opposite direction, following neurophysiological principles that have been investigated in great detail. The sympathetic nervous system becomes excited, the tonus of the vasoconstrictors and of the body musculature increases, the body as a whole becomes tense, the pale face reddens, and the adrenal glands discharge large quantities of epinephrine. As a result of all this, there is an increase in general readiness to respond (the details of which need not concern us here). The inhibition of the central nervous system gives way to general excitation, which among other things leads to release of the instinctive motor patterns of fighting behavior and hence to reciprocation of the slap.

On the other, experiential side of this phenomenon, we can trace a chain of psychological processes, in which each component can perfectly well be regarded as the cause of the next. The person who has received the slap initially experiences great pain and alarm, as a result of which his self-esteem sinks to a minimum. But there then follows a pendulum-like swing on the psychological side as well, affecting his feelings and emotions. The depth of psychological depression is immediately followed by an angry resurgence of self-esteem. This urgently demands self-assertion, which is expressed in the virtually orgiastic pleasure accompanying the release of the inhibited fighting response, literally terminating in sweet revenge.

Tracing of these two causal chains, which we can understandably sketch only in an extremely superficial manner, is entirely legitimate from a methodological-scientific point of view. Indeed, without departing in the slightest from our strictly inductive scientific approach, we can proceed further by a considerable step. We can, in fact, establish a whole series of cross-connections between the individual components of the physiological and psychological causal chains. The experience of alarm undoubtedly occurs in parallel with the first shock effect and coincides with it in time. The depressive decline and the following, contrasting rise in self-esteem quite certainly correspond to the initial fall and subsequent rise in muscle tonus mentioned above. We only need to think of the way in which even in our colloquial expressions the externally visible effects of muscle tonus have become symbols of psychological states. If we say that a person hangs his head or, alternatively, holds his head high, we no longer think at all of the tonus of the muscles of the neck and back that is responsible for these effects. Instead, we think exclusively of the psychological states that represent the experiential side of the same overall phenomenon within the organic entity. Finally, the emotion of rage corresponds equally certainly to disinhibition of the instinctive motor patterns of fighting behavior. Later, in the discussion of the physiology and psychology of enodogenous-automatic motor patterns, we shall present compelling arguments in support of this conclusion.

In the identification and demonstration of such definite correspondences between physical and psychological manifestations of the same organic phenomenon, we can see the most promising method for approaching the mind-body problem through inductive research. But it is precisely for the purposes of that approach that an utterly clear conceptual separation of the physiological and psychological sides of organic processes is the most indispensable precondition. We may never forget that the cross-connections between a physiological process and experience *are not causal connections.* It is, for example, just as nonsensical and epistemologically misleading to state that the emotion of rage is the cause of disinhibition of fighting responses as it is to claim the opposite. One aspect cannot be the *cause* of the other for the simple reason that it represents *the same thing* that happens to be registered from another direction, with a different receptor organ. As has been said, we may not switch from one of the parallel chains to the other, from one side of the two-dimensional metal object to the other.

Up to this stage, the epistemological approach of the inductive scientist is in agreement with Descartes' theory of psychophysical parallelism. But in one feature the analogy of parallel chains, like that of the two-dimensional metal object, begins to break down. It is, in fact, the case that only *one* side of these analogies—as with the theory of psychophysical parallelism—is unconditionally correct. It is doubtless true that every psychological phenomenon, however small, possesses some counterpart in the form of neurophysiological processes. *But is the converse true, namely that for every component in a physiological chain there is a counterpart in a psychological causal chain?* The reader, who has been well prepared for this question, will at once know the answer to this, which constitutes the entire basis for our criticism of the theory of psychophysical parallelism. Because every experiential process is simultaneously a physicophysiological life process, it follows that every psychological process has a counterpart in the material realm. However, as we have already demonstrated quite adequately (p. 161), it is not true that every life process, not even every nervous process, is simultaneously a psychological process. Therefore, it is obviously far from true that every component in the infinitely branching and interwoven causal chains in physiological processes is accompanied by a discretely identifiable phenomenon on the experiential side. The quite peculiar and particular functions of the central nervous system, which alone have counterparts on the experiential side and manifest themselves as experienced qualities in our minds, *are the integrating products of a relatively small number of quite specific mediating agencies.* These mediating agencies "literally turn many things into a single phenomenon," to quote Volkelt's fitting description.*

The physiological, fundamentally physicochemical concatenation of cause and effect in the systemic entity of the organism is almost infinitely complex and the number of elements involved is incalculably large. But, as we have already learned (p. 137 et seq.), the entity is something quite different from a simple sum of these elements. A vast number of subentities is involved in the determination of the holistic quality of every single cell; a vast number of cells serve as subentities of the entity represented by an organ; numerous organs constitute an organ system and a multitude of organ systems combine to form the holistic Gestalt of a living organism. In this hierarchy of entities that are subordinated to one another in a graded series, it is possible to insert an almost unlimited number of additional links. Now the nervous system of a higher animal, particularly that of a human being, is the most advanced and most complex of all organic entities that we know. It consists of a virtually infinite hierarchy of successively subordinate levels, each of which incorporates a large number of adjacent, lower components within its orbit of influence. *Every one of these* levels receives more causal links from below than it transmits to the next higher level. It is precisely this fact that accounts for the *integrating* function of the central nervous

* *Note in the margin here:* Freud's system WBw.

system. I would like to illustrate the nature of this function with a further analogy. A sergeant does not pass on every small detail that comes to his notice from his platoon, for example, that rifleman Meier suffered from stomach acidity this morning. Nevertheless, Meier's stomach disorder may be *included* in a report transmitted upward by the sergeant if several servicemen suffer from the same complaint simultaneously, such that the individual disorder acquires some *significance.* It is this significance that is then transmitted up to the next level in the hierarchy. It is a *fundamental feature* that the message that is communicated *does not contain all of the details that determine its significance.* The clarity of the organization, the entire *integrating* function of its structure *is only possible because details are excluded at every level that is passed.* The company commander in fact only knows that the kitchen used rancid fat. The stomach complaints of Meier and Müller are no longer detectable in his knowledge; they are simply *no longer present,* even though they have contributed as real causes to the quality of that knowledge. In the course of progressing up through the hierarchy, the process of integration continues. The commander of the battalion knows only that the head cook is inefficient, while the regimental commander is simply aware that company X is a so-called "rabble," and so on. The knowledge of any higher state authority has this kind of relationship to the millions of individual processes that represent "elements" within its sphere of influence. The quality of *experience* is related in just the same way to the multitude of elementary physiological processes that are represented in an integrated form in the individual messages that are transmitted upward from subordinate levels. The experiencing subject registers no more than a *general quality.*

It would amount to a complete misunderstanding of the holistic character of the constitutive *integrating function* of the nervous system if one were to expect to find in experience all of the individual elements that can in principle be dissected out on the physiological side of a given phenomenon. In other words, experience only emerges at a specific, relatively high category of levels, namely at those which generate the connections between conditioned positive and negative taxes and remembered Gestalt qualities. We have to come to terms with the fact that experienced qualities are the products of integration processes which themselves lack an experiential side, like the function of the pyloric reflex discussed on p. 161. To our introspection, an experiential quality is something which simply *preexists* and is *fundamentally indivisible* on the psychological plane. Any attempt to identify "elements" within it is doomed to failure from the outset. The fact that such attempts have nevertheless repeatedly been made, and continue to be made, is the result of the one great and fundamental error in the theory of psychophysical parallelism. This is, namely, the erroneous tenet that there must be a corresponding, individually separable element on the psychological plane to match *every* component of the physiological causal chain.

It was necessary to present here with considerable precision the theoretical position of inductive science with respect to the mind-body problem because it is of great significance for the conceptual and practical methods of comparative behavioral research. As was mentioned at the beginning, this discipline sees itself as the natural science of the human species. It would constitute a virtually unpardonable refusal of knowledge if this discipline were to close its eyes to the fact that human beings possess a *mind,* therefore simply excluding introspection, the entire reality of the colorful world of experienced qualities, as a source of information. Comparative behavioral research is fundamentally the science that deals with the *entity* of the living organism and it must therefore necessarily be both a physiological *and* a psychological science. It attempts to approach from both sides the unity that lies behind the two series of phenomena, which is an extrasubjective reality that is certainly *not double.* For this reason, from very early stages in the development of this discipline, the method adopted has been to concentrate research interest particularly on objects that are relatively suitable for an analysis *from both sides. Instinctive motor patterns,* which display endogenous processes of stimulus production on the physiological side and specific qualities of feelings and emotions on the psychological side, provide one such object. *Orienting responses,* which are certainly reflexes, together with the peculiar parallel experience of *insight,* provide a second. Because of its close relationship to the experience of pleasure and aversion, the general phenomenon of purposive behavior is also a focal point of interest. Research into these particular psychophysiological processes has made a major contribution to the elimination of the ancient, idealistic prejudice that any process that can be subjected to physiological causal analysis ipso facto cannot be a process accompanied by experience.

Over the course of recent decades, the natural sciences have become increasingly more careful about pronouncing a definitive, dogmatic "*ignorabimus.*" They have learned from Max Planck's revolutionary findings that the limitations of human a priori forms of thought and intuition do not always represent the boundaries of inductive scientific research. It is precisely *if* we constantly bear in mind that the relationship between the physiological and psychological sides of experiential life processes is *alogical,* and if we set out from the fact that we *lack* an innate form of thought and intuition for this relationship, that there will be some prospect for success in trying to construct in some way the capacity for thought that is not present a priori.*

* *At this point, the sequentially numbered manuscript pages come to an end. (Because section III of the handwritten text begins with p. 1, we do not know what was missing and the length is also unknown.) The last two pages of section II bear the following incompletely formulated sentences:*

The modest routine occupation of inductive research in our discipline is now concentrated on objects that can be registered with both of our receptor organs. The attempt is being made to determine more and more details from both sides and to establish the correlation between the details of the two pictures. With stereoscopic vision it is also the case that the details, the differences between them and, above all, the correlation between the two retinal images give rise to an entirely novel quality of *three-dimensionality*, which was not present in the original images. Like somebody who has been cross-eyed eyes for some time, we lack the coordinating neural apparatus that will integrate the images registered by our two receptor organs and *unite* them. Like a cross-eyed person, we therefore see the inseparable unity of the organic entity as a double image. We see mind and body, whereas in extrasubjective reality there is only a single thing, existing in its own right. If we fixate the entity of the living organism between two precise observational approaches that sharply distinguish the two aspects, we may perhaps in the distant future succeed in seeing behind the two observed phenomena the real unity that they conceal.

To return to an analogy that we have already used, we possess two eyes, each of which perceives only one side of a real object. As if we were cross-eyed, we lack the coordinating neural apparatus that integrates the images registered by these two coordinating neural apparatus that integrates the images registered by these two receptor organs and combines them into *one*. Like a cross-eyed person, we therefore see the inseparable unity of the organic entity as a *double* image. We see mind and body, while in extrasubjective reality there is, in fact, just a single thing existing on in its own right.

The modest routine occupation of inductive research in our discipline is now concentrated on objects that can be registered with both of our receptor organs. The attempt is being made to determine more and more details from both sides and to establish the correlation between the details of the two pictures. With stereoscopic vision it is also the case that the details, the differences between them, and, above all, the correlation between the two retinal images give rise to an entirely novel quality of three-dimensionality, which was not present in the original images. If we fixate the entity of the living organism between two precise observational approaches that sharply distinguish the two aspects, we may perhaps in the distant future succeed in seeing behind the two observed phenomena the real unity that they conceal. With a new . . .

. . . the two images, each of which is generated by an uncoordinated receptor organ, appear separated from one another in our experience.

Let us return to the analogy that we used previously. A cross-eyed person, whose neural apparatus coordinating the directions of the two eyes has ceased to function, sees *two* separate images where in fact only one object is present. In just the same way . . .

III

*Historical Origins and Methods of Comparative Behavioral Research**

**This title could not be determined from the manuscript but is indicated by the text. In fact, this entire page can only be reproduced in fragmentary form because the edge of the original was damaged and the writing was very bleached.*

10

Preconditions

In all biological sciences, the discovery of a new principle is commonly closely associated with the discovery of an object in which that principle is expressed in a particularly simple form. The direction taken by research is extensively determined by the particular nature of that object, TO WHICH[†] it must become adapted, but that direction in itself influences the future course of research. It is very often the case that such "favorable objects," whose discovery can lead to the development of a new research direction, are structures and functions that do not influence the relative FIXITY and invariability of SOME function of an organic entity but are instead influenced by it.[‡]

Understandably, the greater the complexity and HOLIStic character of an organic system and the greater the degree to which research [*a word is illegible here*] accordingly depends on the laborious [*this is the last word on the bleached page of the original manuscript*] and tedious method of analysis on a broad front (Biological Prolegomena), the greater is the influence exerted by the discovery of such an entity-independent component on the progress of research. With the "tough nuts" represented by research into such complicated systems of virtually universal reciprocal causal relationships, the discovery of a relatively entity-independent component constitutes the creation of an entry point for analysis. It provides an extremely welcome possibility for driving a tunnel into the block represented by the entity in a way that will have a decisive influence on the subsequent choice of method and direction.

[†] *Capital letters indicate that the words or letters could not be reliably deciphered.*

[‡] *At this point, there is a sentence that cannot be deciphered in an intelligible form, but which doubtless refers to chapters dealing with the organic entity and* "relatively entity-independent components." *This can be inferred because the latter expression, along with the words* "chapters" *and* "were discussed," *is clearly recognizable* (*see above, p. 146 et seq.*)

The classic example of this kind of origin of a research direction from the discovery of a favorable object is provided by modern genetics. The origin of this field of research coincided with the discovery of the simplest limiting case in which the parents of a hybrid differed in only *one* hereditary character, in just one gene. Right up to the present time, the direction taken by research has been determined by the method imposed by the nature of the most suitable approach of hybridization. With research into the higher functions of the central nervous system and into the behavior of higher organisms, the reflex (or the conditioned reflex) was initially the only relatively entity-independent component that presented itself as a suitable object for analytical penetration. For this reason, it was the exclusive determinant of the direction and method of behavioral research for many decades. Our own field of research, whose presentation is the purpose of this book, similarly owes its origin to the discovery of a quite specific, extensively entity-independent component of animal behavior, namely, the *instinctive motor pattern.* The physiological peculiarity of such innate, species-specific motor patterns resides in the fact that they are not composed of reflexes. Instead, they are based upon a *different* ancient function of the central nervous system. They depend on automatic-rhythmic *processes of stimulus production* like those that have long been known to be involved in the activity of the heart. The history of the origin of comparative ethology is the history of discovery of the instinctive motor pattern and of its physiological peculiarity. In this case, too, the nature of the object has had such a fundamental influence on the course of its research that the history of its origin and the method cannot be separated. For this reason, it is not only *possible* but in fact *obligatory* to discuss them together.

It lies in the nature of the innate, species-specific motor patterns that we refer to as instinctive motor patterns that it was only possible to *recognize* them from a comparative phylogenetic standpoint. This is, without the shadow of a doubt, the primary reason why they were not subjected to inductive scientific investigation until a very *late* stage. By the turn of the century, the theory of evolution and the conceptual and methodological approach arising from it had long become a common factor in all biological science. For this reason, it is at first sight surprising that the phylogenetic approach had not yet been introduced into psychology and behavioral research, especially since Wilhelm Wundt had clearly formulated the theoretical requirement for comparative phylogenetic psychology. But it is most surprising of all that investigation into animal and human behavior from the *neurophysiological* side so completely ignored the phylogenetic approach and its methods despite the fact that other branches of physiology had long been applying them with great success. But, on closer examination, the reasons for this prove to be quite simple.

In the first place, the simplest and most basic requirement for a phylogenetic comparison of animal behavior was lacking because there was no *knowledge* of the behavior patterns of different animal forms. When morphology, the study of the structure of living organisms, first began to generate conclusions regarding phylogenetic relationships from comparisons of similarities and differences in characters, a rich mine of information concerning comparable details was already available. As a natural science, morphology had already passed through the idiographic and systematic stages of development (see p. 28 of the Philosophical Prolegomena) and could justifiably proceed to the nomothetic stage. By contrast, no one was equipped with a basically adequate knowledge of the behavior of animals, or even of the representatives of a single animal group. Alfred Edmund Brehm, the doyen of experts on animal life, and his scientific opponent B[ernhard] Altum both lived at a time that was somewhat too early for them to benefit from the concepts of evolutionary theory. Certain moves were made by human psychologists to establish comparative phylogenetic investigations of animal and human behavior and there is, in fact, even an American journal entitled *Journal of Comparative Psychology!* But the indispensable prerequisite of extensive knowledge of individual behavioral characters of related animal forms was completely lacking. There was also complete absence of the conceptual and practical methods, and the associated strict discipline, that comparative morphology had established over the course of decades of intensive and detailed work. Last but not least, there was a lack of the intuitive feel for phylogenetic relationships that cannot be taught or be extracted from a book but can only be acquired through years of *personal* thorough investigation of the manifold variability within a related group of animals or plants. No human psychologist was equipped with these essential prerequisites and none was in a position to accomplish subsequently the enormous learning task that would have been necessary to develop them.

The great physiologists who tackled the problems posed by reflexes in animal and human behavior also lacked the comprehensive knowledge of behavioral characters of animals that represents the inductive basis and hence the precondition for all phylogenetic research. As will be demonstrated in the discussion of the two great mechanistic research directions in the field of animal and human behavior, behaviorism and the Pavlovian reflex school, both proceeded directly to the nomothetic stage without previously conducting a *hypothesis-free,* idiographic-systematic investigation of animal behavior. Without first finding out "all that there is to see" (Philosophical Prolegomena, p. 27), both raised a *single* phenomenon that had been discovered early on to the status of an exclusive explanatory principle for animal and human behavior. For behaviorists, this phenomenon was learning through trial and error, while for proponents of the Pavlovian school it was the reflex and conditioned reflex. By virtue of their properties as propitious objects and as entity-independent components, these phenomena were extremely important for the progress of behavioral analysis and as a result the research directions concerned led to great accomplishments. At the same time, however, because these entity-independent components are not the *only* constituents of animal and human behavior, these mechanistic research directions could not grow beyond the narrow, specialized zone of these phenomena to generate explanations for the *totality* of behavior.

In addition to the *material* factor represented by the lack of an idiographic-systematic basis and the regrettable error of early restriction to a single explanatory principle, there is a second, purely *intellectual* factor that prevented the great mechanistic directions in behavioral research from identifying phylogenetic comparability in animal and human behavior. This phylogenetic comparability in the behavior of related life forms resides in specific *inherited* and *innate* patterns of movement and response. Like any morphological characters, they are systematically interpretable features of species, genera, orders, and even classes and phyla of the animal kingdom. Explanation of their special form requires phylogenetic consideration in exactly the same way as all historically acquired features in the bodily construction of living organisms. Inherited, species-specific and innately purposive behavior patterns of animals had, in fact, been recognized for some considerable time. Medieval scholars had already considered them and had passed on the extremely dubious heritage of habitual reference to them as *instincts.* The concept of instinct has become established even in colloquial language and indeed conveys the scholastic sense of an *extranatural* factor that is neither explicable nor in need of any explanation. It is called upon as an "explanation" of a pattern of behavior wherever the latter serves a meaningful, survival-promoting function, but its purposive character cannot be explained on the basis of rational processes. Thus, from the very beginning, "instinct" was one of those words that can be called upon in situations where concepts are lacking. Or, to be more precise, the nature of the scholastic concept of instinct resides in its use as an extranatural pseudo-explanation for natural processes. This peculiarity of the concept of instinct, which has had such unfortunate consequences for the development of our own discipline, exerted a major influence with respect to the innate, species-specific behavior patterns that it was used to "explain." At a very early stage of physiological research, these patterns became the focus of a difference of opinion between two opposing schools of thought in natural philosophy, between *vitalism* and *mechanism*. This state of affairs had a decisive effect of the origin of our own direction of research in that it more or less suppressed it for a considerable period of time. For this reason, we must now take a closer look at these two schools of thought and the contrasts between them.

11

Vitalism

On beholding the beauty of organic creation, the spirit of any emotionally responsive and receptive person is filled with an ever-recurring sense of wonderment and awe. Such wonderment and awe is probably felt most deeply and most genuinely by those who talk least about it but devote their entire lives to a penetration into this marvel of the world of organisms. Nothing is more mistaken than the widespread belief that the objectivity of research blinds one to the beauty and wondrousness of nature. It is simply that the research worker generally tends to *talk* very little about his awe. A genuine natural scientist is the opposite of an idealist and views the wondrousness of the extrasubjective world of things as something *real.* Because of this, his awe drives him neither to soliloquy, as in the case of the idealist, nor to verse, as in the case of the poet. Instead of just talking or thinking about it, he feels the insatiable urge to penetrate deeper into this reality and to engage in *active* interaction with it. For the genuine researcher, analysis—the real *explanation* of a previously incomprehensible and therefore "wonderful" natural phenomenon—by no means amounts to a degradation, a profanation of nature. Tracing of a higher, more special principle back to the greater laws of natural science that govern all matter does not lead to any reduction in the wonder and awe evoked by the more special phenomenon. It does not lead to degradation but rather to repeated higher evaluation of matter itself and to an increase in the awe inspired by the possibilities inherent in matter. For this reason, the genuine researcher will never fear that his wonderment will suffer from the progress of research. He will never, ever succumb to the fear that one day, after the final research goal has been attained, nature might lose its reputation as a wizard and stand exposed as a mere conjurer. For this, it is necessary to feel awe when confronted with something that is worthy of wonder and yet *simultaneously regard it as something real* that will not lose one iota of its wondrousness through a natural explanation of it. One must be a *materialist* down to the depths of one's soul but at the same time carry the awe of organic reality in one's heart. Only a few are capable of this and they are natural scientists in the deepest and uttermost sense of this term.

On the other hand, anyone who is an *idealist* in the depth of his soul, for whom everything that is mortal is no more than a parable, necessarily shows a completely different, in fact fundamentally opposed behavior to everything that arouses his wonder and awe. For such a person, everything that is eternal and demanding of awe lies *outside* the world of things and beyond the laws that govern this world. For him, natural and sacred are two concepts that are logically and intrinsically mutually exclusive. For this reason, everything that can be subjected to a natural explanation is, because of this fact alone, devalued, demystified, and divested of its character of wondrousness. That is why such a person, in the depths of his soul, does not *want* to have a naturalistic explanation, at least for anything that inspires his awe and wonderment. It is precisely in this respect that the idealistic dualism of worlds leads on to a disastrous consequence. Consciously or unconsciously, anybody who is trapped by this fallacy *must become an opponent of causal-analytical research.* Indeed, it is worse than this: It is in fact the best among those who are trapped in this age-old fallacy, whose eyes and hearts are *most* open to the awe-inspiring properties of the universe, who are *least* able to desire a causal explanation that will lead to profanation of that which inspires their awe. Many of the great and good have in this way become enemies of materialism and of scientific research. But these great enemies have actually caused less damage than diminutive allies. By far the *worst aspect* of the phenomenon under discussion is that the stunted and short-sighted, the intuitively blind, and those who are incapable of awe have drifted into *our* camp. From that position, they have contributed—and continue to contribute—to reinforcement of the standpoint of idealism. Because of their obvious blindness to great and awe-inspiring relationships, they have consolidated the belief of all idealists in thought and feeling that natural science represents value-blind profanation of the universe. Nobody, not even the greatest idealist, has damaged natural science and materialism so much as the *value-blind,* who exploit the fiction of the value-free objectivity of natural science to conceal the massive deficiency of their molelike souls.

For the reasons explained, those who are endowed with an idealistic orientation to the depth of their souls cannot *desire* that the great, awe-inspiring properties of the universe should be investigated with scientific methods. But they are nevertheless constrained to reach some kind of compromise with natural science, unless they are sufficiently stubborn and detached from the world (one could even say sufficiently logically coherent) to dismiss natural science in toto as the "summit of dogmatic narrow-mindedness", as has been done in eminently consistent fashion by my highly respected friend K[urt] Leider. As a rule, this compromise with natural science follows the example set by Goethe: "To investigate what can be investigated and quietly to revere that which is immune to investigation." Such respectful, self-imposed modesty appears at first sight to be convincing in its correctness. But it conceals a dangerous trap that resides *in the drawing of a distinction between that which can be investigated and that which cannot.* The unconscious desire not to allow the awe-inspiring to be profaned by a natural explanation leads the idealist to an abrupt termination of his research impulse when faced with anything that appears to be sacred. Deep in the subconscious, an irrational, emotionally determined mental barrier is erected and any conscious information that *could* lead to violation of this barrier is "repressed" in the Freudian sense. In other words, the idealist does not investigate that which can be investigated and revere that which is immune to investigation. Instead, he only investigates that which is *not* worthy of awe, while dogmatically equating anything that is awe-inspiring with that which is immune to investigation. In response to the *fear* of profaning anything that is awe-inspiring, the idealist erects a mental barrier in front of it and within this mental barrier lies concealed the age-old scandal of double logic. The scandal of double logic *must* result when somebody on the one hand tries to account for one sector of natural phenomena with a clear impulse to achieve natural explanations, but on the other hand imposes an abrupt termination before another sector of *the same* phenomenal world simply because his awe (in itself entirely justifiable) has led to the erection of a purely emotional mental barrier. From an early age, we all become so accustomed to this double logic of thought, to this measurement against a double standard, that we are simply not conscious of its inherent illogicality, which must truly be regarded as a *scandal* from the standpoint of scientific logic. Even the greatest and most consistent thinkers have failed to achieve conscious recognition of this. It was Kant who coined the delightful sentence: "There are two things that never-endingly fill the human spirit with wonder: the starry sky above and the moral law within." With almost superhuman clarity, he recognized the laws of the planetary system and the laws of human reason and morality. And his awe did not prevent him from attributing a natural, indeed *genetic* explanation to the *former,* tracing the existing lawfulness of the planetary system back to the facts of its historical origin. The principal features of this explanation, the Kant-Laplace theory, have been confirmed by the results of modern astronomy.

But even for this greatest among the great, the laws of human reason and morality were met with a mental barrier born of awe combined with religious and philosophical tradition. He did not make even the slightest attempt to approach the moral law within him with the same approach and the same manner of natural understanding as that with which he explained the laws governing the star-filled sky above him. Indeed, Kant's entire moral theory reflects the same dualistic division of the phenomena of this world into those which are amenable to a natural explanation, and are therefore lacking or neutral in value, and those which are extranatural and fundamentally inexplicable, which alone can possess sacred value. As a consequence, for Kant any human act that derives from a natural *inclination* is ipso facto of neutral value in moral terms. This interpretation was, in fact, soon rejected by Schiller, regarded by Herder as "the brightest of all the Kantians." Consider, for example, this quotation from his wonderful *Xenien:* "I will gladly help a friend, but regrettably from inclination, hence I often find it vexing that I am so lacking in virtue." It is precisely such devaluation of everything that has a natural explanation, which is the inevitable result of every form of idealism, that drives all idealists to erect a mental barrier just at *that point* where a phenomenon in the internal or external worlds appears to possess some sacred value.

For Kant, this unbreachable barrier was situated before the laws of human reason because of the very fact that the latter represented, for him, the most awe-inspiring of all phenomena. For very many significant thinkers, the same barrier lies on the great divide between the inorganic and organic worlds. For anyone acquainted with organic creation, the unique and wonderful phenomenon that we refer to as *life* must indeed represent the most immediately awe-inspiring of all the phenomena in this universe. This is why, even with individuals who refer to themselves as natural scientists, a small residue of idealistic leanings can all too easily lead to the erection of an emotionally charged barrier exactly at the point where the phenomena of life begin. This leads to the conclusion that something *supernatural and simply inexplicable is the essential property of life.* It is highly paradoxical, but at the same time typical of the schism imposed by double logic, that the greatest founders of physiology, Johannes Müller and Claude Bernard, who made major contributions to the natural explanation of so many life processes, should have regarded life itself as fundamentally inexplicable. In order to "account" for life, they postulated the existence of a special, extranatural factor—a "vital force"—that provides direction for individual physicochemical processes and unites them within the overall entity of the organic system. Johannes Müller, in particular, reflecting the schismatic nature of the idealist, simultaneously became the father of causal-analytical physiology and the founder of the disastrous mongrel theory that we refer to as *vitalism* because of the assumed existence of a "vital force."

It has already been adequately demonstrated that, in the quest for a natural explanation, it is rationally and logically simply indefensible to decide on some voluntary limit with respect to any given natural phenomenon. But, in addition to this let us now raise a *different* question: Is it justifiable for a natural scientist, in trying to explain *any* phenomenon, to postulate the existence of an inexplicable and intangible force? Is this perhaps legitimate given the precondition that *all other* explanations are unsuccessful? Of course, anybody is free to believe in something special, higher, and perhaps even supernatural that resides in life and in ourselves. But what one may *not* do is to introduce something supernatural as the *determining factor* in the causal chain of natural phenomena and then to *pretend* that this provides an *explanation* for natural processes. Here as always, for us an explanation is the tracing back of some special unknown phenomenon to a more general familiar basis. A special "factor" conjured out of the air to explain a specific phenomenon is everything but a more general, familiar law. In the Middle Ages, it was customary to be extremely generous with the introduction of such factors. Gases or liquids are sucked in by any vacuum. Why? Why should we rack our brains over such a question. "Nature" simply has an "aversion" to empty spaces, a "*horror vacui*"! A pigeon that is forcibly removed from its loft makes every effort to return home. Why? Because it has a "homing instinct"! Do such "explanations" assist us in reaching a better understanding of the phenomena concerned? Not in the slightest! As the American pragmatist John Dewey clearly stated, a factor introduced for the purpose of explanation is nothing other than the articulation of the already known fact with a new word. The fact that air flows into a vacuum or that a pigeon returns to its loft from a great distance is a *real* phenomenon. To observe, describe, and record such phenomena is certainly the task of natural science, even if it is initially confined to the idiographic stage and is not yet in a position to provide an explanation. By contrast, the "*horror vacui*" and the "homing instinct" are *not* real. They are empty words that, in addition, cause great damage because they fraudulently *pretend to be concepts*, providing an explanation of the process. After all, with the same justification we could introduce such factors for *any* phenomenon that we care to choose and pretend that we have explained them as a result. If a child asks me what makes a train run, I could reply that it is simply brought about by the so-called "locomotive force." I hope that any of *my* children would then go on to ask where this locomotive force came from! But anybody who is *satisfied* with such a pseudo-explanation does *not* ask any further questions, and that is where the danger lies. We are *fundamentally unable to know* how far our causal understanding of life processes may eventually go, or where the boundary lies between that which can still be rationalized and that which is beyond reason. But vitalism dogmatically claims to know where this boundary lies in that quite specific life processes are simply declared to be beyond rationalization. The nature of such dogmatic definition of boundaries often reveals very clearly the correctness of our interpretation of the origin of the mental barrier created by idealism: It is precisely the most wonderful, the most beautiful, and the most awe-inspiring life processes that have been, and continue to be, dogmatically proclaimed by vitalists to be beyond rationalization.

Vitalists have already made repeated errors with such claims. For example, Johannes Müller still believed that chemical compounds involving more than two elements can only be generated by the influence of the life force, hence being beyond rationalization. Because the *principle of energy* was unknown at the time, he invoked the influence of the "vital force" more liberally and more naively than later vitalists. With advances in causal research in physiology, which increasingly generated the ability to analyze processes specific to life and to trace them back to their physicochemical basis, vitalism increasingly withdrew its barrier to rationalization, eventually confining it to the realm of quite specific, extremely complex life processes. These were, above all, processes that were characterized by particularly high degrees of *lawfulness* and *finality.* Particular examples of such phenomena are inheritance, embryonic development, restoration and regeneration, and unfortunately also "instincts," which are of special concern to us here. These phenomena are indeed quite wonderful in the literal sense of the word. Even for the knowledgeable natural scientist, these particular processes represent something that repeatedly evokes a deep sense of wonderment, of ϑαυμάζειν. A tiny sperm cell can transmit not only the endless multitude of species-specific characters, the entire anatomical framework, but also individual traits, including not just morphological features but also highly specialized, personal mental properties. The process by which this is achieved is also a "miracle" for us, in the sense that it evokes our most intensive "marveling." It is equally "wonderful" how the apparently simple material of the egg cell gives rise to the entity of the embryo, which—while continually maintaining its holistic character—undergoes "differentiation" to produce the great multitude of individual body parts. Eventually, this marvelous process produces a creature such as a young lizard or a duckling that emerges from the egg fully prepared for the struggle for existence. Restoration and regeneration represent, if anything, an even greater "*miracle.*" If the arm of a newt is transected at some randomly selected point, its original form is regained through regeneration. In the absence of this intervention, the cells in the remaining stump would have otherwise remained as connective tissue cells in the upper arm, but they now begin to divide and proliferate. How do they "know" *which parts are missing?** How do these connective tissue cells of the upper arm "know" the anatomy of the elbow joint and the number and form of the bones and muscles in the lower arm and hand, which now redevelop with unfailing accuracy?

Finally, the greatest "miracle" of all is presented by the "instincts," by innate behavior patterns. My deepest philosophical wonderment is repeatedly evoked by the sight of a duckling, freshly removed from the incubator, that begins to preen itself with skillful, well-adjusted movements as if it had already done this a thousand times before. When confronted with *this* particular phenomenon, we can almost forgive the vitalist for erecting his mental barrier, precisely because it is a product of a sense of reverence that we share with him. Indeed, this is what has induced us to devote our lives to research into just one of these phenomena. The vitalist is much closer to my heart than any fellow materialist who has only joined the camp of materialism because he cannot see the wonder in these things. Our criticism of the vitalists is not so much inspired by their comprehensible reverence before the wonder of life processes; it arises far more from the fact that their refusal of knowledge also stems from abject resignation, leading them to abandon their research when confronted with the sheer *complexity* of the processes involved.

It is, indeed, extremely difficult to conceive of mechanisms—even at a purely speculative, model-building level—that could potentially provide a causal explanation for these processes. But it is nevertheless instructive to attempt this as a *thought experiment* because it makes us aware of the quite enormous complexity that such a phenomenon must possess *as a minimum.* We only need to consider the tremendous complexity that such a mental model must possess simply to explain the fact, presented in the example of regeneration above, that only missing parts are redeveloped, no more and no less. It is not only necessary for the entire anatomy of the missing body part to be represented in some form. In addition to this, the potential for regeneration must be different at every plane; the residual body must

**Note in the margin here:* Not just cells?

"know" the number and nature of the components in the distal part of the extremity. Insight into the undeniable fact that a causal explanation of life processes is immeasurably complex, so complex that a single human brain will never be able to encompass it, is not in fact something novel. It has already been said that the complexity of extrasubjective reality is in all probability infinite, whereas the capacity of the human intellect for portrayal of this reality is finite (Philosophical Prolegomena, p. 20 et seq.). Human cognition therefore bears the same relationship to the most complex life processes as it does to extrasubjective reality generally. Recognition of the *unfathomability* of an object therefore provides no justification whatsoever for tackling it with anything other than the usual scientific methods. In particular, it does not justify the assumption that the great natural laws governing all matter are subject to an *exception* at some arbitrary point that we may choose. Yet the postulation of any "factor" constitutes precisely such contempt for the general applicability of natural laws. In particular, it is an infringement of the second main principle of physics, the principle of energy.

Modern vitalists, most notably [Hans] Driesch, know how to conceal this fact behind clever dialectics. They never allow their life force to exert a causal effect, however diffuse; instead, as has already been indicated, they simply invoke a "directional" influence on natural causal processes. But the *determination* of a direction means nothing other than the *modification* of a preexisting direction of movement. Even the smallest change in direction of the lightest atom can only result from the influence of a real, physical force. As is well known, in mechanics any change in direction possesses the dimension of an *acceleration*, which is inconceivable without the operation of a *force*. What is really awe-inspiring about the universe, the *one* great miracle, is the fact, that all of the wonderful phenomena of life, including all of the highest achievements of mankind, take place without "small miracles," without *infringements* of the generally applicable great natural laws.

With progress in causal analysis, many of the "small miracles," that were previously postulated by vitalists have been explained one after another on a natural basis. For example, the so-called school of developmental mechanics founded by [Wilhelm] Roux and [Hans] Spemann has achieved extensive causal analysis of an entire series of phenomena in embryonic development that were previously interpreted, initially by Müller and later by Driesch, as direct effects of an extranatural factor. In particular, by means of highly refined manipulations, it has proved possible to bring about nonholistic *errors* in embryonic development experimentally in a *predictable* fashion. This quite convincingly demonstrates that a "holistic" and "directional" factor is simply not present. Modern genetics has produced extraordinarily exact causal analyses of processes which Driesch had just previously firmly attributed to the effects of "entelechy," the name that he gave to the directional life force, and had thus declared to be inexplicable in causal terms.

The same also applies to the phenomena involved in restoration. To take just one example, vitalists had placed particular emphasis on certain especially holistic and purposive regulatory phenomena exhibited in the *coordination of gait sequence* in arthropods possessing many legs, following the loss of individual limbs. After the loss of one leg, a ten-limbed crustacean *immediately,* that is to say, without even once attempting to persist with the previous gait sequence, proceeds so to speak with a 9/9 rhythm. The movement of the remaining extremities is coordinated in the most holistic and purposive fashion imaginable. With all possible combinations of limb removal, the number of which extends into the thousands, a corresponding harmonic regulation is immediately present after their loss. It was argued by the vitalists that it is impossible to assume that the animal possesses a special, harmonic gait mechanism as a kind of emergency reserve for all possible combinations of loss of extremities. Hence, the observed regulation can only be the immediate effect of an extranatural, holistically restorative factor. But in the meantime [Erich] von Holst has discovered through experimental investigation that there is an admittedly somewhat complex but nevertheless comprehensive causal physiological explanation for these regulatory processes. Finally, it is the task of this book, through its presentation of comparative ethology, to demonstrate just how far the "miracle of instinct"—the last great domain of vitalism—can be mastered by the progress of inductive causal analysis. From this, the general implications of what has been said will be seen in individual details. We shall see just how greatly damaging the effect of vitalism can be in *inhibiting research* by dogmatically imposing a barrier between the rationalizable and the nonrationalizable at an arbitrarily selected point where there is, in reality, no such barrier.*

**The manuscript contains another version for the last part of this sentence:* . . . whereas in reality such a barrier fundamentally does not exist, but is practically located at a point that simply cannot be predicted—in line with the principle of asymptotic progress discussed above.

12

Mechanism

The many retractions, large and small, that vitalism has been obliged to accept since the time of Johannes Müller have contributed to a reinforcement of the opposing school of thought, which claims that *all* life processes are fundamentally explicable in physicochemical terms. This school of thought is referred to as *mechanism* and we shall retain this term here, although the linguistically correct word would be *mechanicism.* As far as I am aware, this latter term has been used only by the vitalist physiologist [Gustav von] Bunge. In the field of research into animal and human behavior and psychology, there are three great schools, to some degree interconnected, that are justifiably labeled as mechanistic: Wundt's association psychology, American "behaviorism," and Pavlov's reflexology. Because of their fearless—one is almost tempted to say disrespectful—causal analytical approach to the most complex of problems, indeed because of their conscious simplism, they have achieved major and lasting successes in their research. Seen merely as a working hypothesis, mechanism is not only enormously fertile; it is simply the only legitimate approach available to the research worker. As has already been stated repeatedly, the truth is represented by the working hypothesis that least blocks the way to a better successor. Even if there *are* fundamentally insoluble problems in the realm of life processes, it will be of little detriment to advances in our understanding of the world if a research worker should now and again tackle such an insoluble task in vain. To the contrary, we will thus come somewhat closer to the possibly insoluble problem through improved demarcation. On the other hand, the resignation of vitalists in the face of insoluble problems can be the cause of immeasurable damage, as has, for instance, been the case with the central problem of our own discipline, the question of instinct. It is a *question of belief* whether one feels in one's heart that there is something supernatural that is immune to research. As a *researcher,* however, one *must* be a mechanist. Even the great vitalistic natural scientists, ranging from

Müller and Bernard to Uexküll and Driesch, arrived at their greatest and most enduring achievements in cases where they approached life processes with purely mechanistic working hypotheses. As *researchers,* they too were mechanists! Perhaps the finest articulation of the optimal human and scientific attitude to these questions was produced by Alfred Kühn. He ended a lecture on viruses and the boundaries between the organic and inorganic realms with the citation from Goethe already quoted above: "To investigate what can be investigated and quietly to revere that which is immune to investigation." He hesitated for a moment and then, raising his voice to drown the gathering applause, added quite impulsively: "No! No! Not *quietly,* gentlemen."

In epistemological terms, however, mechanism has the same failing as that attaching to any "-ism" in the natural sciences. As unprejudiced inductive investigators of nature, we can neither claim that there is something fundamentally inexplicable, as vitalists do, nor rule out this possibility, as mechanists do. We have already seen that, *as a working hypothesis,* the mechanistic claim that all life processes are amenable to physicochemical explanation can do no harm and is solely beneficial. From a philosophical and ideological point of view, however, the *dogmatism* attached to this claim can all too easily lead to an equally dogmatic fixation on individual statements about *how* particular life processes are to be explained. Certain such dogmatic explanations are so characteristic of the mechanistic schools that they are immediately thought of in some circles whenever "mechanism" is mentioned. This is particularly the case when a sharp distinction is made between mechanism and materialism, as with the prevailing philosophy of dialectical materialism in the Soviet Union. For this reason, it must be specially emphasized that the sole basis for criticism of the great mechanistic schools by materialistic philosophy and all genuine inductive scientific research lies in their dogmatic fixation on specific individual explanations. It does not concern the fundamental conviction that life processes can be explained in terms of the great laws that govern *all* matter. This conviction is generally materialistic and is not "mechanistic" in the narrower sense that provokes criticism.

We must now take a closer look at the criticism of attempts to explain complex life processes mechanistically, in the narrower sense of the term. As has already been indicated, all three great mechanistic schools suffer from the fact that each has raised a *single* phenomenon, discovered at an early stage, to the *sole* explanatory principle for the entire, complex harmonic system of animal and human behavior patterns. This is an infringement against the methodological principles of "analysis on a broad front," already discussed in detail, which are obligatory in tackling organic systemic entities. Without any doubt, the reason for this lies in the historical fact, briefly introduced above, that all three great mechanistic schools of behavioral research started out from the discovery of a single "entity-independent component," constituting a "favorable object" (see Biological Prolegomena, p. 146 et seq.). The possibilities for immediate detailed analysis that were thus opened up then, as it were, *enticed* investigators to *skip* the idiographic-systematic stage and to proceed too rapidly to the nomothetic stage. With Wundt's "scientific" philosophy at the turn of the century, *association* was the entity-independent component that determined and dominated the entire direction of research in this way. With behaviorism, it was *learning through trial and error* and with Pavlov's school it was the *conditioned reflex.*

In the final years of the last century, Wundt discovered the process of *association,* that is to say the lawful development of *coupling* between two thoughts or two sensory impressions that occurred in direct temporal sequence once or several times. Such coupling is subsequently expressed through the fact that the occurrence of the first sensory impression or thought automatically evokes the second. Wundt investigated the principles governing such thought associations and formulated the *laws of continuity* and *successivity,* which also play an important part in the theory of the conditioned reflex. The first of these two laws states that associations only arise between sensory stimuli or ideas that have occurred in immediate temporal sequence on one or, preferably, several occasions. The second law, on the other hand, states that only the *first* of the two processes can evoke the second through association and not vice versa.

The discovery of association revealed a psychological process that was not only easily accessible to experimentation but also, because of its comparatively simple and fixed mechanism, represented a typical, relatively entity-independent component of an organic system. It permits the attainment of correct results even with a naive, atomistic approach to analysis, that is, one that completely neglects the holistic character of the organism (Biological Prolegomena, p. 145 et seq.). These circumstances led to extremely significant initial results from research into association that were of lasting influence to the extent that they doubtless bore fruit in the origin of behaviorism and hence also in that of Pavlovian reflexology. On the other hand, these great initial successes led to massive overestimation of the range of applicability of the principles that were discovered: Wundt's school arrived at the view that *all* of human psychology can be interpreted as a "bundle of associations!" Association psychology was, in fact, far removed from biology and had no healthy appreciation of the extreme systemic complexity that is exhibited by even the simplest organisms. Still less did it employ the conceptual and practical approach of analysis on a broad front which represents the only means of understanding "systems of universal reciprocal causal relationships" (Biological Prolegomena, p. 137 et seq.) and which was already entirely familiar to the great masters of physiology almost half a century earlier. Indeed, association physiology thrived in the proud conviction that its procedure was particularly "scientific" because it endeavored "strictly in accordance with the example of chemistry and physics" to trace all psychological phenomena back to the "elements" of sensory impressions and associations.

Even though the naive simplism and monistic explanations of association psychologists are almost incomprehensible to modern biologists and Gestalt psychologists, and even though we are inclined to use quotation marks for the adjective in their own chosen label of "scientific" psychology, we should not forget that research into association *really* marked the beginning of penetration of scientific methods into psychology. Even if these methods are subject to gross errors, the mere fact of the introduction of *experimentation* into a science such as psychology is of enormous intellectual significance. Neither naive simplism nor crass oversimplification have been detrimental to research. Indeed, a clearly expressed error that stimulates opposition and more precise investigation has always been far more beneficial to science than cautious refusal to make any statement at all. In fact, the only damaging effect of association psychology was brought about by the *rebound* in opinion that resulted from its eventual failure in the analysis of organic entities. The inability of association psychology to analyze the entity of psychological processes unfortunately had the pronounced effect of discrediting the scientific approach as such in psychological circles. Above all, the fallacious atomism of association psychology was confused and equated with the scientific approach itself. Finally, a rebound effect from the extreme atomism of association psychology was also evident in that *Gestalt psychology*, which definitively banished atomistic (or, more exactly, entity-blind) working hypotheses from psychological research, developed a pronounced *vitalistic* leaning. Unfortunately, the concepts of Gestalt and entity have, for many highly significant authors, more or less taken on the character of "factors," which are regarded as being neither amenable to nor in need of a causal physiological explanation! For these reasons, it cannot be emphasized enough that neither the endeavor to trace phenomena back to known elements nor the mechanistic tendency to press on to the physicochemical level constituted an error that eventually led to the failure of association psychology. The exclusive blame for this is attributable to methodological infringements of the laws of inductive scientific research.

Association psychology lacked a sufficiently broad inductive basis because it had never engaged in hypothesis-free idiographic recording of animal and human behavior. For this very reason, there was no way of knowing where this field of knowledge belonged in the hierarchical system of the natural sciences and it therefore failed to establish contact with the next broader science or even to identify that science. As a result, a vain attempt was made, using fundamentally inappropriate methods (Philosophical Prolegomena, p. 46), to establish a direct link with chemistry and physics. Association psychology was blind to the organism's character as a system of universal reciprocal causal relationships and this gave rise to the erroneous belief that this entity could be resynthesized from the only relatively independent component that had been successfully analyzed. We shall soon see that the other two, more modern mechanistic schools commit a methodological error that is fundamentally the same, albeit less naive and subtler.

Behaviorism is the name given to a research discipline that primarily arose in America and is still flourishing there today. It is totally divorced from the observation of subjective experiential processes and the focus of its investigations is confined to overt *behavior* as the only feature that can be *objectively* grasped. [Edward Lee] Thorndike, one of the most significant founders of this discipline, discovered an objectively accessible process in the form of *learning through trial and error.* This undoubtedly closely corresponds to the objective behavioral manifestation of the *same* psychophysiological process as that which was investigated from the subjective-experiential side in Wundt's association psychology. An animal confronted with some kind of problem, for example, a cat that has been enclosed in a "puzzle box" with a simple and accessible bolting device, initially shows completely *undirected* motor restlessness. It runs around in search of a way out, climbing up the walls, scratching and sniffing here, there, and everywhere. In the course of these initially quite undirected attempts to escape, the animal eventually comes across the bolting device quite by chance. Then, as a result of a movement that is initially completely lacking in insight, the animal succeeds in pushing the bolt back and hence in escaping. If the experiment is repeated, the cat does not abruptly gain insight into the function of the bolt, as would be the case with a human being or a chimpanzee. Instead, the amount of time that the animal spends in undirected attempts to escape decreases quite gradually, following a progressively declining "learning curve." Eventually, when the animal is locked in the box, it will at once turn to the bolt without any superfluous activity and push it back with a purposive movement of the paw. In this way, the animal has *learned* what it has to do through trial and error. Comparable learning processes doubtless play an extremely large and important part in the overall behavior of animals and humans. Particularly with various mammals that can be said to possess "middling" intelligence, a considerable part of their behavior is determined by self-conditioning arising through trial and error.

The discovery of this easily analyzed, entity-independent component accordingly led to great and really significant successes in causal analysis. But these successes also led to the same erroneous conclusions as in the case of association psychology and that would also prove to be the case with reflexology. The explanatory principle that had been discovered was dogmatically raised to the status of an exclusive determinant. In other words, the attempt was made to account for virtually all animal and human behavior on the basis of trial-and-error learning. This early fixation of the entire behaviorist school on a single working hypothesis had the result that a hypothesis-free, idiographic-systematic study of the organic entity of animal behavior was regarded as unnecessary and was simply omitted. Because the school therefore confined itself to conducting experiments dictated by its sole working hypothesis and never engaged in hypothesis-free general observations of animal behavior under natural conditions, the resulting absence of a proper inductive basis simply excluded any possibility for recognizing the inadequacy of that working hypothesis.

On the one hand, the behaviorists achieved great and lasting success through their research into behavior patterns that are actually composed exclusively of conditioned responses acquired through trial and error. On the other hand, they understandably made major errors in cases where their working hypothesis was *not* appropriate. In particular, their great stubbornness did much to distort the facts and completely false explanations were then imposed upon them. This is evident in an especially crude form with attempts that the behaviorists made to explain all innately adaptive, species-specific motor patterns. Their "explanation" resides essentially in the fact that the existence of the phenomenon they sought to explain was effectively denied. [J. B.] Watson, the most outstanding representative of the behaviorists alive today, simply does not accept the existence of long, complex coordinated sequences of innate motor activities. He maintains that all apparently innate motor patterns of young animals are derived from the form of the motor *organs* by means of trial and error. His opinion, as it were, is that young animals try out all possible movements with their wings, legs, and so on and hence through self-conditioning gradually arrive at the end result that wings are suitable for flight and legs for running. Just a few minutes' observation of a chick or duckling freshly removed from the incubator is sufficient to demonstrate conclusively the utter untenability of this hypothesis. The behaviorist school nevertheless doggedly clings to it.

We are obliged to emphasize here these failings of behaviorism and their methodological causes because they have exerted a pronounced disruptive influence on our own field of research. But this criticism by no means amounts to an underestimation of the considerable value of behaviorist research or of the benefits that it has brought. A major advance was made in the exactitude of the approach through taking objective behavior as the exclusive focus of observation. The complete exclusion of subjective internal processes had the welcome effect that it brought about an (albeit extremely radical) separation between investigation from the physiological side and psychophysiological processes. It definitively put an end to the previous utterly typical but logically unacceptable confusion of physiological and psychological causal chains (see Biological Prolegomena, p. 169 et seq.). Today, we can recognize that it is *no longer* necessary to engage in this way in such "psychology without a soul," with its associated refusal of knowledge, because we have learned to maintain an orderly separation between the physiological and experiential sides of psychological processes. But this very methodological orderliness in separating the two incommensurable sides of all neurophysiological processes accompanied by experience, the very method that renders statements about psychological phenomena outside of our own egos at all *possible*, is a heritage of that "psychology without a soul," a heritage of behaviorism! Indeed, one can say that Descartes' recognition of a fundamental duality of all experiential life processes first became clearly established in the minds of psychophysiological research workers when behaviorism broke through the Gordian knot of hopelessly confused psychological and physiological causal chains. It did so by simply excluding all psychological aspects from its realm of investigation! It was precisely this move by the behaviorists, and nobody else, that first opened the way to that observation of the internal world from the exterior, that reflection about reflection, which we have discussed in the Philosophical Prolegomena (p. 12 et seq.), emphasizing its great intellectual significance. The behaviorists, and nobody else, discovered the "other side of the mirror" in the living subject!

Behaviorism experienced a major impetus accompanied by a fertile increase in the precision of its approach through Ivan Petrovich Pavlov's great discovery of the *conditioned reflex.* According to his own autobiographical account, Pavlov was heavily influenced by behaviorism, particularly by

Thorndike, at the beginning of his research. The process of development of conditioned reflexes that he discovered represents without doubt the most important physiological basis both for the learning processes of the behaviorists and for the phenomenon of association. If a morsel of meat is placed in a dog's mouth, it elicits the secretion of saliva. The nervous elements from the organ of taste lead to the center, where they transmit their excitation to other elements leading to the peripheral effector organ, which induces the gland to release its secretion. This constitutes the nature of the simple reflex process, which had already been demonstrated some time previously through the research of Bell and Magendie. These investigators discovered the one-way conducting system of the neurons and were the first to distinguish afferent and efferent nerves. Other workers, including Johannes Müller, had investigated the process further.

This simple salivation reflex takes place in exactly the same way in *any* dog, independently of any prior experience, and hence occurs even in a young animal that has never tasted meat. For this reason, in contrast to the process that is about to be described, it is referred to as an "*unconditioned*" reflex. We can now conduct an experiment in which we repeatedly administer a different, *arbitrarily selected* stimulus immediately *before* the stimulus that elicits the unconditioned reflex. The preceding stimulus, for example, a light signal or the ringing of a bell, is itself of no biological significance and elicits *no* response. The result of this procedure is that a quite unusual central connection is established between the two stimuli. This quite remarkable *acquired* connecting pathway for stimulus conduction has the effect that the first stimulus, which previously elicited no response, henceforth operates as a releaser for the unconditioned response. After such "conditioning," which depends upon the *condition* that the two stimuli must repeatedly occur in regular temporal sequence, the first stimulus, which is in itself of no significance, acquires the capacity to *replace* the stimulus eliciting the unconditioned reflex. In other words, the dog will subsequently salivate in response to the light signal alone. Following Pavlov, we refer to the individual response pattern that is acquired in this way as a *conditioned reflex.* Without a trace of a doubt, the conditioned reflex represents the physiological basis of the phenomenon that Wundt identified as association. Indeed, above and beyond this, it provides the common basis for everything that we refer to as *learning,* hence also providing the foundation and precondition for the principle of "trial and error" studied by the behaviorists.

The entire ability to *adapt* behavior in a purposive fashion to changes in environmental conditions, which is possessed by any individual belonging to a higher animal species, is based on the capacity to develop conditioned reflexes. When a mountain goat learns not to wait for the rain to fall but to seek out a protective cavern as soon as dark clouds appear in the sky, such behavioral plasticity is similarly based on the development of conditioned reflexes. The same applies when a young jackdaw learns to flee from a cat after the parents have *just once* responded to its presence by uttering the unconditionally escape-releasing warning call (see the special part,)* or when a horse learns to accelerate its pace in response to the crack of a whip, without waiting for the unconditional "leg-spurring" stimulus of the whip. With respect to survival value, the function of the capacity to acquire conditioned reflexes always resides in the fact that it permits the organism to *prepare* itself for an expected, biologically significant external stimulus. This then allows it to engage in anticipatory behavior that will in fact permit it to *avoid* the incidence of that unconditioned stimulus. In this, the conditioned reflex exhibits highly significant analogies to the much more complex function of the mental form of causality, which takes place on a much higher plane, and it undoubtedly provides the precondition for it. We shall need to return to these relationships in a special section of this book. Before that, however, the conditioned reflex itself and the special principles applying to it will be discussed in detail in the general part, in the section dealing with "*learning*."

The discovery of the conditioned reflex presented the analytical investigation of animal and human behavior with a tool of a kind that had never existed previously. Thanks to the favorable possibilities for exact experimental design, over the course of a few decades an enormous and, in most cases, highly important body of factual evidence was accumulated. As a result, we are today better informed about the physiological processes involved in conditioned reflexes than about any other similarly complex and significant process in the central nervous system of higher organisms. But the unique intellectual significance of Pavlov's results probably resides primarily in the following: Although he himself, like most members of his school, followed the behaviorist example of distancing himself from any consideration of the processes investigated from the *experiential side*, Pavlov's choice of object and direction of research was without doubt heavily influenced by a quest for knowledge that was directed at the mind-body problem. This is evident from certain autobiographical statements that he made regarding the influence exerted on his own research interests by Descartes and his theory of psychophysical parallelism. Even though Pavlov exclusively investigated the physiological side of quite specific nervous processes, he nevertheless selected as the object of his investigations *particular* processes

**At these points, the author is referring to a projected later section of the book that was never written (see footnote on p. xliii).*

that *also* have a psychological side. We therefore see it as his most important achievement that, through his investigation of the *physiological* side of a phenomenon that is *also psychological,* he radically dispensed with that ancient, deeply rooted prejudice that any process that is *analyzable* in causal physiological terms ipso facto cannot be a psychological process. Even for somebody who is ideologically unprejudiced, experience shows that it is extremely tempting to equate "analyzable in causal physiological terms" with "lacking a psychological dimension." We surely owe it largely to Pavlov that this fallacy, which exerted a major inhibitory influence on research, was eliminated.

The extremely great value that we accordingly attribute both to Pavlov's direct causal-analytical results and to their intellectual effects should not prevent us from engaging in criticism of fundamental methodological errors. This is particularly so in cases where our own specific field of research is directly concerned. As has already been noted in the Biological Prolegomena, in the discussion of "entity-independent components" (p. 146 et seq.), the danger of overestimating the range of applicability of a newly discovered explanatory principle increases with the importance of the discovery and with the range of possibilities it presents for proceeding with further causal analyses. Understandably, the further a researcher has been able to advance on the basis of a research method determined by a specific "favorable object" or "entity-independent component," the more difficult it becomes for him to change his method and return to holistically appropriate but tedious analysis on a broad front when he reaches the limits of the entity-independent component and thus the limits of operation of the associated explanatory principle. In this light, it is easier to understand—if it did not sound presumptuous, one might even say excuse—Pavlov's repetition, albeit in a far less naive form, of the same fundamental error that was made by association psychologists and behaviorists. Even Pavlov undoubtedly *overestimated* the explanatory value of the principle that he had discovered by attempting to explain *the totality* of animal and human behavior on the basis of unconditioned and conditioned reflexes. In a treatise on theory of education published in 1916, Pavlov arrived at the conclusion that "the most important function of the central nervous system proves to be a reflex projection, that is to say, a transfer, a conduction of the stimulus from centripetal to centrifugal pathways" (что главнейшей деятельностью центральной нервной системы является рефлекторная, отображенная . . . переброс раздражения с центростремительлых путеи на центробежные). In various places, one can find quite specific statements that the conditioned reflex represents the basic element of all of the most complex functions of the central nervous system.

Now Pavlov's explanatory monism exerted a marked inhibitory influence on research into the innate, species-specific behavior or animals and humans because of the interpretation that the reflex and the conditioned reflex are the only elements underlying such behavior. In fact, as we have already noted, extremely important innate motor patterns—the so-called *instinctive motor patterns*—are certainly *not* composed of reflexive processes. Instead, they depend upon an entirely *different kind* of elementary function of the central nervous system, on processes of *automatic-rhythmic stimulus production* that are fundamentally similar to the operation of the stimulus-production centers of the heart. It was therefore unavoidable that Pavlov and his school proposed a whole series of quite incorrect explanations precisely in the area that constitutes the central research interest of our discipline, namely in the realm of the "*instincts.*"

In retrospect, it is quite amazing how late in the day even those researchers who thought in precise mechanical terms realized that the *spontaneity* of all "drive-governed" or "instinctive" behavior patterns of humans and animals constitutes a phenomenon that is fundamentally *not* explicable through the principle of the reflex. Even at a time when my own research had made me closely familiar with the phenomena of accumulation of response-specific excitation, the lowering of the threshold for releasing stimuli, and the so-called "vacuum response," to the extent that these expressions of the spontaneity of internal processes of stimulus production had become the sole object of my investigations (see also chapter 20, p. 281 et seq.), I unswervingly clung to the opinion that all of these phenomena could be explained with a *supplementary hypothesis.* In other words, I was well aware that they could not be explained through the principle of the reflex, but did not want to abandon the convenient reflexological explanation that all instinctive motor patterns are chains of reflexes. As is well known, once a new discovery has been made and consolidated, it is cheap and easy to be clever *after the event.* Thus, given the present standpoint of our knowledge about the physiological peculiarity of instinctive motor patterns, it seems almost incomprehensible that at that time I failed to draw the following logical conclusion: The individual invariability of innate, species-specific motor patterns, their very limited adaptability in the face of changed environmental conditions, and their frequent malfunctioning, which is easy to bring about under experimental conditions, clearly demonstrate that they are functions of the central nervous system attached to specific *organic structures.* For this reason, they certainly have nothing to do with a vitalistic "factor," with a "directional instinct," but are undoubtedly amenable to a *physiological* explanation. But this physiological explanation cannot be sought in reflexive processes because the lowering of the threshold for releasing stimuli, which takes place progressively during inactivity, eventually followed by stimulus-free eruption of the entire motor sequence, that is, the entire phenomenon of spontaneity, cannot be explained through the principle of the reflex. It is an inherent feature of the concept of the reflex that it indefinitely sits at the ready like an unused machine and "waits for" the peripheral stimulus without which its activation is simply inconceivable. Accordingly, because the fixed mechanical behavior patterns concerned, which are obviously attached to organic structures in the central nervous system, are *not* reflex chains, they must be based on some *other* structurally defined basic function of the nervous system! Do we know of such a function, in particular one that would be able to account for the stimulus-free eruption of the motor sequence, for its spontaneity? We do indeed know of such a function, from the stimulus-production centers of the heart! The fact that I myself failed to draw this practically self-evident conclusion, despite the clearest delineation of these two premises, is exclusively attributable to the fact that—like other mechanistically thinking researchers—I had stubbornly clung to the erroneous idea that it would be a concession to *vitalism* to depart from the chain reflex theory of the instinctive motor pattern.

The whole of the next chapter is devoted to such research-inhibiting effects of the conflict between mechanism and vitalism. Without any doubt at all, the explanatory monism of the Pavlovian school of reflexology, expressed in the dogma that the reflex is the only "element" from which all more complex functions of the nervous system are constructed, is similarly one of these effects. Here, I shall provide examples of two cases of *spontaneous* behavior, which, as will be demonstrated in detail later, quite certainly owe their spontaneity to endogenous processes of stimulus production, but which Pavlov subjugated to an explanation based on the reflex principle while undeniably ignoring certain facts. That is, they were simply labeled with the *name* of the reflex, which under these circumstances possesses no explanatory value whatsoever. Thus it was that Pavlov still wrote in 1916: "The analysis of the activities of humans and animals leads me to the conclusion that there must be, among the other reflexes, a special goal-oriented reflex (рефлекс цели) that represents a striving to attain a particular, stimulus-transmitting thing." He goes on to state that this goal-oriented reflex is closely linked with the function of two basic unconditioned reflexes, the feeding reflex and the breeding reflex. As we shall soon need to elaborate in detail, spontaneous, drive-governed striving to attain specific releasing stimulus situations is in fact one of the most immediate effects of endogenous processes of stimulus production and has no direct connection with reflex processes. Subsequently, in 1917, Pavlov was confronted with an even more immediate effect of spontaneous stimulus production, in the form of a particularly lively dog that "was unable to keep still" and continuously attempted to escape from the constraints of the experimental condition. He dealt with this by simply introducing a new reflex, the "freedom reflex" (рефлекс свободы), which we regard as a flagrant "*contradiction in terms.*"

These errors, made by one of the greatest natural scientists of all time, are undoubtedly an outcome of the methodological mistake that we have already identified, namely that of dogmatically declaring a *single* explanatory principle that has been discovered to be the only possibility and completely ignoring the potential for completely different kinds of explanation. If we have placed great emphasis on clear definition of these errors here, this is certainly not due to any underestimation of Pavlov and his enormous significance, which also applies to our own field of research. Instead, this has been done for the sole reason that the underlying methodological error had major damaging effects precisely in our particular discipline. Any research worker who is concerned with the investigation of "entities," in the sense of systems with universal reciprocal causal relationships, should constantly and unfailingly remember such errors made by one of the great masters. In this way, one may avoid falling into the trap of overestimating the range of applicability of any explanatory principles that one has discovered. Then, one will always be aware that the capacity to analyze organic entities is directly proportional to the willingness to shift to a different explanatory principle, to leave the previously traced causal thread provisionally unattended, and to pursue a quite different one, as is required by the nature of holistically appropriate analysis on a broad front. The more the object investigated possesses the character of a system of generalized reciprocal causation, that is, the character of an entity, the greater the degree of urgency and obligation to employ this method for research. *Moreover, there is no object in the entire realm of natural science to which this applies more decidedly than the central nervous system of higher organisms and the totality of its function, which we refer to as "behavior."*

13

The Implications of the Conflict Between Vitalism and Mechanism for Behavioral Research

As has been explained in the previous chapter, the errors made by the three great mechanistic schools arise from clearly demonstrable logical and methodological misconstructions because of their blindness to the nature of the organic entity as a system of universal reciprocal causal relationships. But the fact that both behaviorism and Pavlov's school of reflexology arrived at such surprising and fallacious pseudo-explanations with respect to innate, species-specific motor patterns, which are of particular interest to us here, is also attributable to a second factor. This, in fact, was the opposition that existed with respect to the dogma expressed with particular clarity by the *vitalists.* As materialists and natural scientists, we are to a certain extent sympathizers of the mechanistic schools and we are therefore inclined to blame vitalism for a number of their errors. However, it is doubtless more justifiable to observe that, with respect to their statements about innate, species-specific behavior patterns—that is, with respect to "instincts"—*both* of the parties in this conflict of opinion were forced into extreme positions that *neither* of them would have adopted if it had had no knowledge of the rival interpretation. If the vitalists created a supernatural "factor" on the basis of the holistic Gestalt or organic phenomena, in the face of which any attempt at analysis would have been tantamount to sacrilege, the mechanists retreated almost voluntarily into blindness to the entity accompanied by an extreme, methodologically fallacious atomism. The most damaging consequence of this was the emergence of the explanatory monisms discussed above, which distinguish "mechanism" from materialism. If the vitalists identified the purposivity of animal behavior as a miracle, by declaring this to be an immediate product of a supernatural factor, the mechanists avoided taking purposivity into account, ignoring even the important, undeniable fact of simple survival-promoting purposivity. Particularly in the case of certain behaviorist authors, this led to the aberration that *pathological* phenomena were, in an extremely misleading fashion, simply indiscriminately lumped together with physiological, meaningful behavior patterns with a survival-promoting function. While the vitalists identified a supernatural factor—variously labeled as a vital force, a holistic ency, entelechy, instinct, or suchlike—that in the final analysis amounted to a *soul,* the mechanists pursued "psychology without a soul." They did this even in cases where, with proper attention to clear methodology, as discussed in the Biological Prolegomena (p. 201), introspection can be extremely informative. Such exclusion amounts to the worst offence against the spirit of inductive natural science, namely a refusal of knowledge.

But by far the most damaging influence of this reciprocal "radicalizing" interaction between vitalism and mechanism came to bear on the field that is the immediate concern of this book. That is to say, it affected the field of research into innate, species-specific behavior patterns of animals and humans, the field of research into so-called "instincts." Innate, species-specific behavior patterns that are purposive in a survival-promoting sense presented *no problems* for the vitalists. Like other life phenomena characterized by a particularly pronounced harmonic, holistic nature and especially obvious finality—such as inheritance, embryonic development, and restoration—innate, species-specific behavior patterns presented a classic case in which the vitalists could postulate the direct influence of an extranatural life force that is neither amenable to nor in need of an explanation. Johannes Müller, who (as has already been explained) possessed the dual personality of an idealistic natural scientist and could hence become the father of modern reflexology as well as the father of vitalism, clearly regarded "the instincts" *not* as an object of the former but as lying in the domain of the latter! In a passage that I unfortunately know only second hand, he discusses examples of the immediate expression of the "vital force." He recounts how an embryo that is still lacking a central nervous system nevertheless grows to become an entity. He notes how in a butterfly pupa the organs that are later to be used by the imago are developed in a purposive fashion and how the nervous system that is already present becomes reorganized to fit the requirements of the new entity, in that some *pairs of ganglia* in the abdominal neural chord move together and become fused while others disappear. He discusses the metamorphosis from a tadpole to a frog, describing how the spinal cord becomes reduced to match the loss of the tail, and other examples follow. Then, in the same passage, he adds that there is a comparable unconscious, organizational, entity-generating force that is also expressed in the instincts of insects! This grand master of physiological research, whose investigations of reflexes were to provide an important foundation for all later—*mechanistic*—research into the central nervous system, thus included instincts among the wonders that cannot be explained in causal terms. Against that background, how could later vitalists endowed with far more limited analytical talent and far less desire for causal explanation do other than abdicate from an explanation of instincts. In 1940, Bierens de Haan could still write: "We observe instinct, but we do not explain it!"

For the mechanists in the other camp, innate, species-specific behavior patterns, with their simply undeniable harmonic, holistic character and their survival-promoting purposivity, represented a process for which their research approach held out little prospect of success and they therefore provided no stimulus to investigation. Because the vitalists made such a fuss about "the instincts," for mechanists it was almost beyond the pale even to mention them. The few mechanistically oriented investigators of reflexes who go so far as to mention them restrict themselves to the convenient explanation that innate, species-specific behavior patterns consist of chains of unconditioned reflexes. This is an explanation that happens to be infinitely more illuminating and closer to real life than the instinct hypothesis of the behaviorists set out on p. 200, and the author himself remained stubbornly attached to it for many years! However, no reflexologist has made the attempt to consolidate the reflex chain theory with concrete observational facts. Indeed, no behaviorist or reflexologist was ever likely to have the opportunity to witness the performance of a typical, highly differentiated chain of innate, species-specific motor patterns. This is because the research method of both schools was of course limited to experiments that brought about a *change in state* of environmental conditions acting on an animal and recorded the *response* of the organism to this. The preconceived opinion that the reflex and the conditioned reflex are the only elements of behavior determined a quite special, scarcely varying kind of experimental setup in which the central nervous system under investigation had no *opportunity* to show that it was capable of anything other than responding to the influence of external stimuli. In this way, it was quite unavoidable that the opinion *necessarily* developed and became consolidated, that the functioning of the central nervous system is restricted to receiving and responding to external stimuli. Because nobody observed what animals do on their own initiative, it was impossible for anyone to notice *that* they can, in fact, show spontaneous behavior in the absence of any external simulation. The entire phenomenon of *spontaneity*, which is so crucial for recognition of the *physiological* peculiarity of such a large fraction of all behavior, thus remained concealed from the very research workers who wished to learn about the physiology of behavior. Conversely, the vitalistic psychologists, for whom the physiological explanation of behavior was anathema, did actually clearly see the spontaneity of behavior, but they regarded it as a "miracle" that was exclusively used as an argument for the fundamental inexplicability of "instincts." Having become aware of this truly tragic, in fact almost tragicomic, dilemma arising from escalation of the conflict between vitalism and mechanism, one is reminded of a line from Goethe's *Faust:* "What we do not know is precisely what we need, while what we do know, we do not need!" Those who would have been able to derive reasonable physiological conclusions from it did not *see* the spontaneity of behavior, while those who *did* see it were incurably prevented by idealistic prejudices from drawing the correct conclusions!

As a result, the great and fertile field presented to inductive natural science by spontaneous, species-specific behavior patterns of animals and humans lay completely barren and uncultivated as a *no man's land* between the front lines of two opposing dogmas. No wonder that it became "a playground for infertile speculation in the humanities," as [Max] Hartmann once put it! [*See the footnote on p. xxxviii.*] Missing from the picture were investigators who, like the vitalists, were receptive to the complex holistic character, the finality, and above all the spontaneity of behavior of higher organisms, and *simultaneously* tackled it with a causal-analytical approach accompanied by a healthy conviction that these phenomena are amenable to explanation, as in the case of the mechanists. But what was lacking above all else was the first and most indispensable precondition for the development of behavioral research into a genuine inductive natural science; what was lacking was an inductive basis assembled with a hypothesis-free idiographic and systematic procedure! Above all, there was a lack of *research workers,* well versed in the conceptual and practical methods of induction in general and in the holistically appropriate analysis of organic systems in particular, to *carry out* the Herculean task that both the mechanists and the vitalists had left undone. What was needed were investigators prepared to undertake the demanding and important, yet very modest and infantile, task, the dedicated, tedious, and yet infinitely rewarding task, of simply observing everything there is to see. If any one of the great mechanists had ever done this, mechanism would have been delivered of its decisive fallacies!

14

The Inductive Basis of Comparative Ethology

As we have tried to demonstrate in the preceding chapters, both mechanists and vitalists were incurably inhibited by quite specific conceptual errors and prejudices that were magnified by their clash of opinion. This prevented them from initiating research into animal and human behavior at the point where it should have begun, namely with straightforward, unprejudiced *observation* of healthy animals living under normal conditions. They were quite incapable of seeing behavior for what it is, that is, as an extremely complex, organic *systemic entity* consisting of quite different components; one which, like *any* organic system, owes its particular constitution to a quite specific historical process of development. For this reason, in the event it was not behaviorists or reflexologists, not human psychologists or vitalistic instinct theoreticians, who blazed the only genuinely inductive, scientific, and methodologically legitimate trail into the investigation of animal and human behavior. This task fell to quite "simple" zoologists, who quietly engaged in the observation of animals free of any idealistic clash of opinion.

No reflexologist, human psychologist, or behaviorist, but least of all no vitalistic instinct theoretician, would ever have undertaken the extremely tedious and demanding task of developing an acquaintance with *all* of the behavior patterns possessed by even a single animal species, of compiling observations of the *totality* of its vital activities. Seen alongside the really penetrating experiments of the reflexologists and the apparently even "more penetrating" speculations of the vitalists, this task seemed to be quite trivial and immaterial. At a superficial glance, this generally seems to be the case with the first task of research into natural phenomena, the basic idiographic stage. It was [Herbert S.] Jennings who was the first to regard it as a task worthy of a research worker to observe and describe in the finest detail the behavior of animals left to their own devices. Jennings must therefore be regarded as the first genuine *idiographic* investigator of behavior. He was also the first to approach the investigation of the behavior of an animal species in a methodologically impeccable manner as research into an organic *system.* In methodological terms, it must be seen as an invaluable achievement that he defined the *concept* of the *action system* of an animal species as the concept of a *holistic system incorporating all of its behavior patterns.* He established the methodological requirement that this system *as a whole* must be familiar to the research worker, at least in its broad outlines, *before* a single behavior pattern of the species concerned can be subjected to detailed analysis. The unconditional justification for this requirement is sufficiently obvious from the considerations presented in the section dealing with organic entities and analysis on a broad front. Further, it can be clearly seen from our criticism of mechanism that adherence to this seemingly self-evident requirement would have precluded all of the main errors of the great mechanistic schools from the outset.

It is a reflection of his deep insight into the nature of organic systemic entities and into the difficulties inherent in their analysis that Jennings selected the action systems (behavioral repertoires) of the *lowest* organisms as the first objects for his research. He observed thoroughly a whole series of lower animals, almost exclusively protozoans and coelenterates, literally down to the finest detail. He was thus able to establish, as completely as anyone could wish, *inventories* of the behavior patterns exhibited by each individual species. With the animals that Jennings observed, it is scarcely worthwhile to engage in "gleaning." However, he did not confine himself purely to observation. In cases where his demanding criteria for a holistic appreciation of the action system of an animal species had been met, he conducted refined experiments aimed at further analysis of individual behavior patterns. Jennings was the *first mechanist* to direct his attention to the investigation of animal behavior who was not *blind to the nature of the entity.* He was the first to operate with mechanistic working hypotheses while at the same time striving to attain a causal analysis of behavior with methods that, in the best sense of the term, were appropriate for dealing with organic entities.

Jennings never conducted phylogenetic comparisons between the behavior patterns of related animal species. None of his statements includes reference to phylogenetic homologies in the behavior of related species. Nevertheless, despite the fact that he never concerned himself with the *systematic* aspect of behavior in the phylogenetic sense, he can be seen as a pioneer of actual *comparative* behavioral research, as a forerunner of our own discipline. This is because, unaffected by the storm of opinions, he began to tackle the idiographic stage of behavioral research, which provides the essential precondition and the foundation for any systematic approach and hence also for comparative phylogenetic systematics.

But the actual creators of our research discipline were two investigators who were the first *to see the phylogenetic comparability of behavior.* Those investigators were Charles Otis Whitman and Oskar Heinroth, the first a zoologist and the second a medical doctor by training but a zoo manager by profession. As far as the direction of their research interests was concerned, however, both were first and foremost outstanding phylogeneticists and systematists. Quite independently of one another, Whitman and Heinroth discovered the phenomenon of genuine phylogenetic *homology* between the behavior patterns of related animal species. Although Whitman's relevant publication appeared in 1898, it was unknown to Heinroth as he formulated the same ideas in 1910. This is a clear indication of how completely Whitman's extremely important results remained unheeded in both psychological and behaviorist circles. Even today, there is hardly an American psychologist who is even aware of Whitman's name. The extremely close correspondence between the results obtained independently by Whitman and Heinroth provides a very convincing argument in favor of their correctness, such that we can almost be grateful today that Heinroth cannot be seen as Whitman's *disciple.*

Both investigators possessed an invaluable common strength that gave them a head start on any psychological or mechanistic proponent of behavioral research: *both were genuinely acquainted with higher animals and their behavior!* In the discussion of the organic entity and analysis on a broad front, we have already established the principle that the role played by both conscious and unconscious *comprehensive knowledge* of concrete individual facts increases as the object investigated increasingly acquires the character of an *entity.* That is to say, an entity in the sense of a "system of universal reciprocal causal relationships" that even an atomistic mechanist who is entirely blind to the Gestalt quality is obliged to recognize. Everything that was stated in that section was exclusively aimed at the goal of establishing the precondition for a genuinely complete understanding of the following facts: *The central nervous system of higher animals and of humans is undeniably the most complex and, in the sense expressed above, the most holistic system that we know, both with respect to its structure and in terms of the functional entity of its operation that we refer to as behavior. For this reason, comprehensive knowledge of the finest details—in other words, the most intimate familiarity with the object—is more immediately decisive for successful induction than in any other natural science.*

There is, however, a quite peculiar aspect of such a treasure of conscious and unconscious knowledge, which constitutes the precondition for correct induction just as much as for the correctness of intuitive, Gestalt-based perception of governing principles. It can, in fact, only be acquired by someone whose interest is captured by inexhaustible, ever-recurring *pleasure in the object.* In our discussion of the three developmental stages of inductive scientific research (Philosophical Prolegomena, p. 27 et seq.), we have already established how closely a scientist's quest for knowledge is related to that of a young child *inasmuch* as the *pleasure in the Gestalt* and in the *expression of that Gestalt* is the driving force that leads to the collection of known individual facts. This driving force, which enables a young human being to accomplish such immense feats of learning during the first years of life, that pleasure in the Gestalt, is the root of *curiosity.* It is the root of the elementary urge to collect knowledge or to collect *beautiful* objects that express the Gestalt. As has already been said, the systematic natural sciences such as mineralogy, botany, and zoology surely *originated* from the collection of objects. This was not primarily motivated by a conscious search for knowledge but by that far more primitive pleasure in the object for which our colloquial language has no expression other than that of a *pastime.* A pronounced playful interest in a particular object is the unconditional *prerequisite* for collection of a really extensive, rich treasure of knowledge of individual concrete facts. To dismiss as "unscientific" such *hypothesis-free* knowledge, collected because of pleasure in the object, is to scorn the idiographic and systematic stages of science. In the ultimate analysis, such scorn inevitably leads to the kind of failure brought about by hasty advance to the nomothetic stage that we have witnessed with the great mechanistic schools.

By now, it has surely been established sufficiently clearly that, for success to be achieved in research, comparative behavioral research depends more unconditionally than any other science on a broad, rich, comprehensive knowledge of individual facts, gathered in the absence of any hypothesis. Similarly, research into the behavior of living organisms is more dependent than any other science on pleasure in the object, on *playful interest* in an object. It is, of course, true that one can investigate a relatively entity-independent component of animal behavior, such as a reflex or a conditioned reflex, without being an animal lover. One does not need to be a great lover of dogs to engage successfully in surgical experiments on the spinal cord of the dog. But anybody who fails to approach the patterns of his research object with "primeval gusto," with a deep, elementary pleasure, will never succeed in encompassing the *entity* of the behavior of an animal species, be it only the simple action system of a paramecium. Still less will he be able to acquire the far greater treasure of knowledge of small details that is required to achieve real familiarity with the immeasurably more complex system of behavior patterns of a higher bird or mammal. Least of all, however, will he be able to acquire a detailed knowledge of the action systems of *many* related higher animal species. This provides the precondition for the identification of *homologies* between behavior patterns of related animal species and hence for any *comparison* in the phylogenetic sense.

A quite massive investment of time and attention is required, say, to become acquainted with the action systems of 15 to 20 dabbling duck species to the extent that one can memorize their motor patterns down to the finest detail. This is what is required in order to be moved at all to carry out comparisons, just as the comparative morphologist must commit to memory the physical features that constitute his inductive foundation. If playful interest and the primary pleasure in the object did not render observation itself enjoyable, even the most self-sacrificing, patient observer would never succeed in spending months and years gazing for hours everyday with the most concentrated attention at a duckpond or an aquarium. This is what Whitman, Heinroth, and the author himself had to do, for hours on end, and all of us initially believed that we were doing *nothing* other than cultivating a somewhat childish hobby! The relatively enormous *duration of observation* that is necessary in order to establish the inductive basis for phylogenetic comparison in behavioral research simply cannot develop without a pronounced playful interest in living animals.

It is extremely instructive to scour the writings of behavioral scientists of different leanings in search of such pleasure in the object. With typical mechanists, it is always virtually lacking, whereas it is conspicuously present with some vitalists. There are readily comprehensible typological and perceptual psychological reasons for this. Quite understandably, the atomistic "ultradetailed research" of the mechanistic schools is particularly attractive to the type of research worker who possesses little talent for the Gestalt-based recognition of complex entities. In this respect, too, the conflict between vitalism and mechanism undoubtedly had a negative effect. Research workers with a well-developed talent for the recognition of Gestalts were repelled by the atomistic blindness to the Gestalt that characterized the mechanistic schools and were driven into the camp of the vitalists, where any need for causal analysis that they may once have felt was thoroughly suppressed. But even if we must accept that some vitalists derived a certain healthy pleasure from the living animal, from the entity of its harmonic, systemic Gestalt, in them it never reached the intensity that shines forth from the works of Jennings, Whitman, and Heinroth. With none of the great vitalistic philosophers, such as [Herbert] Spencer, [C.] Lloyd Morgan, or [William] McDougall, does it reach such a level that one of them would ever have hit upon the idea of keeping a bunch of living animals in his private apartment and of spending his entire life in direct contact with them. I know of no vitalistic investigator of instinct, no reflexologist, who would have done that and I know of no comparative behavioral investigators who would *not* have done it. One only becomes familiar with higher animals by literally living with them. Otherwise, it is simply impossible to acquire the enormous amount of observation that is needed. There is just no easier way of establishing the inductive basis for comparative behavioral research!

Our research discipline is thus, first and foremost, the *school of animal connoisseurs.* However modest our present achievements may be in evaluation of the nomothetic stage, in advancing to the final abstractions from our young discipline, we are convinced of the enormous importance of the factual foundation that it has already constructed. The possession of this inductive basis also provides our justification for engaging in decisive criticism of developments of opinion that have taken place in the absence of such a basis. In what follows, we shall become acquainted with a whole series of facts that remained unknown to both behaviorists and reflexologists, and necessarily so because of the nature of their research procedure. The interpretation that the reflex and the unconditioned reflex represent the sole element of animal and human behavior is a fallacy that only arose because of the absence of a broad inductive basis, collected in a hypothesis-free manner.

The discoveries of Whitman and Heinroth led to the development of a new direction of research. They came about through a fortunate, but by no means accidental, convergence of a phylogenetic approach and methodology, based on a fundamental training in morphological work, with an extremely rich, comprehensive knowledge of details of animal behavior that could only originate from a playful interest in animals. For both investigators, the comparative investigation of behavior grew out of their shared love of a wonderful object that had already inspired the studies of that grand master of all animal lovers, Alfred Edmund Brehm: love of the world of birds. A bird is one of those typical natural objects that—like certain minerals, flowers, butterflies, and so forth—elicits from people with a pleasure in Gestalts that impulse to "possession" that, in the manner described above, constitutes the root of scientific collection. Right from very early times, people in a wide variety of cultures have kept living birds because of simple pleasure in the Gestalt. Both pioneers of our research discipline were passionately engaged in the hobby of keeping captive birds and from an early age they were experts in their care. In the meantime, *keeping animals* has become an established method of comparative ethology that, as is common with methods generally, has exerted a decisive influence on the progress and direction of our research. For this reason, we must take a closer look at animal keeping as a research method.

15

Animal Keeping as a Research Method

One does not become familiar with animals by enclosing them in a cramped laboratory cage and providing them only with the opportunity to exhibit quite specific responses to stimuli that are determined by the rationale of an experiment. On the other hand, it is also impossible to become fully familiar with animals if, out of exaggerated respect for the entity constituted by the organism and its environment, they are observed exclusively in their natural habitat. In the former case, one is too close to the animal and in the latter, too far removed. In the first situation, one is so close to the animal that only a small fraction of the entity of its behavior patterns is visible. In the second situation, one is so far removed that one can no longer clearly see those "microdetails" of behavior that in fact constitute the "treasure of unconscious knowledge." These details represent, so to speak, the "handholds" with which analysis comes to grips with the object, just as the hand of a mountain climber relies on small projections in the rock face for support. The first basic principle for the scientific observation of animals necessarily takes the following form: *Approach the animal as closely as possible without producing a significant disruption of its behavior that cannot be controlled.* Practical application of this motto in research virtually always entails the need to allow animals to perform their specific behavior patterns under *controllable conditions;* in other words, it requires the *keeping* of animals in captivity.

Observation of an animal species under natural conditions is, of course, the ideal method for behavioral research that is *appropriate to the entity.* Above all, it will always be indispensable at the *beginning* of any research into an animal species that is new to the investigator. Here, the first task is *necessarily* that of achieving an *overview* of the highest functional entity constituted by an organism and its environment. Observation in the field is equally indispensable as a *control* for results obtained in captivity, particularly with respect to answering the question: "normal or pathological?" We shall soon see just how important a correct answer to this question can be. It is, however, an unattainable ideal to aim for reasonably complete recording of the action system of the behavior of an animal species, let alone a more penetrating analysis of it, exclusively through field observation. Simply in terms of the *investment of time* required, the richness and complexity of the behavioral system of a higher bird or mammalian species would place demands on the investigator that could only be fulfilled in exceptional cases. In cases where individual research workers, such as Selous, Vervey, Kortlandt, Goethe, and others, were able to achieve a really penetrating knowledge of the behavior patterns of an animal, it was necessary to devote months or even years to observation of a single species. And even these field observations, which are the most complete among those ever conducted on birds, were confined to reproductive biology and were only possible when brooding behavior or courtship of the species concerned took place at a single site. Only in extremely rare, exceptional cases does the observation of free-living animals permit us to follow the life histories of *individuals,* which is of the greatest importance for more penetrating behavioral analyses. With the American song sparrow (*Melospiza melodia*), [Mrs.] M. Morse Nice followed the lives of individuals for several years and thus carried out what is probably the most complete investigation of the behavior of a higher animal species to be achieved through field observation. But even this success was only possible because the investigator lived for many years in the area where these birds, which are characterized by extreme site attachment, had their breeding territories. She literally lived with her animals for year after year. It is truly the case that one can only become familiar with higher animals by *living with them.*

A further, very important limitation on the effectiveness of field observations lies in the restricted possibilities for conducting experiments. With higher animals that show a marked escape response to humans, the observer is forced to remain in a hiding place. The flight of the animals that is provoked every time the observer leaves the hiding place constitutes a disturbing factor whose effect on their overall behavior is difficult to assess and to take into account in examining the results. The situation is different with invertebrates, particularly with insects, which only flee from human beings when they move or approach too closely. With such organisms, there are no limits to experimentation, even in the natural environment. In cases where the species concerned show a reasonable degree of *site attachment,* it has been possible to conduct complete behavioral analyses with a combination of field observations and experiments. The studies of Tinbergen, Baerends, Molitor, and others bear witness to this.

It would, however, be a quite fundamental error to believe that the essential contribution of animal keeping as a method lies only in elimination of the drawbacks and limitations involved in observation under natural conditions. By the *keeping* of animals, we do not simply mean the endeavor to keep an animal alive as long as possible in captivity, as is the case with the average bird lover or zoo. Instead, we define the method of keeping animals as the attempt to *allow an animal species to perform before our eyes as much as possible of its entire action system under known, controllable conditions.*

For many important experiments—as in all other areas of biological research—the keeping of animals is a *precondition* to the extent that only the *constancy of controllable and hence artificial environmental conditions* can provide a guarantee that an experimental change in conditions is really the cause of the resulting response of the organism. Given the multitude of factors that determine the behavior of a higher animal, it is in many cases of quite decisive importance to be able to exclude uncontrollable incidental stimuli as potential additional influences.

But the second, most crucial strength of animal keeping as a method resides in the fact that it encompasses the organism and its environment *simultaneously*, including the reciprocal relationships between them. After all, an animal kept in captivity will only exhibit its complete system of behavior patterns *if* we succeed in replicating all of the conditions that are necessary for the continuation of the species. But such conditions include all of those *stimuli* that elicit *unconditioned* reflexes in the broadest sense. Hence, keeping an animal in captivity necessarily requires *a complete reconstruction of its "environment,"* taking [Hermann] Weber's concept of the environment as the sum of all conditions that are necessary *as a minimum* to keep that animal. True success in the keeping of an animal species therefore always requires successful *reproduction* as well. Real success in keeping an animal in fact always requires prior completion of penetrating analysis of the unconditioned responses of the species concerned and simultaneously an extremely precise grasp of the stimuli that are needed to elicit those responses. It is precisely this simultaneous grasp of the organism and its environment that elevates the keeping of animals to a research method that is *appropriate to the entity* in the most genuine sense. This fact, in turn, leads to important consequences that we must now examine in detail.

Like quite a number of other biological methods, the keeping of animals is a great *art* that demands quite specific and special talents of the person responsible. Many very intelligent research workers never manage to master this task, however hard they may try. In my opinion, this is attributable to a particular psychological problem in that the animal keeper must *simultaneously* possess a great gift for the observation of tiny details and an ability to see the Gestalt of the entity concerned. These are capacities that are rarely found combined in the personality of a single investigator. Quite often, a truly "detective-like" analytical talent is required to determine the needs of captive animals from tiny details of their behavior, or to infer from the nature of the *disruption* of a behavior pattern which environmental conditions are lacking or which captive conditions prevent its normal performance. On the other hand, a well-developed ability to perceive organic entities and a pronounced gift for creative *synthesis* are required to replace the conditions that are lacking or to exclude the disruptive factors, within the limited scope of animal management. It is precisely this interaction of extremely precise, analytical observation of the animal with creative imagination in the structuring of the artificial environment in which it is kept that constitute the art, the pleasure, and the cognitive value of scientific animal keeping. In illustration of this, let us now consider three concrete examples that demonstrate how the elimination of a behavioral disturbance can simultaneously shed light on peculiarities in the stimulus physiology of a captive animal species and on properties of its natural environment.

First example: One day, a male North American blackbanded sunfish (*Mesogonistius chaetodon*) begins to glide around in small circles, while performing peculiar lateral tail-lashing movements, at a specific place on the bottom of the aquarium that is free of sand. The entire success of maintaining this species, the answer to the question whether the animal keeper will now gain greater insight into the action system of the species or not, depends on whether or not he will arrive at the correct interpretation of this behavioral dysfunction. He must first of all realize that the activity shown by the fish is an instinctive motor pattern that is "intended" for the excavation of a shallow trough, or redd, for spawn. He must appreciate secondly that the substrate in the aquarium is too coarse for this special nest-building pattern, and thirdly that the exposed zinc floor of the aquarium "is thought to be fine sand" by the fish, which "erroneously" performs the motor pattern "designed" for fine sand at this particular spot. Here, I am deliberately using the naive form of expression found in my schoolbooks, which is far from precise but readily understandable. On close examination, the correct interpretation and elimination of this minor behavioral disturbance turns out to involve a number of analytical feats. To start with, it is indispensable to recognize the fixed immutability of the instinctive motor pattern, which is performed in just the same way on a smooth metal substrate as when sweeping away fine sand. But it is also necessary to have some idea of the fact that the fish responds to the smooth metal surface "as if to fine sand" and prefers it as a *substitute object* to the coarse sand elsewhere on the bottom of the aquarium. Above all, however, it already involves a truly scientific attitude to "the instinct." This is by no means regarded as a miracle but is treated just like any *organ* that is either fitting or unfitting, such that the animal keeper must ingeniously adjust the maintenance conditions to match its mode of operation.

Second example: A pair of blackcaps (*Sylvia atricapilla*) kept in an aviary repeatedly breed successfully, but the nestlings are regularly ejected from the nest on the third or fourth day of life. Closer observation reveals that the parents return to the nest over and over again with *food*, but that the nestlings fail to gape to take the food. When the nestlings are removed and hand-reared, they turn out to be completely healthy. I must admit that at the time, as a schoolboy, I did not manage to find the reason for this misfunction, whereas Heinroth promptly hit upon the explanation when he was confronted with the same behavioral disturbance with his shamas. Ejection of the nestlings is due to the fact that in captivity the provisioning parents can find unlimited amounts of food far too easily. Because the intensity of the provisioning drive is adapted to conditions in which finding food is far more difficult and time-consuming, the parents in the aviary provide food so often that the nestlings are quickly satiated and subsequently fail to show the gaping response. Now many insectivorous songbirds exhibit a special, species-specific response that is adapted for the removal of ailing or dead nestlings, which is elicited by any offspring that does not gape. Satiation of the nestlings simply does not occur under natural conditions and this possibility is therefore not "foreseen" in the "construction" of the action system of the species concerned. Thus it was that my blackcaps and Heinroth's shamas responded to their satiated nestlings with the "response to dead offspring" and threw them overboard! Removal of the feeding bowls and provision of the food in small portions at once eliminated this behavioral disturbance for good. In this case, too, it is clear that relatively deep analytical penetration into the response patterns of animals is a prerequisite for the elimination of behavioral disturbances.

Third example: At the Amsterdam zoo, Portielje kept herring gulls (*Larus argentatus* Pontopp.) with the aim of establishing a breeding colony of these birds for behavioral studies. Most of the birds had been hand-reared from an early age and they were in perfect health, utterly free, and fully able to fly. They soon came into breeding condition, formed pairs in the normal way, and went on to lay many eggs. Nevertheless, breeding was at first always unsuccessful because every bird's clutch was always devoured by a neighboring pair. Even for an expert, it initially seems most likely that a *pathological* phenomenon is involved, that is, some effect of captivity that is expressed in the organism itself. We shall soon be looking at such effects in some detail. Accordingly, it must have required a particularly refined talent for observation and combination for the investigator to hit on the solution that external factors in the artificial environment were, in fact, responsible for the behavioral disturbance. The flat area of grassland on which the herring gulls were breeding presented too few optical landmarks to permit the establishment between the pairs of territorial boundaries, which are usually quite fixed. A few digs with a spade were enough to turn the flat area into an irregular, uneven surface and thus to prevent the pairs from launching attacks on neighboring territories, definitively eliminating the behavioral disturbance.

These three examples suffice to show how much the animal keeper can and must learn, through practical management alone and prior to any deliberate attempt at analytical experiments, about the action system of a species and its physiological adaptation to the stimuli of its natural habitat, if its maintenance in captivity is to be fully successful. All three examples involve behavioral disturbances that were exclusively provoked by conditions in the artificial environment that released completely normal response patterns from the animals investigated.

The acumen of the scientific animal keeper is, however, presented with even more difficult challenges because of the fact that behavioral disturbances that are only indirectly due to the conditions in captivity can very easily emerge. They result directly from abnormal responses of the animal and the cause is therefore to be sought in the organism itself and not in the stimuli to which it is exposed. For this reason, with every behavioral disturbance the fundamental question arises as to whether it represents a *pathological* response of the organism under investigation. Although the concept of the pathological, the abnormal, is self-evident to anyone, it is surprisingly difficult to provide a clearly demarcated definition of it. I believe that I can best and most briefly provide the reader with an idea of what is involved by recounting a naive illustration from my school days. My friend B. Hellmann and I had the habit of asking the following question when confronted with behavior patterns of our animals that were initially incomprehensible because their survival-promoting function was not immediately apparent: "Is that how the designer planned it?" To refer to something that is normal and nonpathological as something that "conforms with the plan of the designer" is an extremely crude definition. However, as far as the core of the concept is concerned, it is a very appropriate and perfectly satisfactory graphic definition.

Response patterns that "do not conform with the plan of the designer" are extremely common with captive animals. Minor health problems and minimal defects in the overall constitution of higher organisms can give rise to quite major disturbances in the system of species-specific behavior patterns. Animals that, in a medical sense, are far from being ill but are merely slightly disrupted, perhaps having some minor complaint because of inappropriate rearing (e.g., those suffering from a minor vitamin deficiency or from some other imbalance in feeding conditions), almost always show great defects in their response patterns. Because these mild health problems are brought about by inappropriate conditions in the environment that we provide for animals in captivity, there is really no clear boundary between behavioral disturbances that are due directly to environmental conditions, as discussed above, and those due to the pathological effects of captivity.

Here, too, an example can be provided: large fish of the family Cichlidae are animals that exhibit, in association with highly differentiated patterns of brood care and family defense, highly intensive *fighting responses.* For this reason, they show a high level of incompatibility. In this respect, they are very reminiscent of birds belonging to the duck family Casarcinae (see the special part) [*see the footnote on p. xliii*]. Old males, in particular, are ferocious combatants. If such an individual fish has lived *alone* for a long period of time, it is usually difficult to pair it with a new partner because it will respond exclusively to *any* conspecific that appears on the scene, regardless of its sex, with frenzied fighting responses. Even as a schoolboy, I obviously must have suspected that a kind of "damming" of the fighting drive was involved in such cases. In any event, I had already devised at that stage a trick that I used before making any new attempt to pair off such old champions. I first let such a male "fight to a draw" by providing it with an equally matched opponent in the form of its own mirror image. After the male's fighting responses had been exhausted to the point of unresponsiveness after hours of vigorous attacks on its mirror image, he was usually restored to a state in which he could recognize a sexual partner in the female that was then introduced to him. This case accordingly involves a behavioral disturbance that concerns the normal responses of the organism itself. Although there is an abnormal lowering of the threshold value for the stimuli that elicit fighting responses, this lowering of the threshold can scarcely be described as "pathological." In fact, it is directly generated by environmental conditions, namely the continuous lack of a fighting partner, and it immediately disappears when the corresponding environmental stimuli are replicated.

Such behavioral disturbances are much more clearly pathological in character in cases where the cause also lies in inappropriate environmental conditions but where the effects are longer-lasting and sometimes permanent, as can happen with disruptive influences during rearing. With this kind of behavioral disturbance, *deficiencies* or at least marked attenuation of certain highly particular, species-specific drives and inhibitions are especially common. A pair of little bitterns (*Ardetta minuta*) is in perfect health and breeds every year, but the nestlings are always devoured immediately after hatching. A female leopard gives birth to a healthy cub every year but kills it on each occasion. This is because of a behavioral defect expressed in the fact that the mother consumes not only the placenta and part of the umbilical cord after the birth but "stops eating too late" and goes on to consume part of the cub's abdominal wall and liver along with the rest of the umbilical cord. A hen pheasant breeds normally and hatches healthy chicks, but she then allows them to freeze to death during the first night because she flies up to roost in the trees, as in the nonbreeding season, instead of spending the night on the ground to keep her offspring warm. All three examples involve the omission or disruption of a specific *inhibition* connected with highly specific processes involving unconditioned reflexes, so-called innate schemata. Perhaps even more common are behavioral disturbances due to a quantitative *impairment of individual instinctive motor patterns,* which in extreme cases can lead to the complete absence of part of an action chain. A song thrush repeatedly takes a round, smooth object into its beak and strikes it against a solid surface, such as the floor of its cage or a large, smooth stone. In terms of survival value, the response is designed for smashing the shells of snails. With caged thrushes that are not in peak condition, the intensity of this motor pattern is often reduced to such an extent that the bird never really succeeds in breaking the shell of a living snail, despite the fact that it spends a great deal of time performing the behavior concerned. Weaverbirds, which construct their marvellous nests even when not engaged in breeding, also devote a great deal of time to nest-building activities in captivity. However, the nests that are built are in very many cases incomplete because the intensity of the nest-building patterns never reaches the required level. For this reason, in zoos one commonly sees in aviaries containing weaverbirds the initial stages of nests, but it is very rare to find a completed nest. Countless birds in captivity will begin to engage in breeding, but only a small fraction of them will proceed to really successful reproduction. Some will perhaps carry a few straws to a nest site, but will then fail to complete construction of a nest; others build a nest in the species-specific fashion and then show such a low intensity of brooding behavior that the clutch dies; yet others allow their offspring to starve. In a nutshell, a rupture can occur at any point in the chain of behavior patterns because the animal "fails to put enough effort into it."

It is directly obvious from the examples provided that it can be very difficult in any individual case to decide whether a particular behavioral disturbance is brought about by *external* or *internal* factors, to determine whether environmental conditions or pathological changes in the organism itself are responsible. A great deal of the time, this question can only be answered through the manner in which the behavioral disturbance is eventually eliminated, that is, through a diagnosis "*ex iuvantibus*" as a medical doctor would say. It is equally apparent from the examples provided *just how important* the correct answer to the question "normal or pathological?" is for the great cognitive value of animal keeping. Both the reconstruction of the normal action system of the species investigated and the degree of exactitude with which we succeed in replicating the natural environment through controllable artificial housing conditions are directly dependent upon this. Accordingly, a finely developed sensitivity to the delicate and *often diffuse* boundary between the not-quite-normal and the already pathological is perhaps the most important talent that a scientific animal keeper must possess! On the other hand, however, the keeping of animals in our sense is an apprenticeship that renders one's feel for the pathological just as acute as actual training in a medical clinic. It is surely no coincidence that Heinroth, the outstanding master of animal keeping as a scientific method, was, like the writer of this text, initially trained as a *medical doctor.*

An animal keeper is, of course, not exactly pleased when he has to deal with behavioral disturbances in the course of his daily work. He should never forget, however, that he is not keeping the animals simply for the purposes of propagation, like some animal lover or breeder. Instead, he aims to penetrate into their action system, and any disturbance in the performance of behavior is the most valuable tool available to assist him! This applies to the exogenous behavioral disturbances that are provoked by environmental conditions just as much as it does to endogenous, pathological disturbances. We have already seen with respect to the former that elimination of behavioral disturbances is the most important source of knowledge for identifying the appropriate environmental conditions. But endogenous, pathological behavioral disturbances also provide a direct source of knowledge for the ethologist. A large part of our knowledge about the physiology of the central nervous system is derived from the study of *functional deficiencies* arising from accidental, pathological, or experimentally induced ablation of specific parts. In exactly the same way, a large part of the evidence for our analysis of the species-specific action systems of higher animals is actually derived from the explicitly pathological behavioral disturbances that are so very common with captive animals. Here, it is possible to begin by citing a quite fundamental example: No animal keeper who, as in the examples provided, is really familiar with such behavioral dysfunction from his daily work would ever think of postulating an "infallible instinct" or some such supernatural factor as the explanation for the normal performance of behavior. The quantitative defects in certain behavior patterns that have just been described were the source from which I derived my first clear impressions of the nature of "instinctive behavior patterns"!

When a novice first *begins* to keep a particular animal species, for example, a bird species of some kind, the captive conditions provided are, of course, initially somewhat inadequate. As a result, all of the finer, more differentiated behavior patterns, especially those involved in reproduction, are initially seen *only* in such an incomplete form because the intensity of response-specific arousal is insufficient. In spring, the males briefly display courtship toward the females and one or more of them will occasionally carry a straw in its beak or perform a couple of nest-hollowing movements on a suitably forked branch, but *nothing* more occurs this year. But, to the extent that the animal keeper learns to improve the conditions provided for his animals, other, previously unseen behavior patterns "emerge" and those that were first seen increase in vigor and intensity: carrying of straws and performance of nest-hollowing movements become more intensive and other nest-building movements appear as well. It can still happen that a rupture will occur at some arbitrary point in the sequence, as explained on p. 228. But *if* the animal keeper succeeds in bringing the animals to full health and in providing all of the necessary environmental conditions, the mosaic of individual instinctive motor patterns and other response patterns will gradually come to constitute a meaningful entity "as the designer intended." The initial pathological disruption followed by the progressive reconstitution of the entity of the action system of his animals will, as it were, parade before the animal keeper's eyes a *real* analysis and resynthesis. This is probably the most impressive and instructive lesson that can be provided by animal observations! The animal keeper is compelled to realize the sense in which the system of behavior patterns is a harmonic entity. At the same time, the innumerable behavioral defects shown by his animals will sufficiently knock it into his head that they have no inkling of the survival-promoting purpose of their own behavior patterns. He will realize that they do not, in fact, consciously pursue the goal of fulfilling the survival-promoting function and that no "holistic factor" tells them what they have to do in order to attain that goal. Instead, it will become patently obvious to the observer that the animal "discharges" individual behavior patterns in a completely blind fashion. It is only because of the "design" behind such completely fixed responses that, when each of them is performed at full intensity, they combine to form a *mosaic* of relatively entity-independent components that expresses the survival-promoting *purpose* of the overall behavior. The observer is taught in the most convincing fashion that "the instinct" quite simply *does not exist* as a "*factor.*"

Any animal keeper who has succeeded, in the manner described, in observing the action system of a species, at first in fragments and then as a cohesive entity, would have to be analytically blind to fail to see a fact that is of fundamental significance for all scientific investigation of behavior: *there are functional parallels between species-specific behavior patterns and organs.* He is obliged to realize that the species-specific behavior patterns of an animal species are related to the environmental conditions with which they interact *in the same way as organs.* They either *fit* or *do not fit* and they are just as incapable of adapting to the "wrong" conditions of the captive environment as is the form of organs. If we want to keep a mole in captivity, the structure of the artificial environment that we provide must be adapted to the structure of its digging organs. In exactly the same way, in our example of the sunfish we must provide environmental conditions that are suitable for the *motor pattern* of excavation. Just as the mole has no digging organ *other* than the structures that are typical of the species, the sunfish has no *other* possibility for excavating a nest other than that provided by the instinctive motor pattern developed by the species to serve this function over the course of its evolutionary history. Indeed, *no* organism can perform behavior patterns *other* than those for which the potential is provided in structures of the central nervous system that the species concerned simply *possesses.* There are no exceptions to this principle, even in the highest mental achievements of the human species, as has already been demonstrated at length in the Philosophical Prolegomena (p. 15 et seq.). There, we attributed great significance to the *structural dependency* of psychological functions. *Here,* it can be stated that recognition of this fact is ultimately *derived* from the observation and analysis of behavioral disturbances of animals kept in captivity. The *structural dependency of behavior* is the great lesson that captive management, and particularly the resulting disturbances of behavior, have to offer for anyone who has eyes to see.

Anybody who has completely absorbed this lesson that behavior is also an *organ function* will also understand the inevitable consequence that the *essential problem* of species-specific behavior and its species-preserving purposivity, like that of any organ function, resides not in the relatively easily analyzed mechanism of its operation *here and now.* Instead, it resides far more in an inquiry into the *source* of such purposive design in these highly complicated mechanisms. The question concerning the "designer," so often invoked in the mechanical parallels of the diaries from my school days, in fact amounts to nothing other than an inquiry into *the totality of factors in the evolutionary modification of species that led both the organs and the action system of every animal species to develop in a particular way in the course of a unique historical process.* In this way, the attention of the investigator is unerringly directed toward the question of the *phylogenetic origin* of species-specific behavior patterns.

In summary, we can draw the following conclusions: The great strength and the main value of keeping animals as a method lie in the fact that it *obliges* the investigator to see the organism and the environment simultaneously as an entity. As we have just seen, this *holistic appropriateness* of the method automatically reveals the species-specific behavior pattern as an *organ* of the animal species concerned. This, in turn, reveals its structural dependency, its character as the *function* of an organ, namely that of the central nervous system. As a result, the approach and methods of *evolutionary biology,* which have long been applied as a matter of course to *organs,* are extended to include the *behavior* of organisms. Animal keeping inevitably leads the ethologist to recognize the necessity for a phylogenetic approach!

In addition to this, however, the keeping of animals also provides special opportunities for comparisons between animal species that are, to varying degrees, related. It is, of course, only with captive animals that one can observe the behavior of *closely related* species in immediate proximity. As is well known, under natural conditions such species are always mutually exclusive at any given site. Quite generally, field studies rarely present an opportunity to become so closely acquainted with a large number of systematically related forms that one is stimulated to compare their behavior. It would, for example, be quite impossible to conduct field studies of the behavior of different representatives of a single genus (e.g., *Anas*) distributed in South Africa, Madagascar, India, Japan, the entire Palearctic region and North America. By contrast, *the pursuit of a hobby,* with its systematically oriented *urge to collect,* which makes such a great contribution to the motives underlying both scientific and nonscientific animal keeping, has the immediate effect that many investigators devote particular attention to a quite specific group of related species. Such specialization on a particular order, family, or genus is typically present even among amateur animal keepers with no scientific interests. One fancier of "ornamental fish" will breed only cyprinodonts, while another will be interested only in "tetras" (Tetragon-opteridae). One bird lover will only keep waxbills, while another will concentrate on softbills, and so on. Both Whitman and Heinroth also had a special fascination for a particular group of related species. For Whitman, the members of the pigeon family were his favorite animals, while for Heinroth (as for me) it was members of the duck group that from an early age exerted a remarkable attraction that is hard to define. With both Whitman and Heinroth, the passion for a particular groups of animals led primarily to a "fine systematic" exploration of the relationships within those groups. Pleasure in the object led both of them, in the manner described in the chapter dealing with the developmental stages of inductive scientific research [chapter 14], to collect objects and to study their *systematics* in fine detail. Such a profound passion for a specific zoological group in fact provides the best foundation for a thorough examination of phylogenetic relationships. Let us return to what was said about the relative taxonomic respectability of individual characters in the chapter on *evolution* in Biological Prolegomena [chapter 6]. The best inductive basis for the inference of phylogenetic relationships is possessed not by someone who traces *one organ, one character* in all of its different manifestations across a wide range of animals, but by someone who is familiar with a *relatively restricted* group of related animals, encompassing all of the characters that can be included, and, if possible, covering all of the members of that group.

16

The Origin of the Comparative Phylogenetic Approach in Behavioral Research

From a historical point of view, it is extremely interesting that it was the very endeavor to encompass an ever-increasing number of characters that led to the attempt to find such characters in innate, species-specific *behavior* as well. With both investigators [*Whitman and Heinroth*], a passionate interest in a particular group of related animals in the first instance led to keeping them in captivity and to collecting as many representatives as possible. With the detailed observation and methodologically first-class comparative phylogenetic investigation that Whitman conducted on pigeons and doves and Heinroth did on ducks, the discovery of genuine evolutionary *homology* between innate motor patterns of different species was simply inevitable.

Let us consider some examples of homologous behavior patterns as characters of restricted or extensive groups of related animals: The mallard (*Anas platyrhynchos*), the Indian spotbill (*Anas poecilorhyncha*), the Japanese spotbill (*Anas poecilorhyncha zonorhyncha*), the African yellowbill (*Anas undulata*), the Madagascan Meller's duck (*Anas melleri*), the American black duck (*Anas obscura*) and a number of other *Anas* species all exhibit a number of utterly identical innate behavior patterns that characterize the entire group. In particular, the highly differentiated courtship patterns of the males are exactly the same. In the special part, we shall provide a detailed description of these motor patterns along with illustrations of them. The species concerned differ greatly from one another in their coloration patterns and in a number of other morphological features. For this reason, in traditional systematic schemes they were not united in a single, closely knit group corresponding to their undoubtedly close degree of phylogenetic relationship. The close genetic relationship between all of the species in this group that is indicated by study of their innate behavior patterns can, in fact, be confirmed from an entirely different direction. All species in this group will hybridize with one another without the slightest reduction in fertility. As these ducks all "regard one another as being the same" because of the similarity in their vocalizations and display patterns, they will mate with one another almost indiscriminately and it is exceedingly difficult to maintain purebred species in zoo management practice. For example, in the Berlin Tiergarten and its neighborhood, despite regular culling of hybrids, there were many ducks flying around with the blood of virtually all of the above-named species flowing in their veins. Accordingly, on the basis of the genetic compatibility of these forms, it is well established that it is entirely correct to infer that they are extremely closely related on the basis of the similarity in their species-specific behavior patterns. The forms concerned have, in fact, not quite reached the level that we would expect of "good species!" I would not go so far as to interpret them as *subspecies* of the mallard, expressing this with a triple scientific name. Nevertheless, it is necessary to find some way of distinguishing them from other dabbling ducks, which in current nomenclature are also labeled, quite misleadingly, with the generic name *Anas.*

Instead of taking narrowly defined motor patterns with a special form as a group characteristic, as we have done with the group just considered, let us take a more *widely* distributed character. An example is provided by the peculiar motor pattern of "incitement" over the shoulder, which plays a very important part in the courtship behavior of the *female* duck. This character is found in *all* true dabbling ducks, with the exception of the whistling ducks, which have undoubtedly secondarily lost the pattern. In addition, it also occurs in a few representatives of *other families,* for example, in the Carolina wood duck (*Lampronessa sponsa*) and in the mandarin (*Aix galericulata*), both members of the Cairininae, and in the few species of the Casarcinae, such as the shelduck (*Tadorna tadorna*). In the special part, we shall take a closer look at the relationships between these forms and the Anatinae.

Let us now take a motor pattern that is even more widely distributed, such as the pattern of *neck dipping* as a prelude to mating. In this case, we find that the pattern characterizes a group containing the true geese (Anserinae), the swans (Cygninae), the tree ducks (Dendrocygninae), and the Casarcinae. Such a broad distribution itself provides eloquent testimony that this particular motor pattern must be quite ancient in phylogenetic terms! Tree ducks, swans, and geese are birds that are quite different in external appearance and it would quite probably be very difficult to identify *morphological* characters that would permit such *simple* diagnosis of a group. It must be emphasized, however, that the birds concerned are undoubtedly *also* related in morphological terms. The extremely isolated genus *Coscoroba,* which is usually erroneously classified with the swans, actually represents an independent lineage roughly intermediate between tree ducks and swans, and it provides a link between them.

Such widely distributed and obviously ancient motor patterns can serve as extraordinarily useful characters for zoological "diagnosis," for the *identification key* to a group. Birds belonging to the pigeon order (Columbae) certainly form a cohesive and well-defined group of related species. Nevertheless, it is exceedingly difficult to provide in purely morphological terms a succinct diagnosis of the order that applies to all its members and excludes all nonmembers. A diagnosis taken from the best modern textbook of zoology reads as follows: carinate altricial birds with an inflated, bubblelike base to the bill, relatively flat feet adapted for perching or grasping, and pointed wing tips, with the third primary being the longest. Although this diagnosis incorporates five or even six morphological features, it by no means characterizes all members of the order. To take just two examples, the bill of the tooth-billed pigeon (*Didunculus*) has a totally different form, while the Victoria crowned pigeon (*Goura*) is not really altricial; it has long legs like a chicken and rounded wings which similarly resemble those of a chicken in structure and function. By contrast, one can characterize members of the order with a *single* distinctive motor pattern. Members of the pigeon order drink water by suction, using pumping movements of the esophagus, whereas all other birds drink water by scooping it up with the lower mandible and then tipping up the head. No member of the order provides an exception to this and there is not a single bird belonging to another group that would be erroneously classified with the pigeons on the basis of this "single-character diagnosis!" Only the so-called sand grouse (Pteroclidae) exhibit the same drinking pattern as the pigeons. In fact, authoritative systematists have allocated these birds to the pigeon group as a second suborder *without* knowledge of the drinking pattern, so this character provides further evidence of the inferred relationship.

These examples will suffice for the time being to demonstrate one crucial fact to anyone who has no prior background in comparative morphology and whose knowledge is limited to what has been said about evolution in the Biological Prolegomena: Such innate, species-specific motor patterns represent characters *that must have behaved like morphological characters in the course of evolutionary history.* Indeed, they must have behaved like *particularly conservative* characters.

The extremely close similarity between the phylogenetic investigation and evaluation of species-specific behavior patterns and that of morphological characters is neatly demonstrated by the *expressions* that Whitman, Heinroth, and later systematists of animal behavior used to describe them. None of these animal observers engaged in phylogenetic comparisons ever thought of saying, for example, that a given species "is accustomed" to behaving in a particular fashion or to performing particular motor patterns. Although they were actually concerned with *activities* of animals, they never used *verbs* but always *nouns* in their terminology. Once a particular motor pattern had been described in detail, a noun was used to name it, just as early morphologists had to invent specific terms for newly discovered structural characters. Just as no morphologist doubts that *all* representatives of a newly studied species possess the typical characters that have been determined for an adequate number of reference specimens, no comparative ethologist ever doubts that *all* members of a species possess the same behavioral characters as those that he has observed with his captive specimens, provided that sufficient care is exercised (see pp. 221–233). A species-specific behavior pattern is not something that animals may or may not perform; it is a character that the systematic group concerned simply *has*, just as it has claws, bills, or wings of a particular form! Even the first comparative descriptions of species-specific behavior, which were as yet quite naive, always employed sentences containing the verb *has* and nouns coined to characterize individual motor patterns: tree ducks *have* the same "neck dipping" as geese and swans; lizards of the family Lacertidae *have* the pattern of "treading" with the forelimbs, whereas lizards of the family Agamidae lack it; all members of the duck and goose group *have* the same "back-transfer motion" for nest building; and so on. It is now very common to find that the noun used for a given motor pattern is coupled with the name of the investigator who first described it, as was the case in the early development of morphology.

All of these facts, which were handled quite correctly with naive intuition in the early beginnings of comparative ethological research, inevitably lead us to the conclusion that species-specific behavior patterns are extremely *organlike*. Their special form and function can only be successfully investigated using the approach to which we owe our understanding of the special form and function of animal and human *organs*, namely the comparative phylogenetic approach. In 1898, Whitman had already encapsulated recognition of this in the sentence that has been taken in the first lines of this book as our guiding motto: *Instincts and organs are to be studied from the common viewpoint of phyletic descent.**

This discovery was revolutionary to the extent that, with one fell swoop, it removed the foundation for a whole series of widely accepted interpreta-

**See the footnote on p. xxv.*

tions. If highly complicated behavior patterns are reliable, phylogenetically interpretable characters of species, genera, and orders, like any morphological characters, then this fact alone is enough to demonstrate that these behavior patterns cannot undergo substantial modification through individual experience, as had been assumed by [Herbert] Spencer, Lloyd Morgan, and others. It thus became evident that, contrary to McDougall's belief, the constant feature in instincts is not their "goal," their survival-promoting end effect, but the *motor pattern* as such. Presumably, both Whitman and Heinroth were initially largely unaware of the theoretical consequences of their discovery and treated it as something that was simply self-evident. Both of them attached more weight to the phylogenetic relationships that they determined on the basis of behavior patterns than to the significance of the fact that such conclusions can indeed be based on similarities and dissimilarities in behavior patterns. For them, comparative behavioral research was initially just a means to an end; it in fact arose in the *service* of phylogenetic reconstruction!

17

The First Steps in the Nomothetic Stage

Through the comparative systematic behavioral studies of Whitman and Heinroth that have just been outlined, the constancy of motor patterns as characters of species, genera, and taxonomic groups in general was first of all *discovered* as a phenomenon. The phylogenetic implications of this were then subjected to extensive research. But the manner in which comparative, purely systematic description of species-specific behavior patterns subsequently led to recognition of the *special physiological laws governing the instinctive motor pattern* provides an extremely elegant and typical example of the way in which the idiographic-systematic stage of a young science leads to the nomothetic stage (Philosophical Prolegomena, p. 29).

We now know that species-specific behavior patterns, which provided the object and the foundation of the phylogenetic investigations conducted by Heinroth and Whitman, possess quite peculiar physiological properties. As has already been said in anticipation at the beginning of this section on the developmental history and methods of comparative behavioral research, instinctive motor patterns are not based on unconditioned and conditioned reflexes like all other behavior patterns of animals and humans. Instead, they are founded on *another basic function of the central nervous system, namely on endogenous, automatic-rhythmic processes of stimulus production,* of the kind that have long been known in the form of the stimulus production centers of the heart. At this point, it must be added that *coordination* of the motor patterns that is brought about by central, automatic stimulus production takes place without the participation of peripheral nerves, that is to say once again, without any involvement of reflexes. Thus, the *entire* neural phenomenon—up to the stage where the already fully patterned motor impulses are directly conducted to the musculature—takes place *within* the central organ itself!

Whitman and Heinroth were as yet unaware of the special laws governing instinctive motor patterns that are briefly introduced here and which will be discussed later in some detail. Whitman referred to these patterns simply as "instincts," while Heinroth avoided the loaded term *instinct* and referred instead to "species-specific drive-governed patterns." Understandably, neither author distinguished them from *other* innate behavior patterns that do not possess this endogenous-automatic character but are based on reflexes, most notably *orienting responses* or *taxes*. But while they simply recorded the behavior patterns of their pigeon and duck species in a straightforward, unprejudiced fashion and classified them on the basis of their similarities, the intuitive sensitivity of these two investigators forestalled the explicit knowledge of their discipline by some decades. Their systematic finesse resulted in the following quite astonishing achievement: Of course, genuine phylogenetic homology also exists between innate behavioral elements with a *reflex* nature, such as taxes, innate schemata, and individual reflex chains. Because of their dependence on external stimuli, however, all of these *reflexive* motor patterns understandably never constitute such "*impeccable*," constant, and unchanging characters of species and groups as do the "instinctive motor patterns" that are generated by endogenous, centrally coordinated automatic processes! The basic physiological properties of the instinctive motor pattern have been outlined in an anticipatory fashion above simply in order to highlight the great *independence* of these endogenous automatisms and the necessarily fixed, unchanging overt performance of these motor sequences. They exist, so to speak, as "emancipated" sequences of impulses generated by the central nervous system that are independent of all *external* stimuli and contingencies. But it was these very properties that attracted the attention of Whitman and Heinroth as they diligently sought behavioral characters that were clearly comparable. From the wealth of behavior patterns of different kinds, they were induced to *pick out none other than the endogenous-automatic instinctive motor patterns as phylogenetic characters!** In this way, the purely systematic arrangement of systems of behavior patterns led not only to the *discovery* of the instinctive motor pattern as a phenomenon. It was so sharply distinguished from the background of behavior patterns of other kinds that the discovery of the special physiological laws applying to it was simply a matter of time.

Although Heinroth himself at no point articulated recognition of these special laws in an abstract verbal form, he doubtless came close to it, as we shall see more clearly in the next chapter when discussing his concept of "moods." In fact, his reference to "drive-governed" activities already im-

**Note in the margin here:* Which ones? Not those superimposed by taxes! Therefore, more details are needed concerning courtship and other releasers.

plies recognition of the *spontaneous, eruptive* discharge of such behavior, which is so characteristic of all endogenous instinctive motor patterns and which Heinroth himself portrayed most illustratively in many different places. Thus it was that the pioneers of comparative behavioral research, most particularly Heinroth, had already laid a substantial foundation for recognition of the physiological peculiarity of the instinctive motor pattern that was subsequently clearly established by Wallace Craig, by myself, and especially by Erich von Holst. Before we proceed to a step-by-step presentation of our discipline, it is first of all necessary to say something about the other findings of Whitman and Heinroth and about these two investigators themselves.

18

The Research Personalities of Whitman and Heinroth and Their Findings

Discovery of the existence of phylogenetic homologies in behavior patterns and systematic, descriptive identification of instinctive motor patterns, which brought us so close to an understanding of their physiological nature, doubtless represent the most important accomplishments of Whitman and Heinroth, designating them as the founders of a new field of research. But the significance of their work is by no means confined to these two extremely important findings. For an understanding of the further development of comparative behavioral research, it is therefore necessary to provide an additional commentary on the research personalities, philosophical views, and other findings of these two pioneers.

Although these two investigators were so extremely similar with respect to their inductive methodology and their purely scientific results that I have so far been able to discuss them together in a *single* account, their personalities and philosophical leanings were as different as could possibly be. Whitman possessed an extensive knowledge of philosophy and had a very high opinion of the humanities in general and of Spencerian philosophy in particular. Despite the fact that his crystal-clear gaze as a scientific observer remained unsullied by idealistic prejudices, in his final conclusions he nevertheless leaned toward Spencerian dogmas that now appear to be completely untenable. Unfortunately, this applies to one of Whitman's most important findings, which we must now examine closely.

One of the dogmatic tenets of Spencer's philosophy concerning instinct was that "the instinct" is a phylogenetic precursor of intelligence. In the absence of any inductive factual evidence, Lloyd Morgan, in particular, provided a detailed discussion of the manner in which gradual higher development and differentiation of "the instincts" could lead in a smooth progression to the origin of insightful behavior. Any reasonably exact investigation of animal behavior at once reveals the utter implausibility of this purely deductive proposition. Animals that are *richly* equipped with highly differentiated innate behavior patterns consistently show only very limited modifiability of the behavior through learning and instinct. "Instinct specialists" are always "dumb" animals in that they exhibit only a limited capacity for developing conditioned reflexes and for showing directly insightful behavior. On the contrary, those birds and mammals that show peak achievements in learning and instinct are typically *poorly equipped* with complex innate behavior patterns. Once one has some understanding of the action systems of higher animal species, "instinct" and "intelligence" prove to be *alternatives.* In other words, one can provide a functional substitute for the other and—if highly developed—they are to a large extent mutually exclusive. Whitman had already recognized all of this with great clarity. In his fundamental text *Instinct,** he comes to the conclusion, which is in our view completely correct, that disintegration of instinctive behavior through a *process of reduction* is a precondition for higher development of intelligence in an animal species.

In this context, Whitman achieves a remarkable feat that can be seen virtually as a stroke of genius from the standpoint of our modern, somewhat further developed factual knowledge. He draws a connection between the origin of specifically *human* intelligence and *deficits* in instincts that are shown by domestic animals as a result of the *process of domestication.* This interpretation was based on a number of fine examples of cases in which domesticated animals proved to be capable of solving a particular problem that the corresponding wild form failed to solve. This was because the *loss* of a fixed, instinctive response pattern creates degrees of freedom in the behavior of a domesticated animal, while the wild ancestral form cannot diverge from the fixed tracks of its innate responses. We shall return to the enormous importance of these phenomena in the last part of this book, in discussing domestication and its relationship to the problem of human emergence. At the time, Whitman himself had already drawn the following remarkable conclusion: "These faults of instinct [. . .] are [. . .] not intelligence but [. . .] the open door, through which the great educator, experience, [comes in and works every wonder of intelligence]."

We regard it as an intellectually highly significant achievement by Whitman that he thus clearly demonstrated the utter implausibility of the Spencer–Lloyd Morgan interpretation of "instinct" as a phylogenetic precursor of "intelligence." Indeed, he showed that the exact opposite of this interpretation is true! It is therefore all the more regrettable that, in the summary of his book cited above, he tries to reach a compromise with that dogma

**This is, in fact, a reference to Whitman's* "Animal Behavior."

and thus sacrifices an *important* scientific perception to the authority of a sage in the humanities. As disciples of Whitman and trustees of his intellectual estate, it is our contention that, using inductive scientific methods, he clarified the real reciprocal relationships between instinctive behavior and the higher development of intelligence in phylogeny. He deserves our veneration quite particularly because, with the foresight of a true genius, he established at such an early stage the connection between *domestication-induced* deficits in instinctive behavior and the great leap involved in the origin and higher development of *human* intelligence!

Whereas Whitman also had one foot in the humanities, Heinroth was a pure scientist, a materialist to the core, and an inveterate satirist of things intellectual. He always regarded philosophy as "empty speculation on the basis of inadequate scientific knowledge," and because such speculation was of no interest to him, he always discarded a philosophical book after reading only a few pages. In an apparently joking fashion, but in reality with a quite serious undertone, Heinroth habitually referred to philosophy as the "pathological idling of innate human functions intended for an understanding of nature." His entire lifestyle was decisively tempered by the most fundamental conviction that the observed fact, the inductive basis, is *everything*, whereas the individual conclusions that are drawn from individual details are quite trivial. I owe it to him as a teacher that he also infused me with this conviction at an early stage, even if this involved a sometimes painful constraint on my youthful joy in hypothesizing. The *broadest possible* inductive basis combined with *maximal caution* in the formulation of hypotheses was always his first guiding principle, although in one sense this was without doubt overdone. In fact, in almost all of Heinroth's writings one notices a certain tendency to present to the reader, as far as possible, *only* the factual evidence that is significant with respect to a particular conclusion, while leaving the most important part, the conclusion itself, buried between the lines. A typical example of this approach is provided by his behavior at the International Ornithological Congress in Oxford in 1934. On that occasion, Heinroth gave an extremely informative lecture on the maturation of species-specific drive activities in young birds. The time that he had available was extremely limited, but he did not allow this to have the slightest influence on the exactitude of his description of his splendidly observed facts. But just as he arrived at the actual point of his presentation, which was his conclusion that all of the developmental processes in such species-specific behavior are *processes of maturation* that are completely independent of learning and experience, he looked at the clock and realized that his time had elapsed. Then, instead of trying to summarize in a compressed form the implications of the facts presented, he simply said, without the slightest trace of regret in his voice: "Well, now you can work out the rest for yourselves." It is this principle of "work it out for yourselves" that Heinroth doubtless carried too far. In this way, he left very important conclusions buried between the lines, where they could only be discovered by his closest fellow scientists, because his own inductive basis still did not seem to be broad enough to justify formulation of the conclusion that he had long since clearly recognized. Even in ripe old age, his own work was aimed exclusively at the accumulation of an ever-increasing body of concrete factual evidence. His major treatise on doves, which was eventually published during the war, provides eloquent testimony to this.

Heinroth was modest in the evaluation of his own results to the point of exaggerated self-criticism and suffered from a mild impediment in developing his own hypotheses. He also demanded the same standards of others with respect to the breadth of the inductive basis and exactitude in evaluation. With an uncompromising stance that came close to making him a scientific outsider, Heinroth dismissed any study that did not seem to fulfill these demands. For this reason, he established little connection with the school of human psychology, having no sympathy for its leanings toward the humanities. Only somebody who knew him very well could comprehend just how far his research interests were motivated by the goal of achieving a better understanding of *humans* through the study of animal behavior. In contrast to the behaviorists and the reflexologists, Heinroth was not averse to a certain *psychologization* of animal behavior, but the manner in which he did this is characterized by particular methodological caution. On the basis of his precise knowledge of the "instinctive" behavior of animals and *humans,* he draws the undoubtedly correct conclusion that the specifically human *feelings* and *passions*—"moods" in his terminology—represent the *subjective side of innate, species-specific response patterns.* The only psychological extrapolation from his own experience to that of animals resides in the fact that he also attributes to them a fundamentally similar experience of feelings and passions. But he never engages in any unjustified, speculative statement of the nature and quality of these feelings and passions. His customary response to the old mechanistic proposition that animals are machines was vastly more correct and closer to life: Animals are, to the contrary "people with strong feelings but very little intelligence."

The view that feelings and passions represent the experiential side of "instincts" was expressed independently of Heinroth by the vitalistic philosopher McDougall. The latter also came to the conclusion that the specific, qualitatively distinguishable feelings and passions of human beings correspond to the experiential side of an equal number of "instincts." The term *emotions* as employed by him was a somewhat broader concept simultaneously covering both feelings and passions. Now it is extraordinarily instructive to take a closer look at the conclusions that the vitalist McDougall and the natural scientist Heinroth drew from this common and undoubtedly correct inference. It is scarcely necessary to add that Heinroth had never read McDougall's books and certainly had not been influenced by him.

McDougall fully grasped the extremely important fact that the specific feelings and passions of human beings, such as hunger, fear, love, hate, and so forth, *are the experiential side of "instinctive" responses.* But, as a vitalist and hence in essence an idealist, he follows the old idealistic route from the "certain" internal realm to the "uncertain" external world. This is a classic illustration of the comment made by Bunge, cited above. Purely introspectively, he analyzes human "emotions" that can be reliably distinguished in qualitative terms and infers from them the number and nature of instincts that exist, arriving at a total of exactly 13. Given the current state of our knowledge, we would hardly dare to make such a definitive statement about the number and nature of the response-specific arousal qualities that characterize human beings. This is particularly so because the externally observable instinctive *motor patterns,* which for us provide the most important and only objectively definable *indicators* of response-specific arousal [in animals], have been subject to extensive *reduction.* Although we would therefore reproach McDougall for showing a certain recklessness in arriving at his hypothesis, we must nevertheless admit that his approach to "emotions" and instincts, *insofar as it is applied to human beings,* is not only extensively correct but also reveals a certain genius in its intuitive grasp! But then comes McDougall's great idealistic step from the interior to the exterior, which we simply do not accept. These 13 instincts, which, as we have just seen, were inferred exclusively from the internal processes of *humans,* represent for McDougall *all* of the instincts that can and do exist. On this basis, he allocates all of the innate behavior patterns of all animals to these 13 instincts. Of course, this generalized transfer of results obtained from humans to "the animals" amounts to a thoroughly unjustifiable anthropomorphism. If we accept the correctness of McDougall's initial proposition that each specific instinctive behavior pattern corresponds to a specific "emotion" and vice versa, we must necessarily attribute to any animal that has a *much larger number* of instinctive behavior patterns than a human a correspondingly *larger* number of qualitatively distinguishable experiential qualities of feelings and passions. On the other hand, we are fundamentally unable to state the nature of those qualities.

Without knowing anything of McDougall's work, Heinroth emphasized very clearly in a number of different places that the terms that are available for human feelings and passions are unsuitable for the naming of experiential processes in animals for the simple reason that they are insufficient in number. For example, in discussing the escape responses of the Burmese jungle fowl (*Gallus bankiva*), the ancestral form of our domestic chicken, he states quite precisely that the presence of *two* quite different escape responses in these birds makes it highly unlikely that the assumption of a *single* emotion of fear would be justified. Burmese jungle fowl, just like other galliform birds, react to the sight of a mammalian carnivore with a quite *different* kind of response-specific arousal as when a flying raptor appears on the scene. In the first situation, they utter a quite specific alarm call, which is well known to anyone as the egg-laying cackle of the domestic chicken. The neck is extended and, as the response intensity increases, they show *flight intention movements.* At an even higher intensity, they take to the air and will usually fly into the nearest tree. In the second situation, when a flying predator comes into sight, they utter a quite *different* alarm call, a soft, drawn out "rrrrr." As the response intensity increases, the body posture becomes increasingly crouched. Eventually, they abruptly show a "negatively phototactic" response and flee to take cover in the nearest dark place, where they then remain crouched and immobile. Although it is fundamentally impossible for us to know *what* takes place in the subjective experience of these animals during these responses, we would have no justification in assuming that *one and the same* emotion represents the correlate of these two completely separate and independent response patterns.

Following recognition of a correlation between specific "moods" and particular, species-specific response patterns, Heinroth's interpretation is totally different from McDougall's. Taking this insight derived from human beings, he makes just *one* generalization to other organisms, namely, that *there are internal conditions of readiness to perform particular, species-specific behavior patterns.* When, in discussing a particular animal species, he refers to flight mood, escape mood, attack mood, and so on, these terms can always be transposed into a purely objective context. Each indicates an internal psychophysiological state that constitutes the *precondition* for the performance of a specific motor pattern. All of the terms that Heinroth coined using the word "mood" were defined in tandem with the designation of a *motor pattern!* Recognition of the fact that a specific "mood" corresponds to an upsurge in a particular *response-specific* arousal quality, which was admittedly inferred primarily from human experience, quite logically led to the concept of the *response intensity* of a particular behavior pattern. The *same* quality of response-specific arousal can lead, at different degrees of intensity, not only to differential levels of performance of a motor pattern but also to the performance of different motor patterns, each of which corresponds to a particular degree of intensity of the motivation concerned. To take just one example: A flock of graylag geese that are grazing in the late afternoon some way from their habitual sleeping area do not possess the psychological capacity to "decide" on the appropriate time to fly home. Instead, the birds gradually develop *flight motivation,* beginning to do so long before they actually fly away. They stop grazing, their necks become increasingly extended and slender, and the multisyllable contact call that accompanies grazing gradually and smoothly gives way to the two-syllable flight signal (see also the special part). Simultaneously with this latter call, repeated lateral shaking of the bill is performed. Once this motor display pattern has appeared, it is usually only a fraction of a minute before the birds remove their wings from their wing pouches. They then raise their necks, flatten their feathers, lower the front ends of their bodies, and, immediately afterward, take off while uttering the loud, trumpeting flight call and rise whirring into the sky. Heinroth's *objective* discovery, which incorporates the complete underpinning of his concept of moods, resides in the following: A *scale* of increasing intensity of a given motor pattern, indeed in some cases a graded succession of different motor patterns, is activated in a strictly governed sequence by the upsurge in *a single* kind of response-specific arousal. As we shall see later, this sequence corresponds to a series of increasing threshold values. The sequence of the motor patterns is purely "mechanical." At any stage in the process, the specific arousal can begin to ebb away and "peter out." Stated in psychological terms: the animals literally do not "know" what they want, but simply, quite blindly, obey their surging and ebbing "emotions." This is precisely what Heinroth meant by "people with strong feelings but very little intelligence." In our example of the flight motivation of the graylag goose, it very often happens that such a buildup to takeoff proves to be "abortive" in the manner described. The animals then "have to" start again from the beginning. In other words, they begin with the motor pattern corresponding to the lowest intensity of the specific arousal quality. Or, to put it more precisely, the renewed gradual upsurge in arousal is expressed in the same regular sequence of motor patterns with different thresholds as before.

Any presentation of Heinroth's theory of moods would be incomplete and would fail to convey full understanding of its importance without some discussion of the phylogenetic origin of the function of *motor display patterns,* on which it sheds such significant light. From the first weak indications of instinctive motor patterns, of the kind that arise at the lowest intensity levels of response-specific arousal, one can clearly recognize the direction which the motor activity of the animal is likely to take. Because it is therefore possible, so to speak, to infer from them the intentions of the animal, Heinroth referred to them as *intention movements.* In its original form, an intention movement represents nothing more than a small, incomplete fragment of a particular instinctive motor pattern. In early spring, a night heron is perching in the branches of a tree. It gradually adopts a peculiar body posture unlike its normal, extremely erect carriage in that the rear end of the body is raised aloft. It then grasps a nearby twig and performs with it a single, unusual shaking and lateral thrusting movement. Immediately afterward, it sinks back into its previous, erect resting posture. Anybody who is familiar with the nest-building patterns of this species at their full intensity can recognize from this single twig-shaking movement an indication of the primary nest-building activities and knows that "nest-building motivation" is beginning to build up within the bird. Indeed, it is even possible to recognize this if the bird simply stands for a while with the rear end of the body raised "as if in the nest" and momentarily stares with both eyes at a twig beneath it. As we have already seen from a number of examples, the converse of an "all-or-nothing law" applies to the overall performance of an instinctive motor pattern. There are all conceivable intermediates between the barely detectable intention movement and performance of the pattern at its highest intensity, at which it serves its species-preserving function.

In their original form, such intention movements are quite undoubtedly *inconsequential byproducts* of the instinctive motor patterns as far as survival of the species is concerned. At the most, they may serve some kind of function in the sense of self-stimulation. But in a great many vertebrates these incompletely performed motor patterns, which are in themselves inconsequential, have acquired a new, extremely important function in that they have been converted into a "means of communication." This highly interesting phylogenetic process, for which we shall later present a number of detailed examples from cichlid fish, lizards, and birds, probably originated through the development in socially living species of an "understanding" of the intention movements of conspecifics. Just as the observer learns to infer from the intention movements of an animal what motor patterns it is likely to perform, many species have developed, in the course of their phylogenetic history, *innate* responses to the intention movements of conspecifics. In other words, they have developed unconditioned reflex response patterns that ensure an adaptive response to the stimuli presented by the intention movements of a conspecific. A jackdaw that perceives the flight intention movements of conspecifics likewise develops flight motivation. A horse that sees escape intention movements in a conspecific similarly develops escape motivation. A chicken "understands" the mildest intention movements of attack in a higher-ranking flock member and immediately gives way, and so on.

We know of numerous examples in which such "primary" intention movements, which have not been specially modified in any way, have thus acquired a function as social "signals." In an even larger number of cases, however, the *new* adaptive function that the intention movement has acquired because of its *comprehensibility* to conspecifics has led to a quite specific, independent further differentiation of the motor pattern concerned. Understandably, it has been the *visually* effective aspects of instinctive motor patterns that have developed an adaptive function as "signals." It is therefore entirely comprehensible in functional terms that further differentiation as a signaling adaptation has repeatedly and in a wide variety of different cases led to emphasis and exaggeration of the *visual* effectiveness of an intention movement. This culminates in the phenomenon that we refer to as the "demonstrative enhancement" of an intention movement. We can take as one example of this the head-nodding movements that many birds show as an intention movement of departure. In many different duck species, mothers that are leading offspring display an almost grotesque "demonstrative" exaggeration of this movement. The ducklings not only respond to such demonstratively enhanced head-nodding by beginning to move; they are also guided to the left or the right by the *direction* of the maternal intention movements, even if the mother is already several steps ahead of them. In a quite remarkable case of convergence, *swimming movements* have undergone a fully analogous process of demonstrative enhancement for guiding offspring in cichlid fish. Conspicuous, but in fact only weakly propulsive intention movements of swimming performed by the guiding parents provoke the swarm of young fish to move forward actively. Further, marked "accentuation" of these intention movements to the right or the left induces the swarm of offspring to veer to the corresponding side, even if they happen to be *in front of* the guiding parents. The young fish hence seem to behave "obediently," although in reality they are responding with unconditioned reflexes. (See also the section of the special part dealing with cichlid fish.)

None other than Charles Darwin was the first to realize that the motor display patterns that serve as external indicators of feelings and passions in *human beings* are to be interpreted as *incipient activities*. These activities *as such* may have lost their function and may even have disappeared in the species

Homo sapiens, while only the intention movements that are *visually* effective as signals of emotional state have been retained. In his classic book, [*The*] *Expression of the Emotions* [*in Man and Animals*], Darwin explains how the expression of extreme anger, with deeply knitted eyebrows, wrinkled nose, and pronounced retraction of the upper lip, can be interpreted as a *preparation for biting.* It is hence a homologue of the menacing facial expression of all mammalian predators, which all show a basically similar innervation of the facial musculature, that is accompanied by threatening growls.

Heinroth's exact phylogenetic demonstration of many such highly differentiated intention movements, specialized for their visual effect, has established an extraordinarily solid underpinning for these doubtlessly correct interpretations made by Darwin. In fact, Heinroth's discovery that the entire intensity range of instinctive motor patterns—from the mild intention movement to the full-intensity performance—has a *common connection to a single quality of response-specific arousal* confirms the correctness of the assumption that a motor display pattern derived from an intention movement is an *indicator* of the corresponding quality of feelings and passions. It indicates the corresponding "*mood*" in Heinroth's sense. The methodological distinctiveness and the *value* of Heinroth's concept of motivation resides in the fact that it is, on the one hand, amenable to a purely objective behavioristic approach, while on the other hand it is completely applicable to *human* experience, without any mixture or confusion of the different aspects! It allows us to engage in a combined parallel observation of animal and human psychology that is far, far more revealing than mere analogies. Examples of animal and human intention movements and motor display patterns have been discussed in some detail here for the very reason that in this case the parallels between human and animals are not mere analogies. The animal and human motor patterns concerned are, instead, *genuine phylogenetic homologues.*

Experience shows that, in the eyes of the uninitiated and particularly for behaviorists and reflexologists, the parallels drawn by Heinroth between the "moods" of animals and those of humans arouse the impression of a certain *anthropomorphization* of animal behavior. But if Heinroth has no qualms about saying that a young graylag gander "falls in love" with a particular goose or is "jealous" of another gander, this by no means constitutes an unjustified extrapolation from humans to animals. Instead, the real state of affairs is exactly *opposite.* Christoleit, a Protestant theologian well acquainted with Heinroth's writings, has far more justifiably—from his point of view—accused Heinroth of inadmissibly "*theromorphizing*" humans, of extrapolating from animals to ourselves! In fact, on the basis of an extremely broad foundation of factual knowledge, Heinroth does conclude his work on ducks by stating that the study of the behavior of higher animals leads us inevitably to a particular interpretation. This is that much of human behavior that we generally regard as the product of human reason and morality, our behavior toward our families and friends, has an innate, purely drive-governed basis and is rooted in much more primitive responses than is generally assumed. In our account of the social behavior of birds, and of the one mammalian species that we dare to describe here as the only representative of its group, it will be shown in detail just how extensive are the parallels between animal and human behavior with respect to all that is "instinctive." They are so extensive that it would be completely absurd and artificial to contrive the invention of new objectifying terms for behavioral patterns of social animals that are utterly identical to those of human beings, instead of using the generally understandable expressions for aspects of human mentality that exhibit such far-reaching analogies and are, indeed, to some extent phylogenetically *homologous.* As has been said, this does not amount to extrapolation from humans to animals, to anthropomorphization of animal behavior. Quite to the contrary, it expresses the conviction that the emotionally based human behavior patterns concerned *depend upon neurophysiological processes that are fundamentally the same.* In short, in both phylogenetic and physiological terms, they are "*the same*" as the corresponding behavior patterns of higher animals!

Heinroth primarily uses his investigation of the phenomenon of moods to make statements about the phylogenetic and physiological nature of certain *human* behavior patterns, while he of course exercises great caution and methodological clarity with respect to the psychologization of *animal* behavior. In fact, he merely states his conviction that particular highly specific feelings and passions of animals accompany the performance of instinctive behavior patterns, as they do in human beings. He refrains from making any comment about the *quality* of such experiential correlates, and—in contrast to McDougall—he quite generally uses terms derived from the human context for animal feelings and passions only in exceptional cases. As a rule, he coins new terms that combine the name of the instinctive motor pattern concerned with the word "mood." This is precisely why his concepts are so unassailable from the behavioristic side and yet are of such great utility in *psychological* terms. With exemplary care, he maintains a clear separation between the physiological and experiential sides in his investigations of life processes. With almost somnambulistic assurance, he *avoids* repetition of all of the errors of the theory of psychophysiological parallelism discussed on p. 157 et seq. and all of those other methodological and logical mistakes that others before him had made on the treacherous terrain of the alogical, parallel duality of psychophysiological phenomena. All of this is all the more surprising and remarkable when one takes into account Heinroth's total, conscious, and voluntary unfamiliarity with philosophy. Perhaps it was his very freedom from this encumbrance that allowed him to keep to his straight and narrow path.

Heinroth's theory of motivations was not only a major achievement in methodological, philosophical, and psychological terms. It was also an enormous step forward across the stony ground of purely *physiological* investigation of innate motor patterns. The strict lawful connection between the different intensity levels of motor activity—from the weak intention movement up to complete performance; the rise and fall of motivations that is always expressed in an almost mechanical fashion in the ascent and descent of the entire scale from mild to intensive instinctive motor patterns—for any observer with some bent for analysis, all of this has such an *inevitable-mechanical* character than it almost seems to *invite* causal physiological analysis. In particular, all of these phenomena in themselves suggest the notion of a *material,* to a certain extent hormonal, basis for the physiological processes involved. This is a notion that is, in principle, already conveyed by the repeated use of allegorical expressions such as "rise" and "upsurge" and "fall" and "decline" for motivations.

Recognition of the fact that there are *general* and *more special* motivations, with the emergence of one of the latter always being *conditional* upon the prior existence of a specific representative of the former, is also already implicit in Heinroth's work. For example, Heinroth refers to a bird as showing a general "reproductive mood," which provides the basis and the precondition for the emergence of specific, more special moods, such as courtship mood, nest-building mood, and so on. Tinbergen—who must be regarded as being just as much a direct disciple of Heinroth as I am—and his student Baerends have conducted extremely fruitful experiments with the three-spine stickleback (*Gasterosteus aculeatus*) and the sand wasp (*Ammophila campestris*), investigating the hierarchical organization of such broader, more general and narrower, more spatial behavioral motivations. The results that they have obtained provide strong support for the assumption that there is a material, in a wider literal sense *hormonal,* basis for the motivation of an organism to perform specific behavior patterns. In the general part, we shall consider this "hierarchy of motivations" in great detail.

Above and beyond this, however, the descriptions that Heinroth himself provided of the rise and fall of moods within the organism itself implicitly anticipate, in several places, a fact that eventually proved to be of decisive significance for recognition of the special physiological laws governing the instinctive motor pattern. At several points, Heinroth describes in full detail the process of *completely stimulus-independent discharge* of instinctive motor patterns. It was precisely this phenomenon that later led to the discovery of automatic-rhythmic stimulus *production,* which constitutes the physiological basis of all instinctive motor patterns. Heinroth devoted particular attention to the accumulation of an internal motivation to perform *escape responses* in tame, hand-reared birds that were never exposed to escape-releasing stimuli and thus had never had the opportunity to discharge the escape activities concerned. He provides an especially graphic account of how a hand-reared black grouse (*Tetrao tetrix*) gradually began to display vigilance and to develop escape motivation in response to ever-decreasing external stimuli. "When is that rascal finally going to show up?" These are the words that Heinroth, in jokingly anthropomorphic fashion, liked to use to portray the motivation of his bird following such "*damming*" of undischarged escape patterns. Subsequently, we objectified this by referring to *lowering of the threshold for releasing stimuli,* which provided the actual basis for inferring the physiological peculiarity of the instinctive motor patterns, namely endogenous, automatic-rhythmic stimulus production.

This summary account of Heinroth's theory of moods (motivations) should not mislead the reader into believing that he himself set it out in an abstract form. What one can find in Heinroth's writings is exclusively a wealth of concrete facts derived from observation along with his combined terms, including the word "mood," which he uses without further comment. "Well, now you can work out the rest for yourselves!" This most strictly inductive of all natural scientists also remained true to this slogan in his treatments of "moods." Despite this absence of abstract conclusions, we regard Heinroth's treatment of motivations as a major intellectual achievement. The behaviorists were the first to observe the processes of human inner life from the outside and they thus established the new approach, that reflection about reflection, whose great intellectual significance has been mentioned in the Philosophical Prolegomena. Pavlov's reflexology school was the first to conduct consistent investigation of a decidedly *psychological* process, namely learning, from the *physiological* side. That investigation set out from "the other side of the mirror," with a conscious methodological refusal to consider the psychological, experiential side at the same time. Now, what is methodologically and intellectually novel about Heinroth's investigations of the phenomenon of "moods" resides in the following: he studied a "psychophysiological" life process, that is, one *accompanied by experience,* as an *entity.* In a methodologically impeccable manner, without any confusion of the two incommensurable aspects, his investigations *simultaneously* covered both the objective physiological *and* the experiential aspects. Further investigation of the particular life process that Heinroth studied—namely, the *instinctive motor pattern*—has shown just how fruitful such a methodologically controlled treatment of one and the same phenomenon simultaneously from the objective physiological and the psychological sides can be. As a result, such simultaneous application of behavioristic and psychological approaches to one and the same phenomenon has become one of the most fundamental standard techniques in comparative behavioral research. The further penetration into the nature of the instinctive motor pattern that is described in the next chapter provides only *one* example of this procedure. We shall encounter another example in the chapter dealing with taxes and insight.

19

The Discovery of "Appetitive Behavior" by Wallace Craig

As we have seen, Heinroth extensively prepared the ground for recognition of the physiological peculiarity of the instinctive motor pattern. We have also seen how far *psychological* consideration of "motivation," in the sense of an internal readiness to perform specific motor patterns, helped this investigator to develop correct concepts of the *physiological* character of the life processes that he studied. A further major cognitive advance in the same direction was made by Wallace Craig, a direct disciple of Whitman. Craig set out more from the *psychological* than the *physiological* side of behavior and concerned himself primarily with the problems of *spontaneity* and the *subjective purpose* of behavior. The vitalistic philosopher McDougall, with whom we have already become acquainted, had also intensively investigated these particular phenomena and had—in polemical attacks against mechanistic explanations of instinctive behavior (p. 249 et seq.)—made a number of entirely correct points. He coined the famous aphorism "The healthy animal is up and doing" in opposition to the reflex machine theory of the mechanists. We have already teased out above the contrasts that existed between the vitalist McDougall and the inductive scientist Heinroth with respect to their treatment of "instincts" and "emotions." It is just as instructive to single out the extensively analogous contrasts that exist between McDougall and Craig with respect to their treatment of the phenomena of spontaneity and purpose in animal and human behavior.

A healthy raven or a healthy young dog is, indeed, a very active creature and for anybody who is acquainted only with animals possessing this kind of action system the reflex bundle theory justifiably seems quite nonsensical. However, we must make a reservation at this point. Like all statements that make general reference to "the animal," McDougall's above-mentioned aphorism is not entirely correct. A healthy ant lion (*Myrmeleon*) or a healthy pike (*Esox*) is by no means "up and doing!" If an ant lion continuously moves around the vivarium, instead of sitting immobile in its funnel trap, or if a pike swims to and fro along the glass panel at the front of the aquarium instead of hovering motionless between the water plants, this indicates to the animal keeper that there is some problem either with the organism itself or with the replicated environmental conditions. As long as it is healthy, a dog or a raven is, indeed, usually active, whereas a pike or an ant lion remains immobile and vigilant. That is to say, it waits for releasing stimuli. The action system of the former contains many spontaneous, automatic-rhythmic instinctive motor patterns, while that of the latter does *really* consist very largely of reflexive responses to external stimuli. Accordingly, for somebody who is, for example, familiar only with the ant lion from personal observations, the reflex machine theory of the mechanists would certainly not seem so absurd. This reservation has been presented in order to make it clear to the reader that one should never construct sentences with "the animal" as the subject. Heinroth used to interrupt such sentences with the mild and friendly interjection: "Are you referring to an amoeba or a chimpanzee?"

We must therefore correct McDougall's sentence to make the following observation: Healthy animals of species whose action system contains many endogenous instinctive motor patterns are continuously engaged in *activity* that is not released by specific external stimuli. To take just one graphic example: A tame raven lands outside on my windowsill. After extensive visual scanning of the room, which is relatively unfamiliar and therefore something "novel," at the same time attractive and disquieting, it makes a bouncing hop from the windowsill onto my desk. I am well aware that the raven's presence will make it impossible for me to continue working, but at first I allow it to do as it likes. After further scanning of the surroundings has confirmed that there is no danger, it begins its activity. It by no means performs pointless, undirected movements; instead, it has quite definite intentions. It pecks and probes inquisitively at all kinds of objects in my room. First, it examines the head of a wooden statue with loud, resonating pecks to test its resilience; then, using a powerful distending movement of its bill that is also shown by nutcracker's and starlings, it raises the edge of the lid of my typewriter and, with eyes bulging, peers between its gaping jaws into the gap. Now it switches its attention to my manuscript. From previous experience, it knows full well how actively I will pursue any pages that are pilfered. In the raven's eyes, my response to this in fact increases the value of the prey because of exactly the same psychological mechanism that first makes an object interesting for a dog when one attempts to take it away. Inconspicuously, I place an already corrected, superfluous page in such a way that the raven can easily reach it and I pretend to make a lunge in pursuit as it takes it away. It then screws up the page and tears it into pieces, but these are immediately abandoned when I fail to follow up the chase. But then, quite against my wishes, it manages to grap my eraser. This is first pecked about, then stowed away in the raven's throat pouch and finally concealed in the gap between the arm and the seat of the sofa. Although the "prey" is no longer to be seen, the raven then stuffs some of the now uninteresting scraps of paper on top of it. This process of concealment takes place behind my back and I can only see it in the mirror. If I were to turn around to look, the raven would immediately break off its activity and seek out a new hiding place. Finally—because I cannot chase the raven away without permanently damaging our relationship—I am obliged to sacrifice the cheese from my breakfast sandwich to entice it out of the window. This is because it has now become attracted by the twitching movements of the ribbon in my typewriter and is fully intent on dismantling it.

Without thinking about it, in this consciously naive and straightforward description of one of the animals with which I am most familiar, I have used the word "intent." On seeing such behavior patterns in higher animals, no impartial observer can escape the impression that the *animal as a subject* has *intentions.* But for a behaviorist who objects to this kind of claim concerning an animal subject, we can define such an intention quite objectively as the pursuit of a subjectively determined *purpose:* Following the very appropriate and useful definition of E. C. Tolman, we can define a *purpose* as existing wherever an organism *maintains a specific goal* and continues to *modify* its behavior until this end result has been achieved. Viewed objectively, this is also precisely what a human being with a quite specific intention does.

McDougall—in contrast to mechanistic behavioral investigators—saw this *intention* of higher organisms quite correctly. He saw the primevally powerful, *spontaneous* drive that wells up within the organism and induces it to engage in an elementary striving after specific goals. Above all, he also saw what the mechanists did not see or did not want to see, namely the wonderful, species-preserving purposivity that underlies such intentions and striving. But because he was an idealist and a vitalist, for him both the powerful striving to attain specific goals and the species-preserving purposivity of such striving were consequences of *one and the same* "factor," namely "instinct," and hence a single thing. For McDougall, the *adaptive function* and the *goal sought* by the organism *as a subject* are as one!

Is this equalization correct? With drive-governed activities that are performed smoothly and without disruption by animals under natural conditions, the illusion is almost created that this is the case. Here, too, we can consider an example: A raven is flying over a snow-covered forest and begins to circle above a small clearing. It has spotted something that was not there yesterday. It is the carcass of a cat that the forester had caught yesterday in a trap and had killed, skinned, and left there as a friendly gesture toward this now rare bird. The crows have already been at the corpse and not much meat is left. The raven circles lower and eventually lands in the dry branches in the crown of an oak tree on the edge of the forest. After spending some time carefully scanning the carcass, which is simultaneously attractive and frightening as a "novelty," the raven slowly descends from branch to branch with bouncing hops. Having gradually become convinced of the safety of the surroundings, it glides from one of the lower, projecting branches into the clearing. Still continuously prepared for flight, progressing only by sideways movements, it hops cautiously toward the carcass. While it is still a few yards away, one of the crows that had previously eaten there arrives. Being far less cautious than its larger relative, the crow lands directly on the carcass and immediately begins to eat. For the raven, this competition increases the attractiveness of the prey and dispels its caution. It covers the last few yards on the wing and, with the feathers on its head and back ruffled and uttering a deep, throaty call, it chases away the crow. The raven first pecks and hacks inquisitively at various parts of the hard, frozen carcass. Then it tests the resilience of the skull with resonant thrusts of its beak. It finds a fissure in the sidewall of the braincase produced by a blow from the forester's axe and succeeds in enlarging it. Using a powerful distending movement of its beak, it widens the fissure. It then begins to eat the brain, but does not seem to be very hungry as it hides a large part of it in its throat pouch and then flies off into the crown of the half-withered oak tree. There, it approaches the trunk and is just about to conceal the morsel in a cavity when it notices that a crow has also landed in the branches behind and is looking on. With an angry grumbling call, the raven chases the crow away. Only when the latter is out of sight does the raven conceal the lump of brain in the cavity, carefully covering it by pressing in pieces of bark after it.

When the innate species-specific instinctive motor patterns of a species are seen in this way *only* in their functionally adaptive form, fully revealing their survival value, they do indeed seem to be so "purposive" that—even for an observer who is not inclined toward anthropomorphization of animal behavior—it is quite easy to believe that the animal *knows* why it is performing all of these individual behavior patterns. *Nothing* in the behavior of our raven in the forest clearing would lead us to suspect that it is *not* consciously pursuing the goal of sating its hunger and even providing a food store for the future. If one were to rely only on this field observation, one could really believe at first that, for the animal subject, the individual motor patterns described are simply *means* to attain the *goal* of obtaining food. One could believe that the biological *function* of the behavior patterns shown is the *goal* that the subject is striving to attain and is *identical* with it. This is precisely what McDougall thought. According to his interpretation, the raven pursues, by means of *variable* behavior patterns, the *constant* purpose of satisfying its "feeding instinct." In his view, a dog hunts only in order to sate its hunger and a bird builds its nest with the subjective goal of using it for brooding and rearing its offspring. By the same token—if one will excuse my stooping to a reductio ad absurdum in the guise of a counterargument—a young man seeks to make an impression on a beautiful young lady in order to become the father of a lively, bawling newborn as soon as possible.

The reader who has followed attentively will by now presumably have arrived at the same discovery as that first made by Craig. The findings made by Whitman and Heinroth must surely have led the reader to reject McDougall's proposition that our raven in the clearing is striving to attain a constant subjective purpose, namely satisfaction of the "feeding instinct," by means of *variable* behavior patterns. We who are familiar with ravens know that the species simply "*has*" all of the motor patterns described, just as it has claws and a beak. And we have seen that the tame raven in my workroom does *not* owe its living to these behavior patterns and quite certainly does not use them as a *means* of attaining some superior purpose. Yet, with photographic fidelity, it performs the *same* motor sequences as its free-living counterpart in the forest clearing and does so with the *same intensity* and the same zeal! A raven kept in even closer confinement, in a small cage, will still perform the same motor sequences with undiminished energy. However, under these conditions it will display them with the most "unsuitable" substitute objects imaginable. Now, this independence of the drive from any superior purpose and its property of bringing about the performance of quite specific motor sequences is characteristic of *all* instinctive motor patterns! As any dog lover knows, the hunting passion of any dog—as long as it is healthy—is completely independent of the food that is provided. Unfortunately, however rich and tasty the food provided, it is quite impossible to suppress the hunting drive or even to reduce it. Further, in the characters of individual dogs, there is not the slightest proportionality between the hunting drive and voracity. Yet this should most definitely be present if dogs hunted in order to eat. The most passionate and successful hunter among all the dogs that I have personally known, an old hound from my "wild dog backcrosses" by the name of Wolf, was chronically lacking in appetite and a fastidious eater. He would not eat raw meat at all and never touched the carcass of any prey that he killed. My red-billed queleas built wonderful nests year after year in their cage, but in spite of all of my efforts they did not "want" to proceed to reproduction. And the same even applies in the case of the *human.* Even the most child-loving adolescent would, following proper introspection, have to admit that there is something wrong with McDougall's equalization of the species-preserving *end result* of an instinctive behavior pattern and the *drive goal* that the subject strives to attain!

Accordingly, species-preserving finality is obviously *not* the *purpose* that the organism as a subject strives to achieve. So what *is* the designated goal of the drive-governed striving of the animal and human subject, which is equally obviously *present?* What is the goal of that elementary, primevally powerful striving and intention that often even ignores the individual life of the organism and overwhelms all responses of self-preservation? What did the tame raven "want" on that occasion when it came into my workroom? What did my hound Wolf "want," although he was fully aware of the mortal dangers of poaching and despite the fact that repeated attempts were made to shoot him? Again and again, he indulged his passion for hunting and that eventually cost him his life. What did that classic lover Leander want as he swam every night across the Hellespont to Hero, an undertaking that reportedly also cost him his life in the end?

I am absolutely certain that my raven was not seeking food, despite the fact that the species-preserving function of its motor patterns lies in obtaining food. I am absolutely certain that my dog Wolf was not out to eat game, although the species-preserving function of his species-specific hunting patterns quite definitely lies in the consumption of the prey. Finally, it would take a lot to convince me that the intention behind Leander's swimming was to have Hero produce a newborn, although the species-preserving *function* of love surely lies in this end result. So what did all of these individuals *want?* The answer to this question is so simple that it needed a very bright scientist to discover it. After all, as Goethe so neatly put it: "The hardest of all is what you fancy to be so easy, to see with your eyes what lies before them." Or, to take the more graphic approach that Heinroth preferred: "No eggs are harder to lay than the eggs of Columbus."* The raven wanted nothing other than to find objects on which it could discharge *those particular behavior patterns* of splitting, distending, concealing, and so on. My hound Wolf wanted to discharge the innate, species-specific behavior pattern of *shaking the prey to death.* This represents

**Oskar Heinroth (born in 1871) died in 1945.*

the culmination of the instinctive motor patterns in the predatory behavior of many carnivores and it is evidently accompanied by pleasurable sensations that are almost delirious and orgasmic. We can assess the *intensity* of such patterns from the force of the organism's subjective striving, but we shall never be able to determine their *quality*. And even Leander—if one will forgive me such crude realism—wanted to perform with Hero innate, species-specific behavior patterns that are accompanied by pleasurable sensations of an orgasmic nature!

A commonplace? Far from it! Much more a truly Copernican breakthrough, eliminating the old prejudice that "hunger and love" bind the world together, as Friedrich Schiller believed. At the same time, it was a correction of McDougall's error in creating 13 instincts from these two and equating their species-preserving function with the subjectively pursued purpose. It was of immense psychological—and also physiological—significance to recognize that *every* instinctive motor pattern becomes the *exclusive* goal of autonomous, subjective striving if it has remained "undischarged" for some time. We shall soon see the consequences that followed from Wallace Craig's seemingly straightforward recognition of this fact.

In his paper "Appetites and Aversions as Constituents of Instincts," Craig formulates in the clearest way possible the following important facts, which are indisputably founded on inductive observational evidence: When an organism finds itself in an—artificial or natural—environmental situation in which specific innate, species-specific motor patterns are *not* released and thus are not "discharged" for some time, it starts to display *motor restlessness.* This persists and increases in intensity until the organism—initially purely by chance—encounters the stimuli that release the motor pattern concerned, or arrives in an environmental context in which its performance becomes possible. In the more complicated case, particularly following the development of participatory conditioned reflexes, this undirected motor restlessness develops into purposive striving by the organism. The objectively definable goal of this is the *attainment of the stimulus situation which releases the motor pattern concerned, such that it is performed until the drive has been completely "satisfied," which means, in objective terms, until the readiness to respond has disappeared.* As an "instinct-reduction animal," a human being experiences such striving after the discharge of behavior patterns only with the two sole remaining intact instinctive motor patterns of eating and mating. For this reason, Schiller's little verse about "hunger and love" and Pavlov's interpretations of the connection of the "goal reflex" with the unconditioned reflexes of feeding and reproduction apply only to humans. Extrapolation of these relationships to animals is a purely anthropomorphic error. We know from the case of the human that purposive behavior, whose objectively definable goal resides in the performance of instinctive motor patterns, is on the *subjective* side quite undoubtedly aimed at *sensual feelings of pleasure.* These already emerge when the *releasing stimulus situation is perceived* but reach their highest intensity through *proprioceptive perception of the movements of one's own body.* Both of these conditions apply equally to the two genuine, highly differentiated instinctive motor patterns of humans: eating and mating. There is quite certainly no impermissible anthropomorphism involved in assuming that all genuine instinctive motor patterns of higher animals are accompanied by analogous sensual feelings of pleasure, even though there are many more of them than in humans. This is true even though we cannot have any idea of the *quality* of such animal experience, except for the few instinctive motor patterns of animals that are true phylogenetic homologues of human instinctive motor patterns.

Now, Craig refers to the purposive behavior by means of which both animal and human subjects strive after the performance of a particular instinctive motor pattern as *appetitive behavior,* while he refers to the drive as *appetite.* Because of the narrower meaning of the German word *Appetit,* which exclusively concerns the consumption of food, I have translated Craig's expressions as *Appetenzverhalten* and as *Appetenz,* respectively. As has already been said, *any* endogenous-automatic instinctive motor pattern can potentially become the goal of its own autonomous appetitive behavior. Each animal possesses at least as many mutually independent behavioral *motivations* as the number of instinctive motor patterns in its action system. "Everyday" or "subsidiary" instinctive motor patterns, such as *locomotion,* are used so extensively under normal circumstances that "damming" of their endogenous stimulus production virtually never occurs, with the result that there is usually no noticeable striving to attain their release. Yet, if their performance is prevented even for a very short space of time under experimental conditions, they immediately become the objects of special appetitive behavior that is clearly directed at them. In the general part of this book, we shall need to take a closer look at such behavior of so-called "instrumental responses."

According to Craig's terminology, in all action chains in which genuine appetitive behavior plays a part, the *last* element in the chain of activities, the purpose for which they are performed, is *always a purely instinctive motor pattern.* Craig as yet had no knowledge of the particular physiological laws governing the instinctive motor pattern. Therefore, as reflected by the title of his paper "Appetites and Aversions as *Constituents* of Instincts," he interpreted appetitive behavior as a *component* of instinctive behavior. Because of this, he was obliged to use another term to refer to the purely endogenous-automatic activity that *we* alone label the "instinctive motor pattern." He refers to this motor pattern, which is the subjective purpose of appetitive behavior, as the "*consummatory action*." But this concept is, in fact, almost identical to that of the instinctive motor pattern used here.

Another important point recognized by Craig is of considerable importance here. The *higher* the degree of development of the psychological capacities of an organism becomes, the *longer* the pathway may be to the goal after which it will strive by means of purposive appetitive behavior. In other words, the lower the intelligence of the animal, the *smaller* the biological scope that can be "expected" of its appetitive behavior. In the simplest case, therefore, appetitive behavior is limited to undirected motor restlessness as the drive for a specific motor pattern becomes augmented and to a few unconditioned reflexive orienting movements when the organism eventually encounters the stimuli sought with such undirected searching behavior. With somewhat more advanced organisms, such blind searching is supplemented, following the principle of trial and error, by development of a number of *conditioned reflexes* that provide for a certain degree of species-preserving, purposive *adaptability* of the appetitive behavior. But the main part of the work to be performed with respect to preservation of the species remains the instinctive *final element* in the action chain, the "consummatory action" itself. In the feeding behavior of a sparrow hawk, for example, appetitive behavior is limited to largely undirected searching, at the most reinforced by a few conditioned responses, primarily in the form of negative conditioning. The main part of its hunting behavior consists of the marvellously differentiated and fixed innate instinctive motor patterns of chasing, striking, and killing its prey. Incidentally, it is directly obvious to any impartial observer on seeing the behavior of such a prey-striking raptor that the purpose for which the *subject* is striving lies in the motor patterns of the animal, striking and killing, which are "*laced*" with orgasmic sensual pleasure. Contrary to the simple anthropomorphic interpretation, this role is not played by the subsequent consumption of the prey! A striking predator finds itself in an exceptional state of maximal arousal that is difficult to describe. Somebody familiar with dogs with a passion for hunting would most easily be able to imagine this. Immediately after striking its prey, the bird shows the same degree of abreaction as a human being—if you will pardon this use of the only graphic analogy—directly after orgasm. Far from greedily beginning to devour its prey, the raptor—even if it is very hungry—will first sit still for several minutes on its prey. Then, with an absent-minded air, it will tug out a small feather and discard it and then embark on the slow, laborious process of plucking its prey, as though half-asleep. Even when the raptor finally begins to eat, it does so in a "dispassionate," mechanical manner, as though not quite conscious. Craig emphasized recognition of the following point: The *sought-after* and most *pleasurable* part of the action chain is located *at the beginning* of the chain of actions consisting of instinctive motor patterns. Apparently, the *first* instinctive motor pattern to follow the appetitive behavior, constituting the immediate goal, is often that which is the most baited with accompanying "sensual pleasure," as is the case with the motor pattern of "shaking to death" that is exhibited by mammalian carnivores.

By contrast, the *higher* the intelligence of the organism, the *further* the distance may be to the goal in the purposive, plastic appetitive behavior that now includes greater behavioral capacities. Eventually, the "consummatory action" that constitutes the goal can lie at the far end of a long sequence of behavior patterns. Any human being engaged in an unrewarding job will spend several hours a day performing highly complicated learned and insightful activities with the sole goal of having something to "get his teeth into!" With the fine touch that is sometimes found in colloquial language, this expression in fact anticipates the principle recognized by Craig that concerns us here and which he summarized with the following sentence: "*The end of the chain action is always instinctive.*" Indeed, it must be emphasized that this is also true in cases where the appetitive part of the chain includes refined psychological achievements. In humans, as "instinct-reduction creatures," all that is left of instinctive motor patterns serving as goals are, indeed, "hunger and love," as was quite correctly seen by Schiller and Pavlov, admittedly two very different thinkers. On the other hand, extrapolation of such reduced, drive-governed behavioral motivation of human beings to animals is crass anthropomorphism. Every animal species has as many autonomous appetites as the number of response-specific qualities in its action system.

As is sufficiently obvious from what has been said, Craig's approach to research and interpretation is actually more that of a psychologist than that of a physiologist. In his schema of the "instinctive" action, the endogenous instinctive motor pattern is not conceptually separated from other, quite different types of behavior patterns on the basis of the particular *physiological* laws that govern it. All of the functions incorporated into appetitive behavior, however complex and distinctive they may be, are for him just other *constituents* "of instinct." Delimitation of the concept that *we* understand under the term *instinctive motor pattern* is found in an abstract form in Craig's work *only from the psychological side.* It is expressed in his statement, acknowledged above, that the "consummatory action," the motor pattern that constitutes the goal and satisfies the drive, is always instinctive. But Craig's definition of "instinct" is expanded approximately twofold in that he interprets not only "appetites" but also *aversions* as constituents of instinct. In Craig's conceptual system, an aversion is a *purposive* form of behavior that, instead of striving after particular stimuli as in the case of an appetite, *is directed at avoiding or eliminating them.* However, purposive behavior patterns of an organism that remain active until a specific external stimulus is eliminated *always prove on closer examination to be reflexive.* Indeed, they turn out to be what we refer to as *orienting responses* or *taxes,* following Alfred Kühn. If a green crab (*Carcinus maenas*) or a house mouse (*Mus musculus*) is placed in the light, it will energetically attempt to return to darkness. In the simplest case, this is achieved with very primitive taxes which are certainly of an unconditioned reflexive and reflex-controlled nature and which result in avoidance of light. In the more complex case, much more highly differentiated action chains are involved. In these, taxes in fact still play a decisive and literally directional role, but there is additional participation of higher functions, above all conditioned responses in the form of learned pathways. A house mouse, for example, is perfectly capable of learning pathways "into darkness" that initially lead toward the light, in *opposition* to the unconditioned reflex tendency, thereby following a detour into protective cover.

Now there is undoubtedly a certain psychological justification for referring to such a behavior pattern as an *aversion.* After all, *aversio* in Latin means *turning away* and the organism does, in fact, turn away from the releasing stimulus. In the majority of cases that we can psychologize from human behavior, it is, indeed, justifiable to speak of turning away from the stimulus and not of turning toward the stimulus-free side. This is true to the extent that the stimulus itself is associated with subjective *feelings of aversion.* Accordingly, "striving after the resting condition," which constitutes the nature of aversion in Craig's sense, is essentially an avoidance of the stimulus rather than a striving after the condition of freedom from stimulation. However, this situation is not always reliably present. From introspection alone, we know that even with quite simple taxes, for example those directed at a particular *optimal temperature,* there may be no feelings of aversion and equally well there may be sensual feelings of pleasure. If there is a heavy frost, we are by no means influenced only by our "aversion" to cold. Instead, we are also affected by the sensual feelings of pleasure that the warm stove awakes in us. Thus, as far as the psychologization of animal behavior is concerned, it is never possible with any processes involving taxes to state unequivocally that the "whip" of aversion or the attractive "bait" of pleasure represents the *subjective* motivation of the objectively purposive behavior. Hence the assertion regarding the experiential side of behavior conveyed by use of the concept of "aversion" by no means carries the same probabilistic certainty as the other assumption made by Craig, namely that the attraction of sensual pleasure is the subjective motivation for all appetitive behavior directed at the performance of instinctive motor patterns. For this reason, we now use Craig's concept of appetite in its original sense, but instead of the concept of aversion we use that of purposive behavior directed at the *resting condition.* The latter is a concept derived from the Swiss ethologist Monika Holzapfel, who has made a substantial contribution to the clarification of the subjective side of taxes and purposive behavior.

The great significance of Wallace Craig's work surely lies in the fact that he was the first to develop a *general theory of action* that is unassailably correct both from the objectifying, physiological side and from the viewpoint of human experience. His clear separation between the concepts of the species-preserving *function* and the subjectively desired *purpose* of a behavior pattern radically dispensed with a conceptual confusion that had an immeasurably inhibiting effect on research, hampering not only the vitalistic philosophy of instinct but also the entire contemporary undertaking of human psychology. Craig developed a scheme of *motivation* for all animal and human behavior. Despite its graphic simplicity, this is in complete agreement with all known facts derived not only from the field of objective behavioral research but also from human experiential psychology.

Craig's results are also significant, to an extent that can scarcely be overestimated, in *physiological* terms because they brought us a big step closer to recognizing the endogenous-automatic character of the instinctive motor pattern. Craig repeatedly emphasized the fact that it is the instinctive action *itself* that, after a period of damming, renders the organism restless and induces it to seek actively after releasing stimuli. Strictly speaking, this observation alone implies recognition of the *internal production* of stimuli.

20

My Own Contribution to an Understanding of the Instinctive Motor Pattern

The most significant personal contribution that I have been able to make to an understanding of the physiological peculiarity of the instinctive motor pattern resides *primarily* in the investigation of certain *unconditioned reflexive* processes that release instinctive motor patterns. More precise penetration into the mode of operation of these releasing mechanisms, which we now refer to as *innate releasing schemata,* brought with it *quantifiability of the releasing stimulus.* Such quantifiability, in turn, opened up the possibility of deriving from the relationships between the strength of the stimulus and the strength of the response an indication of the *internal responsiveness* of the organism. As a result, this revealed the important phenomenon of continuous *accumulation* of response-specific arousal, which builds up while a specific motor pattern is at rest but is then consumed during its performance. Accumulation of a response-specific arousal quality doubtless provides the physiological basis both for the phenomenon of *motivation* that Heinroth studied in detail and for the properties of *appetitive behavior* investigated by Craig. From a purely methodological point of view, it is interesting to see how the *double quantification* of stimulus strength and responsiveness that is about to be described made it possible to investigate quite simultaneously *two* physiologically very different phenomena such as the innate releasing schema and the instinctive motor pattern, while also examining their close functional relationship. This is a typical example of "analysis on a broad front."

The existence of receptor correlates for quite specific combinations of external stimuli is, in itself, nothing new. McDougall had, in fact, already recognized the phenomenon as such and coined the term "innate perceptory patterns," although he never made any attempt to conduct a causal analysis of them. Further, as we have seen, Heinroth had discovered through his observations of *intention movements* the fact that these motor patterns release appropriate responses from conspecifics by means of a mechanism that evidently involves unconditioned reflexes. To use the somewhat anthropomorphic formulation favored by Heinroth himself, such patterns are "innately understood." In higher animals, particularly in birds and even in certain bony fish that display parental care, it commonly occurs that two or more conspecifics cooperate to achieve *one* common function. Examples of this are provided by the behavior patterns involved in pair formation, in cooperative parental care, and in the reciprocal relationships between parents and offspring; in short, in all of the *social* behavior of animals in the narrower sense. In such cases, the appropriate interaction between the parents is almost exclusively brought about by wonderful systems involving finely coadapted instinctive motor patterns serving as releasers and the innate schemata that respond to them. The releasing behavior patterns and those that are released interact together just as smoothly, and almost as mechanically, as the teeth of two cogwheels. Or, to use an even more expressive metaphor coined by F. H. Herrick: "The instincts of a species fit like lock and key."

Indeed, the problem of the innate releasing schema, of the receptor correlate to a quite specific, biologically significant stimulus context, is undoubtedly the problem of a "*lock!*" There could never be any doubt about the fact that the releasing process involved is entirely equivalent in nature to the phenomenon that Pavlov called the unconditioned reflex. But recognition of this equivalence does not in any way explain the function of the innate schema or even help us on our way. The problem does not reside at all in the reflex process but, so to speak, *before* this, in the receptor that only permits quite specific environmental influences to serve as releasing stimuli. Like some kind of *filter,* it excludes other, very similar influences and thus prevents them from operating as releasing stimuli. As far as I am aware, even Pavlov himself never pursued his analyses of the unconditioned reflex as far as investigation of the receptor, which is, so to speak, the afferent limb of the reflex arc. For example, he nowhere raises the question as to why a dog responds to the chemical stimulus of meat extract in its mouth with secretion of mucus-rich "eating saliva," but responds to the purely mechanical stimulus of dry sand with the copious production of watery "flushing saliva." With such responses, the problem of *selectivity* of the innate releasing schema does not intrude, because *different* stimuli—chemical as opposed to mechanical in our example of the salivation response—*stimulate different receptors* and thus represent the starting points of different reflex arcs.

The situation is different with the complex interaction between releasing and released behavior patterns in social animals. Consider the following example: A jackdaw sitting quietly about a foot away from a conspecific does not give rise to any stimulation. If, however, it adopts the species-specific "threat posture," by markedly ruffling the feathers on its head and back and simultaneously lowering its head and the slightly spread tail, thus leading to a peculiar hunched body posture, any other jackdaw—even a young jackdaw that demonstrably lacks any experience—will respond in an appropriate fashion. It will either flee or adopt the threat posture itself. In the same entirely predictable fashion, any jackdaw will exhibit the "response of social preening" by beginning to preen the head feathers of a conspecific, provided that the latter adopts the "submissive posture." In this, the body is held erect but the ankles are somewhat bent and the feathers of the back are flattened, while the head feathers are maximally ruffled and the bill is turned away from the partner; or, more precisely, the ruffled back of the head is presented. On repeated observation of this response, which is very common and occurs many times a day, anybody who thinks in physiological terms must be confronted with the question as to the causal basis for such *high selectivity* of the reaction to the stimuli concerned. It is undoubtedly the case that *unconditioned reflex* processes are involved, but what are the mechanisms that allow a conspecific that adopts a threat posture to spark off a reflex arc that is *different* from that evoked by a submissive posture? Undoubtedly, both stimuli are transmitted through the retina of the eye. The organ, the receptor in the direct physiological sense, is the *same.* The difference resides only in the images that are received. But even for these, it must be said that the sight of a "hunched" jackdaw is not so very different from that of a jackdaw that bends its ankles and turns its head away. Nevertheless, any completely inexperienced young jackdaw will respond "as if to the press of a button" by fleeing if a previously unstimulating conspecific alongside adopts the former posture, yet "must" preen its head with the same unconditioned reflexive certainty if it displays the second postural pattern with its head and feathers.

But if we now seek a causal explanation for this great selectivity of specific response to stimulation, asking *what* is actually *different* in the objectively existing stimuli that the organism receives, the only possible answer is that it must be the different *retinal images* that evoke such incredibly specific responses. Now it so happens that a great deal is already known about specific responses to retinal images through *acquired,* that is, learned, reaction patterns. Both through the investigations of the Pavlovian reflexology school and through those conducted by *Gestalt psychologists,* we know of innumerable cases in which an organism responds to "Gestalt simulation," to *images,* in an entirely specific manner. In another context, we have already become acquainted with Wolfgang Köhler's important aphorism that an organism is able to develop a conditioned response to a particular stimulus situation only if it is capable of perceiving it as a *Gestalt.* In all stimulus situations in which a specific response is made on the basis of previous *acquisition processes,* the combination of external stimuli that elicits a conditioned behavior pattern is accordingly always perceived as a Gestalt. In other words, it is perceived as a *holistic complex quality* in the highest sense of the term. Within this, the individual stimuli that participate in determining this quality are fused to form an unmistakable experiential entity, as we have already discussed in the Philosophical Prolegomena (p. 28; p. 54 et seq.). These facts were already known regarding the *acquired* receptor correlates that make specific stimulus situations—and *only* these—the releasing factor for quite specific behavior patterns. In view of this, it at first seemed very reasonable to conceive of the response of *unconditioned* reflexive correlates to releasing stimulus situations in an analogous fashion as a response to Gestalts, so to speak as a response to "innately remembered images." Thus it was that Jung, in his theory of the *archetype,* developed the notion that organisms are born into the world with a species-specific heritage of images of specific, biologically significant objects, such as parents, sexual partners, prey, and so on. Jung conceived of these "archetypes" as remembered images of some kind, as "species experiences" that had become differentiated in some way in the course of evolutionary history. He did not in the least think of this naively as the inheritance of acquired learning in the Lamarckian sense, but imagined it as an evolutionary process in the sense of modern phylogenetic theory. Thus, when a living organism responds to a specific biologically significant object or stimulus situation in an appropriate fashion *without* any prior individual experience, according to Jung's interpretation this represents a response to a "species-specific remembered image" that is definitely to be regarded as a holistic Gestalt. Both in Heinroth's work and in my own earlier studies, wherever it is stated that an organism "innately recognizes" its mating partner, its prey, a predator that is a threat to members of its species, and so on, this form of expression is implicitly based on exactly similar notions. In particular, it reflects the opinion that the innate response to such a stimulus situation is based in the same way on the "recognition" of a total image that ipso facto possesses the character of a complex quality in the form of a Gestalt.

Revision of these apparently reasonable interpretations occurred quite gradually because of a steadily increasing number of observations of cases in which animals responded "erroneously"—that is, in a manner that is pointless or even detrimental with respect to survival—*to just a few of the stimuli that are contained in such stimulus situations that release innate responses.* As an initial example of this, one can cite observations involving the so-called "rattling response" of the jackdaw (*Coloeus*). It was this response that first made me aware of the peculiarity of the innate releasing schema that concerns us here and which also pointed the way to a *method* of investigating it by means of *dummy experiments.* Any healthy adult jackdaw, that is, one over the age of about four weeks, responds with reflexive certainty in a specific manner as soon as *a conspecific is grasped in the hand.* With specific body postures and movements, which will be discussed in detail in the special part,* and while uttering a loud rattling call, it attacks the hand that is holding the other bird. For the release of this response, it is entirely immaterial whether the attacking jackdaw "loves" or "hates" the bird that is being held. I had a jackdaw that I had reared in isolation, which responded to me with the most intensive patterns of social behavior but at the same time chased other captive young birds of the same species with equal intensity and tried to kill them. Nevertheless, this jackdaw attacked me "in a blind rage" in the manner described whenever I held one of the young birds in my hand. Because of the unconditionally certain predictability of this response and the utter constancy of the motor patterns and vocalizations, it seemed almost definite from the outset that it represented a species-specific, unconditionally reflexive behavior pattern. It seemed equally likely that the species-preserving function is likely to reside in the defense of a conspecific that has been seized by a predator. In fact, however, a single unforgettable observation revealed to me that the jackdaw does not carry an internal "image" of a conspecific in great danger, hence recognizing the occurrence of this "archetype" in the concrete stimulus situation. Quite by chance, I allowed my tame jackdaws to catch sight of me with wet, black bathing trunks in my hand. The next instant, I was surrounded by a flock of angrily rattling jackdaws, raining pecks on the hand that was holding the bathing trunks! I knew from experience with my first jackdaw that the rattling response is always followed by a long-lasting fearful aversion to the human being that evoked it. As will be demonstrated in detail in the special part in the discussion of the jackdaw, such unconditioned reflexive responses lead extremely rapidly to the development of *conditioned* responses. As a result of these, any human who has elicited the unconditioned response a few times becomes a "conditioned substitute stimulus" (see p. 202 et seq.). The occurrence of such a process with the rattling response would have immediately made my jackdaws wary of me, thus ruling out any further work with them. For this reason, I had to limit myself to conducting further experiments on the releasing conditions for the rattling response with the utmost care,

**See the footnote on p. xliii.*

always allowing a considerable time to elapse between them. Accordingly, I kept my eyes well open for any relevant chance observations.

The results of my observations can be briefly summarized as follows. The response is released at full intensity as soon as a shiny, black, moving, and dangling object is *carried* by another organism. Both the nature of the shiny, black, dangling object and the identity of the organism that is carrying it are utterly *immaterial*. This must be particularly emphasized: the response is by no means *stronger* when a real predator is carrying off a real jackdaw. A human who is not himself an object of fear but is carrying black bathing trunks in his hand evidently presents *all* of the stimuli that are also present in the "normal" situation in which the response serves its species-preserving function. The many *additional* details of this situation that seem to us to be so important for the total *image* exert no influence on the release of the unconditioned reflex. The simplified situation is not only "adequate," that is, functionally equivalent to the normal situation, it is also "optimal" for the releasing effect to the extent that there is no object that will exert a stronger influence. Hence, the receptor correlate—the "lock" of the response—is constructed in such a way that, from the multitude of stimuli that occur in the relevant, biologically significant situation, *just a very few, specific* stimuli are, as it were, picked out. It responds to a relatively very *simple* combination of stimuli and certainly not to the complexity of the overall situation. Despite its simplicity, however, the stimulus combination that is thus established as the key to the unconditioned response is always constituted in such a way that it characterizes the biologically significant situation *sufficiently unambiguously* to render an *erroneous* response of the receptor correlate sufficiently improbable. Our term *innate releasing schema* refers to precisely this function of simplified characterization of the situation.

It is of the utmost importance to make the reader aware of the fundamental difference between such *unconditioned* response-releasing innate schemata and the occurrence of a *conditioned* response to the complexity of an acquired perceptual Gestalt. I believe that the best way of doing this is to contrast the observation of the unconditioned release of the jackdaw's rattling response with a conditioned release of the *same* behavior pattern. Some years later, when my jackdaw colony had become somewhat shy and when it was no longer very important to me to keep the birds tame, I began to band the almost-fledged nestlings on a yearly basis, without taking any special precautions. Understandably, every time I did this it evoked intense rattling responses from all of the breeding birds. In the manner already indicated above, the result was that I myself became the conditioned stimulus that released the complete response. The jackdaws would begin to rattle wherever and whenever they saw me. Although they did not dare to engage in active attack like the tame jackdaws described above, they followed me while rattling so persistently that they made it very unpleasant for me to stay in the garden. Of course, they recognized me *personally* and never showed this response to any other human. At the beginning, they only harassed me when I wore the country clothing that was familiar to them; but they soon learned to recognize me as the same Gestalt when I wore other attire. This "hate" directed at me first began to wane some time after the offspring had fledged, but the older breeding birds maintained the conditioned response throughout the winter and the next year they began to respond to my appearance with the "rattling response" even *before* the time came for banding.

The enormous complexity of these releasing conditions for the conditioned responses, which clearly involve the perception of a complex Gestalt quality, stand in marked contrast to the simplicity of the "key" to the lock of the unconditioned response: black, moving, and carried by another organism. Indeed, it is quite likely that additional dummy experiments, which I did not carry out at that time, might have indicated an even greater degree of simplification!

In addition to the immeasurably great, but purely *quantitative,* difference in the *complexity* of the conditioned and unconditioned response-releasing stimulus conditions, there is a *second* difference. This is, in fact, far more significant with respect to the physiological peculiarity of the innate releasing schema. As will be explained more precisely in the section dealing with Gestalt psychology, among the very large number of individual stimulus properties that are involved in determining the complex quality of a perceived Gestalt, *not a single one may be absent* without leading to *modification* of the overall quality of perception *as a whole,* thus *destroying* the Gestalt as such. Consider, for example, the characteristic profile of the silhouette of a human head, which is immediately *recognizable* to anyone who knows the person concerned. If we alter, for example, the profile of the nose, the silhouette is by no means perceived as the previous physiognomy fitted with a new nose. Instead, it is perceived *as a whole,* as a modified, completely new, and different physiognomy. The memory of the Gestalt image of the previous silhouette *simply fails* to respond. By contrast, from the few stimuli to which the receptor correlate of the innate schema responds, not just one but several may be lacking *without* obliterating the response. The response merely becomes less *intense* as the number of stimuli lacking from the "optimally" releasing stimulus combination increases.

Here, too, an example can be provided from my early observations. The "following reflex" of a fledged young jackdaw, the innate response pattern that induces the offspring to fly behind its parents, is released by an extraordinarily small number of stimuli evoking an unconditioned response. If a young bird that has been artificially reared is placed on the ground and its human caretaker first squats alongside it and then leaps up and runs away, it will fly after him "with unconditioned reflexive predictability." The response is even more intense if one emits at the same time the flight-summoning call of the jackdaw, which is easily imitated. The *upward* movement of the releasing object is at least as important for the effect achieved as the movement *away* from the young jackdaw. If the bird is sitting above human head height, so that the feature "upward" is missing, the following response of the young jackdaw is markedly weaker. A human "jackdaw father" is simply unable to present the *optimal* stimulus situation for the operation of the innate releasing schema that elicits this following response. He is unable to display either the feature "upward" or the feature "away" to the same degree or with the same speed as a "real" jackdaw. Further, one "strong" feature that is very important for release of the response is completely lacking: the sight of the unfolding, beating pair of wings is an extremely effective releasing stimulus.

The observations described here clearly demonstrate the fact, already anticipated above, that not just one but several of the stimuli that constitute the *optimal* object of the innate releasing schema may be missing without abolishing the response. A human being, who lacks black wings and is only able to present the characters "upward" and "away" in a far weaker form than the normal object of the response, can nevertheless release this well enough to provoke a young jackdaw to fly behind him during short bicycle trips! However, this only applies as long as no "choice experiment" occurs between the human caretaker and a real jackdaw, which is an object that will more effectively release the unconditioned response. Otherwise, the plus factor in the unconditioned following response will overpower all of the conditioned responses that bind the young bird to the human being. It will blindly fly after another jackdaw, even after it has repeatedly endured the experience of thus losing contact with the "parental animal" and ending up in an evidently extremely unpleasant situation where it is alone. *Any* one of the characters belonging to an innate releasing schema can *in isolation* serve as the releaser of an—albeit less intense—response. Under certain conditions, a young jackdaw that is sitting at head height will fly after a human being who slowly walks away, such that literally only the *one* character "away" comes into effect. Similarly, the jackdaw may fly toward one if it *only* hears an imitation of the summoning call. What is important here is the reciprocal *interchangeability* of the releasing stimuli, any one of which will elicit motor patterns with the *same* response-specific arousal quality. If, for example, the jackdaw is sitting on the ground and, without emitting the summoning call, one stands up alongside it and walks away at normal speed, it will exhibit a following response of specific, moderate intensity. If, on the other hand, the bird is sitting so high up that the character of upward movement cannot be presented, one has to run away quite rapidly while uttering summoning calls in order to evoke a response of the same intensity. It is possible to provide a large number of illustrations of such equivalence between the releasing effects of different combinations of characters that release the *same* response. Numerous examples of this will be given in the discussion of Seitz's studies of the mouth-brooding fish *Haplochromis (Astatotilapia) strigegena.*

Accordingly, with the innate releasing schema it is the *sum* of the characters that evokes a response and not a *Gestalt* that integrates these characters into a *single,* indivisible complex quality and that serves as an all-or-nothing releaser of a conditioned response. It must, however, be noted at this point that *individual characters* that are registered by an innate schema can *in themselves* operate rather like very simple Gestalts. With the stickleback, the character "red *underneath*" releases fighting responses in dummy experiments, while the rattling response of the jackdaw responds to a character such as "*carried* by a living creature," and so on. Each of the effective characters is hence, in its own right, very closely related to a "stimulus Gestalt." But the individual characters are in no way combined *together* into a functional Gestalt. Instead, they operate purely as the sum of independent individual stimuli, albeit with the same effect.

This *summation* of the releasing stimuli that "unlock" an innate releasing schema understandably facilitates *quantification* of the effects of those stimuli. Even with a very crude experimental setup it is possible to obtain a relatively reliable impression of "strong" as opposed to "weak" characters. In an experiment with a male *Astatotilapia,* for example, parallel orientation of the dummy has a far greater effect in releasing a fighting response than, say, the shiny blue color of the display pattern or the dark fringe on the dorsal and anal fins, or alternatively on the upper and lower margins of the dummy. But really exact quantification of the releasing effect of individual stimuli registered by an innate releasing schema only became possible when the *internal responsiveness* of the organism was *simultaneously* taken into account. All that we can *observe* as a criterion for the effectiveness of an impinging external stimulus, and all that we can record quantitatively, is the *intensity* of the response that is given to this stimulus. But, wherever *instinctive motor patterns* are concerned, the response intensity is determined by *two* factors, namely, the strength of the external stimulus and the internal state of responsiveness. It is the latter that we must now consider. I shall illustrate this with an observational example that at the same time has the advantage that it was the first case that initially made me aware of the problem presented by stimulus intensity and responsiveness. It involves the following response of the fledged young jackdaw, already mentioned above. If such a bird is animated and keen to fly, normal quiet departure is enough to induce it to follow even if it is sitting at head height. But the more *frequently* the following response is elicited, the *stronger* are the stimuli required, until eventually only a combination of *all of them presented together* will exert a releasing effect. Looking back, I have to recognize that it was this particular context that led me to the notion of "stronger" and "weaker" stimuli, of "optimal," "more effective," and "less effective" objects of an innate schema. When my first free-flying jackdaw was following me, such behavior more or less compelled me to identify stimuli of ever-increasing strength, simply in order to return home with this bird after it had become tired as a result of flying after me! On such an excursion, it was quite impossible to overlook the *summation* of the stimuli "away," "upward," and "flight-summoning call."

On the other hand, however, this observation comprises another, fundamentally important fact with respect to the quantification of stimulus effects. *The same* stimulus, the same, objectively constant sum of characters evoking a reaction from the schema, can release responses of *different* intensity, according to whether the organism is fresh or tired. Conversely, with a fresh individual a weak stimulus can release a response of the same intensity as that evoked from a tired individual by the strongest possible combination of stimuli. *The threshold value for elicitation of the response increases with the number of releases that have already taken place!* This, in itself, is by no means special, as it is nothing other that an expression of the long-known phenomenon of *fatigue* in the physiology of stimulation. Nevertheless, there is a peculiarity inherent in this variability of the threshold for instinctive behavior patterns. Such a young jackdaw that can no longer be induced to follow even with the most "refined" presentation of stimuli is, in fact, not at all "tired" in the ordinary sense of the term. Indeed, it is not even "tired of flying," because it will readily fly from tree to tree or generally fly "around." It is only the specific motor pattern of flying along behind "as if attached" that can no longer be released, despite all efforts. A far more impressive example of this completely *specific* fatigue of a single response is provided by the "sham disability display" of many birds. This motor pattern, which is particularly highly differentiated in whitethroats, is released when a predator approaches the nest. Its species-preserving function resides in the illusion that the brooding bird is seriously disabled, thus inducing the predator to follow it away from the nest. Similar, but not homologous, motor patterns are shown by mothers of many duck species when they are leading offspring. In this case, however, they generate an impression not of general disability but of mechanical incapacity to fly. This is done by a convincing "performance" of vain attempts to take off.

Study of the releasing stimuli and of the inner responsiveness of such appropriate objects led to insight into the general lawfulness of instinctive motor patterns, which I regard as the most important result of my lifework. My own earlier observations, along with Heinroth's investigations, had revealed an important fact with which the reader is now familiar: in many animal species, if the adequate object for a given instinctive motor pattern, the "optimal" stimulus situation that exactly fits the requirements of the schema, is continuously absent, it will be performed in response to a "substitute object." The male blackbanded sunfish that lacked fine sand for the construction of its nest hollow discharged the nest-building pattern on a smooth zinc substrate (p. 224). An old male cichlid fish that had been kept in isolation for some time, so that its fighting responses had never been released, eventually responded "erroneously" to a female of its own species as if it were another male (p. 227). Heinroth's hand-reared black grouse, which had never encountered an adequate stimulus situation for release of its escape response, responded with escape attempts to substitute stimuli of ever-decreasing strength (p. 257). Indeed, in an extreme case, such a motor pattern can "go off" quite spontaneously *without* a substitute object of any kind. Two examples can be provided to illustrate this. The first concerns the red African cichlid fish *Hemichromis bimaculatus,* also known as the "jewelfish." A female of the same species is introduced to a male. The latter, visibly sexually aroused and radiating the incredible colors of its display markings, exhibits an extremely peculiar motor pattern. It hovers vertically, head downward, a few centimeters above the bottom of the aquarium with its jaws wide open and performs jerking, downward thrusting movements. Subsequent observation reveals that this is a vacuum performance of a motor pattern that serves to clean the *stone* that has been selected as a spawning site. This also plays a part, in a "demonstratively enhanced" form, as a genuine display pattern in the courtship of these fish. An even better example of such a "vacuum response," to use the name that I had used for this process as early as my late schoolboyhood diaries, was provided by a starling (*Sturnus vulgaris*) that caught flies "in a vacuum!" This bird had its "lookout post" on the head of a bronze statue that stood on the mantlepiece in our dining room and continually looked *upward* with tense and clearly *directed* intention movements. From time to time, it took off and flew up into the free space near the ceiling, where it snapped at something before returning to its roosting place. After landing, it would strike its beak a few times from right to left against the substrate, showing the motor pattern that many insectivorous birds use to kill the prey and prepare it for swallowing. The starling followed this with clear-cut *swallowing movements,* after which it relaxed its feathers in momentary "satisfaction." The bird's behavior was so deceptively similar to real catching and eating that on several occasions I stood on a chair to check whether there were any tiny flying insects in the room. Eventually, I realized that this starling was performing a specific instinctive motor sequence as a "vacuum response" *without an object.* There is an almost unlimited number of examples of such behavior. There is, for instance, the very impressive behavior of weaverbirds. In the absence of the requisite materials, they will perform the extremely complex motor sequences for tying the first straws to the nest stem *without* a straw on a perch in the cage.

The *methodological* novelty of my approach to understanding these phenomena was that I focused my attempts at quantification on the *threshold value* of the instinctive motor pattern that was released. A young jackdaw that has not yet "flown itself out" will follow even in response to a slow departure, whereas after stimulatory fatigue it will only respond to a combination of all the stimuli that have been described.

The method described here, namely that of tackling the quantification of stimulus strength and response intensity at the *threshold* of the response and of quantifying the stimulus and the response *simultaneously,* led to the important insight mentioned above (p. 279 et seq.). The "response-specific fatigue" that is expressed in *elevation* of the threshold value for releasing stimuli is the *negative side of the same physiological process as that which at the other extreme of the phenomenon is responsible for responses to a substitute object and "vacuum activities."* The complete continuity between these two contrasting phenomena is easily demonstrated. As we have seen, the threshold value for the release of all genuine instinctive motor patterns rapidly increases from one experiment to the next. But if we allow the experimental animal to *rest* in an environment that is completely lacking in stimuli relevant to the response investigated, the threshold value will gradually decline again. This lowering of the threshold for releasing stimuli *does not stop at a particular norm* but *proceeds continuously* as long as the motor sequence concerned is not performed (including weak indications thereof). As the duration of rest from the response increases, the animal begins to respond to ever-weaker stimuli with ever-stronger responses until it reaches the limiting threshold value of zero and erupts as a "vacuum activity."

This phenomenon of *lowering of the threshold for releasing stimuli in proportion to the time elapsed since the last discharge of a motor pattern suggests the existence of some form of response-specific arousal energy.* *This would have to be continuously produced by the organism but used up by performance of the motor patterns specific to that response.* For the sake of clarity, I shall illustrate this here with a very mechanistic conceptual model that in fact provides analogies for all of the important details of the overall process and, in particular, neatly symbolizes the need for *double* quantification of the stimulus and responsiveness (figure 7). Let us assume that a fluid is continuously flowing into vessel *A* from the tap *B* located above and to the left. This flow provides an analogy for the continuous generation of response-specific arousal. Tap *C* represents the innate releasing mechanism, which can be "opened" to varying degrees by specific stimuli of different strengths, thus permitting the discharge of the accumulated arousal in the form of the instinctive motor pattern concerned. The instinctive motor pattern itself is represented by the emerging jet of water and its *intensity* is illustrated by the power of the jet, which is reflected by the distance that it reaches along the scale *D*. The model provides a good illustration of the way in which the intensity of the response, represented by the distance over which the jet projects, is dependent upon *two* factors. It depends on the degree to which tap *C* is opened, representing the strength of the stimulus, and on the degree of *damming* of the response, which is represented by the height of the column of fluid in the container. As is also graphically illustrated by the model, this *inner responsiveness,* the height of the "current level" of response-specific arousal, directly depends upon the frequency and duration of performance of the motor patterns concerned during the period immediately beforehand. The focal point of all of our attempts at quantification is always confined to the intensity of the response, the distance over which the jet projects. Whenever we investigate the effectiveness of a stimulus, *we at first do not know* how far "the tap is opened"—to use the symbolism of our model. But if we conduct a second experiment to determine a value for the current level of response-specific energy—in practice, usually by presenting the optimal object and thus completely opening the tap—a quite specific relationship is established between the releasing effect of a partial stimulus and that of the complete set of stimuli in the optimally releasing situation. This relationship is found to be completely *constant,* whatever the value of the current level may be.

**Note in the margin here:* Or a substance?

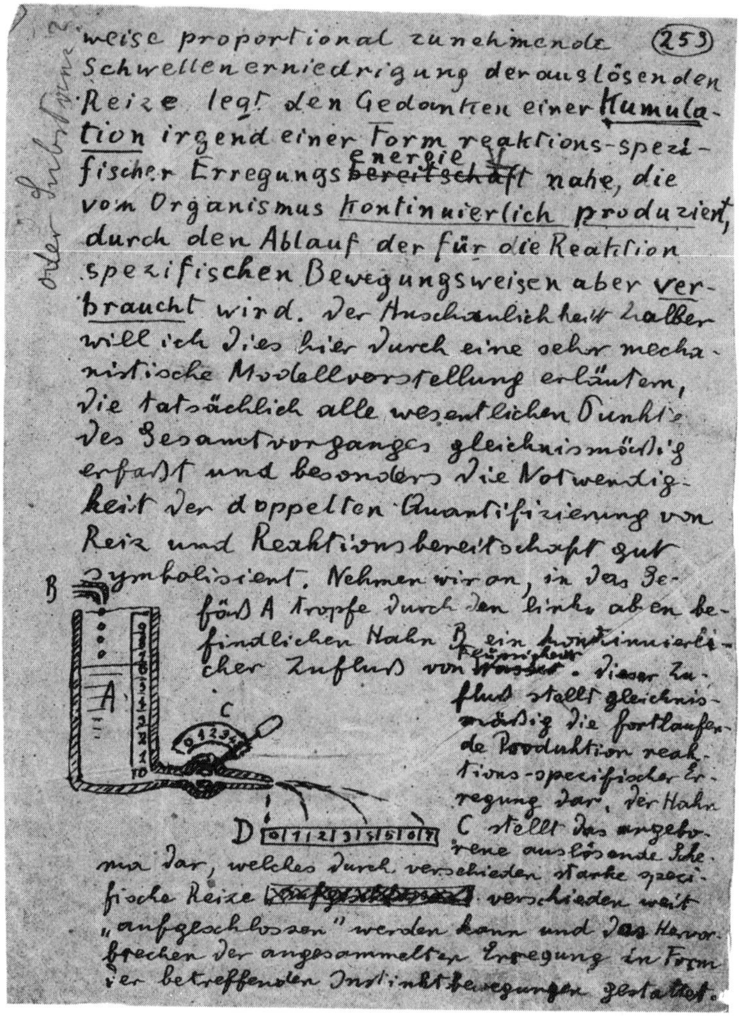

weise proportional zunehmende (253)
Schwellenerniedrigung der auslösenden
Reize legt den Gedanken einer Kumula-
tion irgend einer Form reaktions-spezi-
fischer Erregungsenergie nahe, die
vom Organismus kontinuierlich produziert,
durch den Ablauf der für die Reaktion
spezifischen Bewegungsweisen aber ver-
braucht wird. Der Anschaulichkeit halber
will ich dies hier durch eine sehr mecha-
nistische Modellvorstellung erläutern,
die tatsächlich alle wesentlichen Punkte
des Gesamtvorganges gleichnismäßig
erfaßt und besonders die Notwendig-
keit der doppelten Quantifizierung von
Reiz und Reaktionsbereitschaft gut
symbolisiert. Nehmen wir an, in das Ge-
fäß A tropfe durch den links oben be-
findlichen Hahn B ein kontinuierli-
cher Zufluß von Flüssigkeit. Dieser Zufluß stellt gleichnis-
mäßig die fortlaufen-
de Produktion reak-
tions-spezifischer Er-
regung dar, der Hahn
C stellt das angebo-
rene auslösende Sche-
ma dar, welches durch verschieden starke spezi-
fische Reize verschieden weit
„aufgeschlossen" werden kann und das Hervor-
brechen der angesammelten Erregung in Form
der betreffenden Instinktbewegungen gestattet.

Figure 7

Hence, it has been possible for comparative behavioral investigators, in particular for my student A[lfred] Seitz, to demonstrate *a genuine, exact proportionality between stimulus strength and response intensity.* This is a result that had been vainly sought both by the school of scientific psychology at the turn of the century and by the great mechanistic schools!

Demonstration of an accumulation of response-specific arousal energy undoubtedly provides a physiological foundation both for Heinroth's theory of moods (motivations) and for the phenomena of drives and appetitive behavior investigated by Craig. Without a doubt, the *same* damming processes on the one hand determine the threshold value for releasing stimuli and on the other generate restlessness in the organism as a whole, inducing it to engage in an active search for these stimuli. In turn, all of the phenomena discussed in this chapter are extensively explained in physiological terms by the results obtained by E[rich] von Holst, who demonstrated the presence of automatic-rhythmic processes of stimulus production in the central nervous systems of a variety of animal species.

21

Erich von Holst's Discovery of Automatic Stimulus Production in the Central Nervous System

The discovery of automatic-rhythmic stimulus production in the central nervous system owes its origin to the introduction of a completely new experimental method that represented a fundamental departure from previous tradition. As has been described on p. 211 for the experimental methods of the great mechanistic schools of behavioral research, all of the approaches in invasive experimental physiology were always limited to the *modification* of some condition and recording of some *response* of the central nervous system. In both cases, this kind of approach *necessarily* led to the opinion that the function of the central nervous system is confined to the reception of stimuli and response to them. The fundamental novelty in von Holst's method, representing an extensive parallel to the methods of the first comparative ethologists, was quite simple. He took isolated, surviving parts of the central nervous systems of various animals, such as the ventral nerve cord of the earthworm or the spinal cord of various fish, and *left them to themselves* in order to be able to observe what "they did on their own"! Using this approach, which was as original as it was simple, von Holst immediately discovered an elementary function of the central nervous system that had previously remained virtually unrecognized. Both from the ventral chain of ganglions in the earthworm and from the spinal cords of all of the fish that were investigated, without any influence of external stimuli, there is *continuous* transmission of a well-coordinated sequence of impulses for quite specific, often very complex motor patterns. This applies particularly to certain motor patterns that are used especially *frequently*, such as those involved in *locomotion*. The well-known case of the creeping movement of the earthworm, in which the contraction of the circular or longitudinal musculature of one segment follows from that of its neighbor, such that a "wave" of successive muscle contractions runs along the worm's body, had previously provided the classic example of a reflex chain. It had been assumed that the contraction of one segment provoked the timely contraction of the next in a reflex manner by means of "proprioceptors," nervous elements that respond to internal processes in the body and transmit messages to the center. Neurophysiologists made the same assumption about the swimming movements of fish and, indeed, about all innate patterns of motor coordination. The first experiments conducted by von Holst were sufficient to demonstrate quite decisively the utter indefensibility of this assumption. If the ventral nerve cord of an earthworm is dissected out of the body and suspended in saline solution, it continues without modification to transmit the rhythmic and segmental sequence of impulses for creeping locomotion, as can be demonstrated by electrical recordings. A similar function is also present in the spinal cord of a fish. To demonstrate this, it was necessary (following anesthesia with urethane) to transect the medulla at the level of the tenth cranial nerve in an eel (*Anguilla*) or in a wrasse and then provide artificial respiration. After the fish had recovered from the shock of the operation, the isolated spinal cord began to transmit motor impulses for the typical, well-coordinated swimming movements of the intact fish. Nothing is changed in the coordination of these motor sequences if one transects all of the "afferent" nerve tracts, that is, all of those leading from the periphery to the center, thus reliably excluding any possibility for the occurrence of reflex processes. The eel continues to perform its typical sinuous movements and the wrasse continues to beat its fins in the same way and with the same coordination as in the intact state. A crude illustration of the lack of any involvement of reflexes in the production of these movements is provided by the following experiment: Taking an eel with its spinal cord isolated in the manner described, not only the sensory but also the motor roots of the spinal nerves are transected in the middle section of the body, such that the musculature in this region is paralyzed and rendered inactive (figure 8). A sinuous movement that is initiated at the front end of the body disappears, rather like a train in a tunnel, in the paralyzed section of the body, but then reappears at exactly the right time at the other end of the body. This experiment conclusively demonstrates that the temporally coordinated contraction of one muscle segment is not released in a reflexive manner by the contraction of the preceding segment.

In contrast to an intact organism, however, the isolated ventral nerve cord of an earthworm will continue to "creep" and the isolated spinal cord of an eel or a wrasse will continue to "swim" literally *incessantly* as long as the animal survives. Thus, the endogenous automatism transmits its impulses continuously, but its effects can be suppressed in the intact animal by an

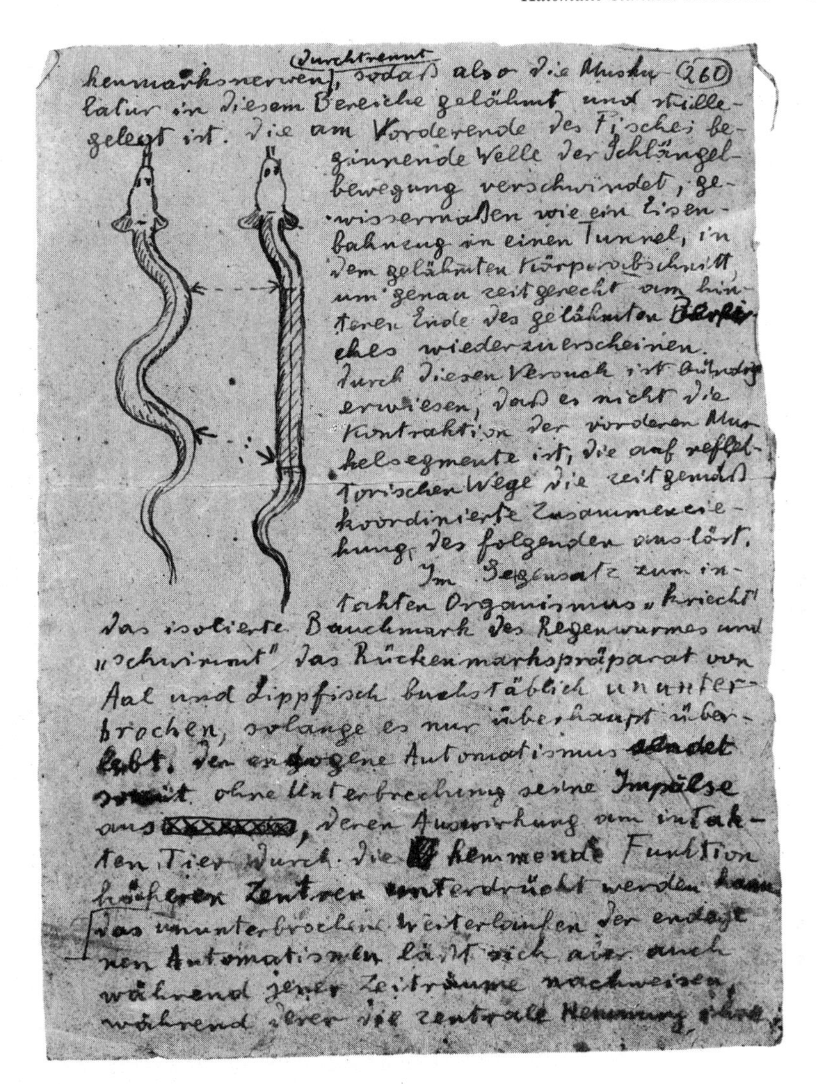

durchtrennt
kenmarksnerven, sodaß also die Musku- (260)
latur in diesem Bereiche gelähmt und stille-
gelegt ist. Die am Vorderende des Fisches be-
ginnende Welle der Schlängel-
bewegung verschwindet, ge-
wissermaßen wie ein Eisen-
bahnzug in einen Tunnel, in
dem gelähmten Körperabschnitt,
um genau zeitgerecht am hin-
teren Ende des gelähmten Berei-
ches wiederzuerscheinen.
Durch diesen Versuch ist bündig
erwiesen, daß es nicht die
Kontraktion der vorderen Mus-
kelsegmente ist, die auf reflek-
torischem Wege die zeitgemäß
koordinierte Zusammenzie-
hung des folgenden auslöst.
Im Gegensatz zum in-
takten Organismus „kriecht"
das isolierte Bauchmark des Regenwurmes und
„schwimmt" das Rückenmarkspräparat von
Aal und Lippfisch buchstäblich ununter-
brochen, solange es nur überhaupt über-
lebt. Der endogene Automatismus sendet
somit ohne Unterbrechung seine Impulse
aus, deren Auswirkung am intak-
ten Tier durch die hemmende Funktion
höherer Zentren unterdrückt werden kann.
Das ununterbrochene Weiterlaufen der endoge-
nen Automatismen läßt sich aber auch
während jener Zeiträume nachweisen,
während derer die zentrale Hemmung ihre

Figure 8

inhibitory function of higher centers. But the uninterrupted activity of endogenous automatisms can also be demonstrated at times when such central inhibition prohibits the occurrence of external motor effects. Demonstration of this can be achieved by exploiting a phenomenon that von Holst refers to as *relative coordination.* In the wrasse (*Labrus*), beating of the pectoral fins and beating of the tail fin are caused by two separate processes of automatic-rhythmic stimulus production, each of which can function independently of the other. There is nevertheless a particular kind of coordination between the beating of the pectoral fins and that of the tail fin. *Without* the participation of any receptors or afferent nerve pathways, that is to say, demonstrably without any involvement of the periphery, the rhythm of the pectoral fins influences the effects of the rhythm of the tail fin. Whenever the impulses from the two unsynchronized stimulus-production rhythms show temporal coincidence in the same sense, the beating of the tail fin is reinforced. Conversely, when there is temporal coincidence in the opposite sense, the beating of the tail fin is attenuated (figure 9). Thus, in kymographic recordings showing the fin beats, the movement of the tail fin shows superimposition of the rhythm of the pectoral fins on its "own rhythm," whereas the tail fin exerts no influence on the beating of the pectoral fins. Accordingly, von Holst* refers to a *dominant* rhythm of the pectoral fins and a *subordinate* rhythm of the tail fin. He refers to the kind of interaction described as *relative coordination.* Now, this phenomenon of relative coordination allows us to follow the activity of the dominant rhythm even in cases where its immediate effect, the movement of the group of muscles that it controls, is temporarily absent. If overt manifestation of the dominant rhythm is blocked by superimposition of a *central inhibition* (figure 10), such that the pectoral fins remain immobile, the rhythm recorded from the still beating tail fin persistently shows the same superimposition of the pattern of the dominant rhythm, although the latter is no longer revealed by beating of the fins! This demonstrates that the same sequences of impulses are being transmitted as before, but in a latent form without any motor consequences.

Production of stimuli by the central automatism also continues unchanged when the locomotor organ that it supplies with impulses is induced in some other way to perform an *additional* movement. The simplest situation is one in which a reflex is involved. The reflexive supplementary movement elicited at *x* by an additional stimulus (figure 11) does *not* become superimposed on the tracing of the dependent rhythm. At this point, in correspondence to the opposite direction of the dominant rhythm, the dependent rhythm shows a particularly small amplitude and certainly shows no reinforcement. In other words, it reflects superimposition of the move-

**Note on figures 9–12: The tracings sketched here are based on diagrams published by von Holst. Interested readers are referred to his publications, particularly to "Vom Wesen der Ordnung im Zentralnervensystem" (1937) and "Die relative Koordination als Phänomen und als Methode zentralnervöser Funktionsanalyse" (1939). In Erich von Holst,* Zur Verhaltenphysiologie bei Tieren und Menschen. Gesammelte Abhandlungen, *Vol. 1 (pp. 3–132). Munich: Piper, 1969.*

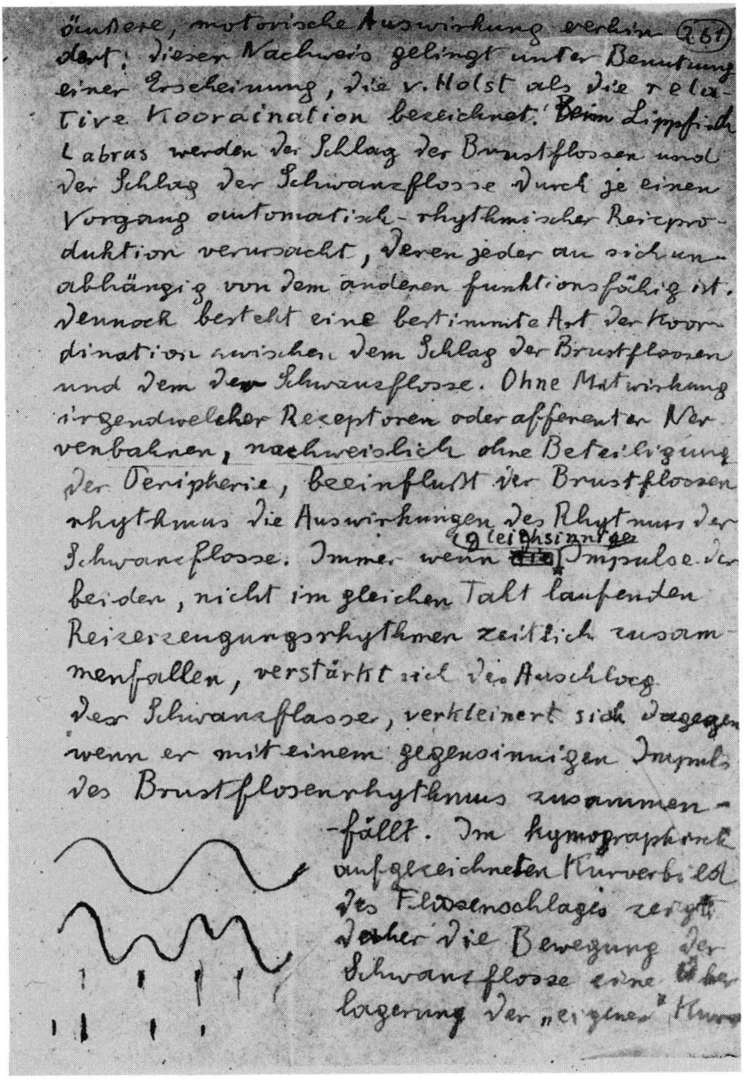

äußere, motorische Auswirkung verhin- (261)
dert; dieser Nachweis gelingt unter Benutzung
einer Erscheinung, die v. Holst als die relative Koordination bezeichnet. Beim Lippfisch
Labrus werden der Schlag der Brustflossen und
der Schlag der Schwanzflosse durch je einen
Vorgang automatisch-rhythmischer Reizpro-
duktion verursacht, deren jeder an sich un-
abhängig von dem anderen funktionsfähig ist.
Dennoch besteht eine bestimmte Art der Koordination zwischen dem Schlag der Brustflossen
und dem ~~der~~ Schwanzflosse. Ohne Mitwirkung
irgendwelcher Rezeptoren oder afferenter Ner-
venbahnen, nachweislich ohne Beteiligung
der Peripherie, beeinflußt der Brustflossen
rhythmus die Auswirkungen des Rhytmus der
Schwanzflosse. Immer wenn gleichsinnige Impulse der
beiden, nicht im gleichen Takt laufenden
Reizerzeugungsrhythmen zeitlich zusam-
menfallen, verstärkt sich der Ausschlag
der Schwanzflosse, verkleinert sich dagegen
wenn er mit einem gegensinnigen Impuls
des Brustflossenrhythmus zusammen-
-fällt. Im kymographisch
aufgezeichneten Kurvenbild
des Flossenschlages zeigt
daher die Bewegung der
Schwanzflosse eine Über-
lagerung der „eigenen" Kurve

Figure 9

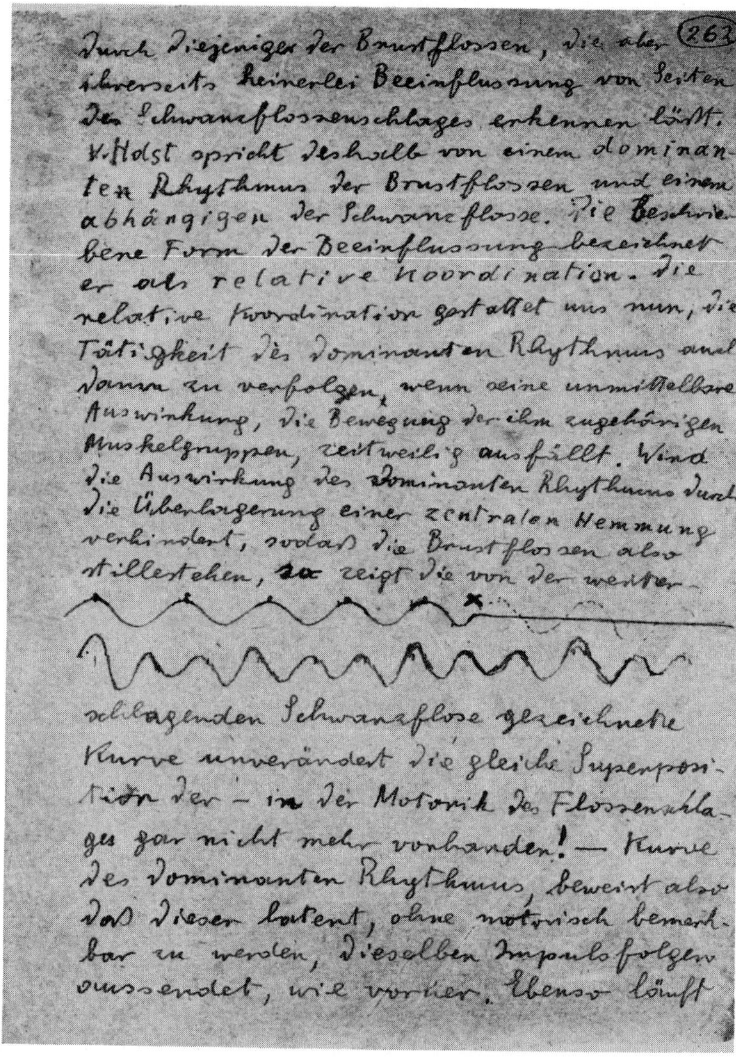

durch diejenigen der Brustflossen, die aber (262)
ihrerseits keinerlei Beeinflussung von Seiten
des Schwanzflossenschlages erkennen läßt.
v. Holst spricht deshalb von einem dominanten Rhythmus der Brustflossen und einem
abhängigen der Schwanzflosse. Die beschrie-
bene Form der Beeinflussung bezeichnet
er als relative Koordination. Die
relative Koordination gestattet uns nun, die
Tätigkeit des dominanten Rhythmus auch
dann zu verfolgen, wenn seine unmittelbare
Auswirkung, die Bewegung der ihm zugehörigen
Muskelgruppen, zeitweilig ausfällt. Wird
die Auswirkung des dominanten Rhythmus durch
die Überlagerung einer zentralen Hemmung
verhindert, sodaß die Brustflossen also
stillstehen, so zeigt die von der weiter-

schlagenden Schwanzflose gezeichnete
Kurve unverändert die gleiche Superposition der – in der Motorik des Flossenschlages gar nicht mehr vorhanden! – Kurve
des dominanten Rhythmus, beweist also
daß dieser latent, ohne motorisch bemerk-
bar zu werden, dieselben Impulsfolgen
aussendet, wie vorher. Ebenso läuft

Figure 10

ment that the pectoral fin *would have* performed *without* the influence of the additional stimulus. On the other hand, the tracing recorded for the dependent rhythm immediately shows an effect if immobility of the pectoral fins is caused by real *extinction* of the dominant rhythm, as commonly occurs when the spinal cord preparation begins to deteriorate after a certain amount of time has elapsed (figure 12). As soon as the dominant rhythm is arrested, the irregularities generated by superimposition in the tracing of

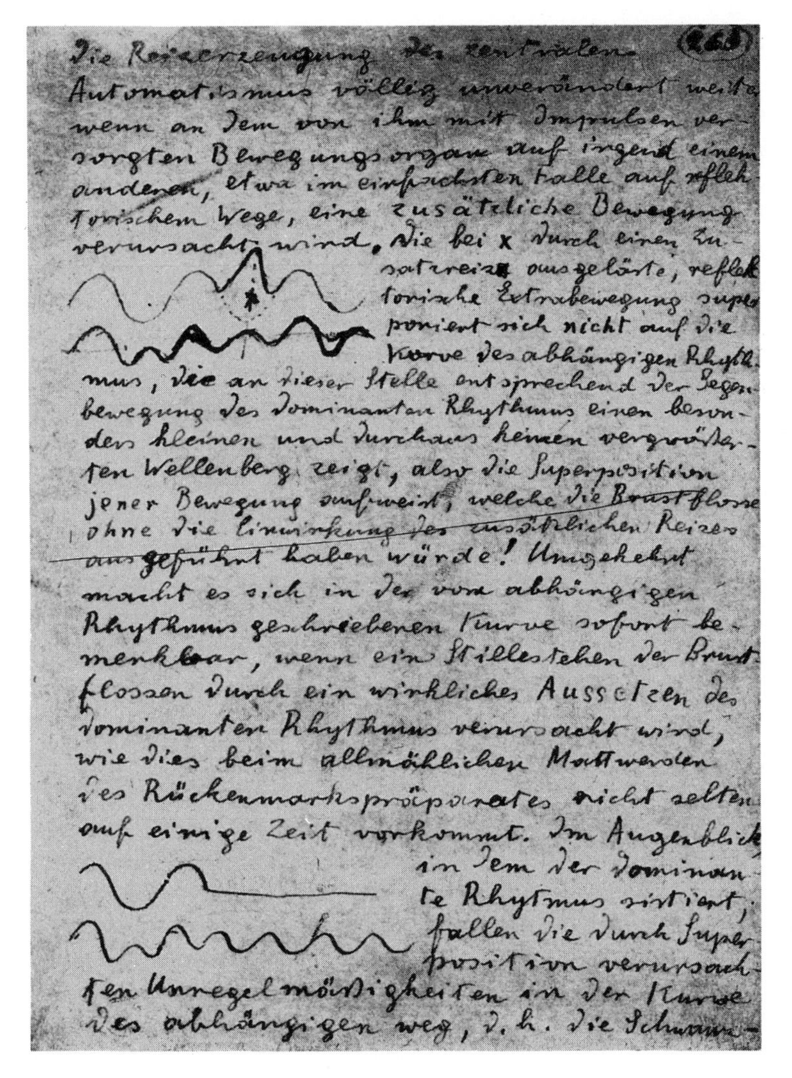
Die Reizerzeugung des zentralen (263)
Automatismus völlig unverändert weiter
wenn an dem von ihm mit Impulsen ver-
sorgten Bewegungsorgan auf irgend einem
anderen, etwa im einfachsten Falle auf reflek-
torischem Wege, eine zusätzliche Bewegung
verursacht wird. Die bei x durch einen Zu-
satzreiz ausgelöste, reflek-
torische Extrabewegung super-
poniert sich nicht auf die
Kurve des abhängigen Rhyth-
mus, die an dieser Stelle entsprechend der Gegen-
bewegung des dominanten Rhythmus einen besonders kleinen und durchaus keinen vergrößer-
ten Wellenberg zeigt, also die Superposition
jener Bewegung aufweist, welche die Brustflosse
ohne die Einwirkung des zusätzlichen Reizes
ausgeführt haben würde! Umgekehrt
macht es sich in der vom abhängigen
Rhythmus geschriebenen Kurve sofort be-
merkbar, wenn ein Stillestehen der Brustflossen durch ein wirkliches Aussetzen des
dominanten Rhythmus verursacht wird,
wie dies beim allmählichen Mattwerden
des Rückenmarkspräparates nicht selten
auf einige Zeit vorkommt. Im Augenblick
in dem der dominan-
te Rhythmus sistiert,
fallen die durch Super-
position verursach-
ten Unregelmäßigkeiten in der Kurve
des abhängigen weg, d. h. die Schwan-

Figures 11 and 12

the dependent rhythm disappear. In other words, from this point onward the tail fin produces an entirely regular tracing, reflecting its now independent rhythm.

For our purposes, one fundamentally important fact can be derived from all of these phenomena. The endogenous, automatic-rhythmic process of stimulus production generates its centrally coordinated sequences of impulses *continuously and without interruption.* In contrast to an intact organism, the ventral nerve cord of the earthworm will continue to "creep" and the spinal cord of an eel will continue to "swim" *uninterruptedly* until the prepared specimen begins to die off. The influence exerted by the governing functions of the central nervous system on such automatisms is obviously limited to *maintaining inhibition of* their effects on the musculature. Only in cases in which the motor pattern concerned is biologically appropriate does *disinhibition* occur, such that the way is opened for the endogenous rhythms to exert their motor effects. The brain or any governing center has no influence of any kind over the *form and coordination* of the centrally coordinated sequences of impulses and is unable *to change them in any way.* We shall shortly return to the significance of these conclusions. In the general part of this book, a considerable amount of space will have to be devoted to von Holst's findings. In particular, we shall need to discuss in detail the conclusions that led von Holst to infer the existence of some *material* form of response-specific arousal quality that is continuously generated by the endogenous automatism and is consumed by the performance of the motor pattern itself. This inference is completely identical to that which I had derived from the cumulative properties of complex instinctive motor patterns and provides a precise physiological basis for it. From what has already been said, one can appreciate the extent to which von Holst's investigations offer genuine physiological *explanations* for all of the properties and characteristics of instinctive motor patterns that have been discussed in previous chapters.

In the first place, a plausible explanation is provided for the conspicuous *inflexibility* and *individual constancy* of the instinctive motor pattern, which so many authors with vitalistic leanings repeatedly attempted to question and which, on the other hand, seduced so many mechanistically thinking investigators into interpreting instinctive motor patterns as chain reflexes. Demonstration of endogenous automatisms that are *coordinated in the center itself*—which von Holst provided for very many, extremely complex motor patterns of various animals such as the entrenching movements of the anglerfish (*Lophius piscatorius*)—can readily provide an explanation for a whole series of facts that clearly conflict with the reflex chain theory! The fact that a motor pattern is sometimes performed with a natural object and sometimes with a quite inappropriate substitute object (or even as a vacuum activity), but in both cases shows *photographically identical* coordination, becomes self-evident if these coordination patterns are already delivered in *finished form* by the central nervous system transmitting the impulses. If, by contrast, the motor sequence were a chain reflex, it would be barely credible that the component reflexes would not include at least some that were influenced by the object. To take just one example: With a weaverbird, the entire, complex motor sequence with which a plant fiber is firmly tied to a branch is performed in precisely the same way *with* or *without* an object. This clearly shows that *within* the motor sequence itself there can be *no reflex* that is *controlled by stimuli emanating from the object.* Only the release—or, as we can now say, the *central disinhibition*—of the instinctive motor pattern as a *whole* is in many cases brought about by unconditioned reflexive processes, as we have already seen in the discussion of the innate releasing schema. For anybody who, from oft-repeated observations, is familiar with the absolute "photographic" fidelity with which the normal performance matches the vacuum discharge of an instinctive motor pattern, and for anybody who has racked his brains for long enough over this extremely peculiar phenomenon, von Holst's demonstration of the fact that *receptors are not involved* in determining the form and coordination of the motor patterns is a veritable revelation. It provides an immediate and utterly convincing explanation for a previously quite puzzling phenomenon.

The assumption that instinctive motor patterns are based on endogenous processes of stimulus production also explains in an equally plausible fashion the entire set of phenomena associated with *spontaneity.* This is true both with respect to drives and appetite on the subjective psychological side and with respect to threshold lowering and the eventual eruption of the motor pattern as a vacuum activity on the objective physiological side. In this case, too, a phenomenon that previously posed a paradox and that many simply did not want to believe abruptly becomes self-evident. Indeed, if it were not already known as a phenomenon, its existence would be virtually a theoretical necessity! The higher centers exert only *inhibitions* on the motor effects of central automatisms and releasing mechanisms represent only the *elimination* of these inhibitions. With progressive damming, internal accumulation of the continuously generated arousal that is specific to a particular motor pattern makes it *increasingly difficult* for the governing centers to *maintain the inhibition!* It is therefore hardly surprising if the threshold value for the releasing stimuli decreases with the duration of internal accumulation and if the arousal eventually breaks through all the dams to be discharged as a vacuum activity.

The fact of endogenous stimulus production can also provide certain explanations with respect to the subjective phenomena of purposive behavior. The assumption of a *material* accumulation of response-specific arousal energy makes it comprehensible that, following an extended period of damming, even a quite special behavioral component that in itself certainly does not reflect activity of the *entire* nervous system will lead to arousal of the organism *as a whole* and then *drive* it to search actively for stimuli. Precisely such a "driving" effect of certain substances has long been known to occur with many endocrine glands. Indeed, the so-called *hormones* (from the Greek ὁρμάω, I drive; ὁρμων, the driver) owe their name to this particular function. In the general part, we shall need to examine in more detail the functional parallels between hormones and response-specific arousal qualities that are accumulated through endogenous processes of stimulus production in the central nervous system. This will be done in the discussion of the "motivational hierarchy" and its investigation by Tinbergen and Baerends.

Undoubtedly, however, the greatest explanatory power, and hence the greatest support for the assumption that there is a fundamental identity in nature between instinctive motor patterns and the automatic processes of stimulus production discovered by von Holst is provided by the correspondence between the two that we shall now discuss. Down to the very last detail, the entire set of related phenomena—such as the intensity scale of instinctive motor patterns belonging to one quality of response-specific arousal, the process of accumulation of response-specific energy, and its consumption through the performance of the motor pattern—corresponds to the phenomenon of *spinal contrast* discovered by Sherrington and, in a way, rediscovered by von Holst. Let us take a closer look at this phenomenon, which—as the name suggests—is located in the spinal cord. With fish such as the wrasse, perch, and the like, which are almost always freely suspended in the water and must therefore beat their fins day and night almost without interruption, the spinal cord must be able to deliver the endogenous impulses for these movements in an indefatigable manner. The same applies to the swimming movements of fish that, at least on occasion, must have the capacity for major *feats of endurance* in swimming, as is the case with the eel. Thus it is that an isolated spinal cord preparation of such species will perform the movements concerned incessantly. By contrast, with fish such as the sea horse (*Hippocampus antiquorum*) that remain immobile virtually continuously and in their normal lives perform swimming movements for only a few minutes every day, an isolated spinal cord preparation will generally *not* perform these movements *at all.* After transection of the medulla, the pectoral and dorsal fins, which are the only locomotor organs of this diminutive fish, remain continuously immobile. However, von Holst observed a small but significant difference between the immobility of the fins in such a preparation and their resting condition in the intact, immobile fish. Whereas the dorsal fin is *tightly folded* against the back of the intact, resting fish, in the preparation it is kept in an *intermediate position,* halfway between this resting position and the maximally unfolded position with which the wavelike swimming movements are performed. Now, through the application of *additional stimuli* that evidently have a similar effect to the central inhibition that is lacking from the preparation, it is possible to induce the fin to adopt the same, completely folded resting position as is found with the resting, intact fish. The simplest way of bringing this about is to apply moderate pressure to the "neck region" of the fish's body. If such a "fin-folding" stimulus is allowed to take effect for a certain time and is then removed, the dorsal fin is extended *further* than the intermediate position that was previously maintained. This is why Sherrington used the term *contrast.*

If the inhibitory stimulus that leads to folding of the fin is now allowed to take effect for an extended period, once it is removed the fin not only unfolds to its full extent but also begins to perform the typical wavelike swimming movements for a short time. After this movement has ceased, the fin gradually returns, initially more rapidly but then more and more slowly, to its "half-mast position." The explanation that von Holst provided for this phenomenon must also be easily understandable to the reader on the basis of the ideas that we have developed, with quite different objects and by means of quite different experimental approaches, concerning the now-familiar phenomena of accumulation and disinhibition of response-specific arousal. The *raising* of the fin and its *wavelike beating* are nothing other than two instinctive motor patterns with different thresholds that are activated by the same quality of action-specific arousal, which they also *consume.* In this, they are like the patterns of fin spreading, gill flap spreading, and tail beating of a fighting male *Astatotilapia.* Corresponding to the restricted "need for swimming" that the sea horse normally has, the continuous production of stimuli is so limited that it is barely sufficient to cover the consumption required for a few minutes of swimming per day. If the generation of arousal is *directly* translated into movement, that is, without any accumulation, it only "feeds" moderate raising of the fin. In a fish in which the spinal cord has been separated from the brain, the entire production of stimuli "leaks out" in the absence of any inhibition. As a result, the "current level" of response-specific energy can never be dammed up sufficiently to generate the response intensity corresponding to the wavelike beating movements of the fin, even with complete disinhibition. The level of spreading that the dorsal fin maintains in the fish operated on corresponds to a quite specific degree of intensity of the response, namely, that at which the *consumption* of response-specific energy exactly *balances* its continuous *production!* Only if the central inhibition that operates in the intact fish is substituted by an additional stimulus is "leakage" of the response-specific energy prevented. This energy is then *stored* so that even with complete disinhibition (corresponding to full "opening of the tap" in the analogy presented on p. 283), current values are reached that correspond to the intensity level of the wavelike beating movements of the fin. When the tap is fully opened, the jet projects to reach the mark for this motor pattern instead of merely dripping as before. However, such projection of the jet is very short-lived because the limited continuous supply is by no means sufficient to balance consumption at this level.

Accordingly, even quite rare motor patterns that occur extremely *discontinuously* in the intact organism can be based on *continuous* endogenous stimulus production. Recognition of this provides a ready explanation for all of the phenomena of response-specific fatigue that we have discussed on p. 280. But, as can be seen, the entire set of phenomena connected with spinal contrast is simply *identical* with the set of phenomena connected with accumulation and consumption of response-specific arousal with instinctive motor patterns. We are confronted with an extremely rare case in which a phenomenon demonstrated with invasive methods on an isolated part of the nervous system is completely identical in nature to the behavior of the intact organism. Our conviction concerning the factual and causal identity of the processes discussed here is greatly reinforced by an important historical fact. I developed the ideas presented on p. 271 et seq. concerning the accumulation and consumption of response-specific energy without any knowledge of von Holst's results. Conversely, at the time when he was conducting the studies of spinal contrast reported here, von Holst was also unaware of my work. The completely independent way in which the results were obtained surely makes their fully consistent agreement particularly valuable!

A further circumstance that provides extremely convincing support for identity in nature between instinctive motor patterns and endogenous-automatic, centrally coordinated rhythms lies in the complete correspondence of their reciprocal relationships with hierarchically higher functions of the central nervous system. We have already become acquainted with one part of these relationships in the discussion of phenomena surrounding *spontaneity* (p. 295 et seq.). The limited influence that the hierarchically higher center exerts on both the automatism itself and on its motor expression explains a number of constitutive properties of the instinctive motor pattern. It is necessary to have a clear grasp of these relationships in order to understand what we refer to as the *refractoriness* of animal and human instinctive motor patterns. Like a mechanic operating a steam engine, the hierarchically higher function of the central nervous system can, so to speak, do no more than control the "tap" of the response. A mechanic cannot make any modification to the form and coordination of the movements performed by the engine; he can only lower or raise the supply of steam to the cylinders. Indeed, he can only exert a limited influence on the production of steam by the boiler. He cannot switch it off at will, nor can he increase it without constraint. In exactly the same way, the central functions of the nervous system exclusively determine whether and to what extent an automatism is *disinhibited*. They cannot increase or decrease the continuous production of response-specific arousal, nor can they produce the slightest modification in the form and coordination of the endogenous impulses. Even in cases where the disinhibition mechanisms of instinctive motor patterns are directly governed by the highest centers of all, which we refer to collectively as "will" because they are provisionally not amenable to analysis, these centers are not able to compensate for any deficit in the current level of response-specific arousal. However, most of the instinctive motor patterns that are directly governed by will, among which locomotor patterns are particularly prominent, are so richly supplied with internal stimulus production that this limitation of will hardly ever manifests itself. As with the wrasse and the eel, the rhythms concerned generate such a rich supply of impulses that the motor pattern is immediately "available" *at any time* that the "will" may "decide" to disinhibit it.

On the other hand, however, there are also numerous instinctive motor patterns that are "voluntarily" disinhibited, including some involved in locomotion, for which stimulus production generates only limited quantities of the relevant arousal quality. In such cases, a relatively long period of *damming* is an indispensable requirement for temporally restricted performance of the motor pattern concerned, as we have already seen with the example of the beating of the dorsal fin of the sea horse. With such instinctive motor patterns, it can occur even in the intact organism that the motor pattern concerned fails to "fire" despite complete disinhibition from the center. In other words, anthropomorphically speaking, the animal concerned "wants to but cannot"! This contrast between continuously available motor patterns and those that are only possible after a given period of accumulation can be found with the *same,* homologous instinctive motor pattern of closely related animals, such as the flight motor patterns of various birds. Small birds that fly frequently always have the motor pattern available in an unlimited quantity, such that it really appears to be completely voluntary. By contrast, large birds that fly less often, such as ducks and geese, are noticeably dependent upon the current level of response-specific energy. One can conduct an experiment in which such a bird is tested immediately after it has flown for an extended period. In the test, the bird is required to fly over an obstacle, for example, by placing it in a small unroofed enclosure, from which it will immediately try to escape. The bird's behavior clearly indicates that it has full insightful control of the situation. It will immediately exhibit directed intention movements for flight, but in some cases it can literally take hours before the bird has built up the necessary arousal intensity for it really to fly up and out of the enclosure. It is precisely in such cases that one gains the impression that intention movements have a certain stimulatory effect. A fresh bird of the same species that is not "flown-out" will fly over the obstacle at once, just like a small bird with "fully voluntary" (i.e., always available) flight motor patterns would do. This form of restriction on "voluntary" behavior is quite common among animal motor patterns and the boundary between voluntary and involuntary movement is by no means clear-cut. The only ready explanation for this phenomenon resides in the relationship discussed above between the automatism and the governing inhibitory centers. For this reason, it provides a further, very convincing argument for an absolute identity in nature between the automatism and the instinctive motor pattern.

The outline presentation of Erich von Holst's findings presented here is quite sufficient to demonstrate to the reader their simply enormous importance for comparative behavioral research. All that remains, therefore, is to point out the inestimable importance of the discoveries of endogenous, automatic-rhythmic stimulus production and central coordination in a much more *general* context. They require a quite fundamental upheaval in the notions that previously dominated, and to some extent still dominate, not only in physiology and its special subdisciplines but also in materialistic approaches to psychology, with respect to the functioning of the central nervous system. This is because a *completely new, previously unrecognized elementary function* of the nervous system has now emerged alongside the *reflex process* as an entirely equivalent and equally significant phenomenon. Previously, the latter had been seen unequivocally and without contradiction as the *only* basic function of the brain and spinal cord. All reflexological hypotheses that have to do with the central nervous system are affected by von Holst's discoveries down to their deepest foundations and therefore require certain important amendments. So far, the instinctive motor pattern is, admittedly, the only process previously interpreted as reflexive in nature whose explanation has been shifted to an entirely new basis through the discovery of automatic stimulus production rhythms. However, it is to be expected that progress in physiological analysis of the more complex functions of the brain that are accompanied by experiential processes will reveal a much *wider* occurrence of stimulus production processes. In particular, the highest conscious and mental processes of human beings show such a high degree of *spontaneity* that it seems to me quite likely—on a purely speculative basis at present—that processes of endogenous stimulus production play a decisive role in them as well. In the next chapter, we discuss the extent to which endogenous processes of stimulus production play a part in simpler mechanisms that have so far been interpreted as reflex processes. We shall also examine how their discovery calls for a further analysis of the phenomena concerned.

22

Implications for the Analysis of Related Phenomena

To round off this outline to comparative behavioral research, centered around an account of its historical origins and methods, it only remains for us to take a brief look at certain implications of the findings that have been presented. Recognition of the endogenous-automatic nature of instinctive motor patterns, in particular, has been of direct relevance to the analysis of various functionally related phenomena. In the course of holistically appropriate analysis on a broad front, every new finding, every correction of a previous mistake always *indirectly* necessitates revision of preexisting notions regarding many other phenomena involved in the system. This is true even when the finding *directly* affects only a *single,* sharply defined phenomenon within the complex system of the organic entity. This is not very different from the situation involved in solving a crossword puzzle. The correction of a single word that has been entered entails the correction of others that are directly or indirectly connected with it. A small advance in knowledge concerning just a single component of the organic entity frequently has additional effects on our understanding of many other components and opens up unexpected possibilities for progress with analysis in apparently quite distant areas.

The problem that most urgently required clarification following the discovery of the endogenous-automatic nature of instinctive motor patterns was the question of its relationships to the *reflex.* Of course, discovery of a new explanatory principle should not mislead us into *underestimating* the importance of previously known laws, despite the fact that their range of application has been *curtailed.* There are two processes that are undoubtedly constructed from reflexes, in particular, that have a very close functional connection to instinctive motor patterns. Indeed, without their participation there would be scarcely a single endogenous automatism able to serve its species-preserving function. These processes are, firstly, *the innate releasing schema,* and secondly, those mechanisms for spatial orientation that are referred to as *taxes,* following Alfred Kühn.

Let us first of all consider the innate schema. Recognition of the automatic nature of the instinctive motor patterns has contributed in several ways to a deeper understanding of its operation. Apart from the few cases in which the disinhibition of genuine automatisms is directly governed by "will" (p. 299), it is almost always brought about by genuine innate releasing mechanisms whose unconditioned reflexive nature was obvious from the outset (p. 272). However, as long as the instinctive motor pattern itself was regarded as a chain of unconditioned reflexes, there was no reason to make a conceptual distinction between the releasing process and the rest of the behavioral sequence, as it was quite obviously just the first link. In the naive form of expression used in my first publications, this view is reflected in terms such as the "rattling reflex" and the "following reflex" of the jackdaw, and suchlike. Long *before* the automatic nature of the instinctive motor pattern was clearly recognized, double quantification of the stimulus and responsiveness, as discussed in chapter 20, had led to a clearer conceptual separation between the released *motor pattern* and its "lock" located in the afferent arm of an unconditioned reflex. The notion of a quite independent releasing mechanism, sharply distinguished in causal physiological terms from the motor pattern, that I envisaged in the concept of the innate releasing schema was understandably considerably clarified by von Holst's findings on the relationship between the automatism and the inhibitory/disinhibitory center that governs it. The *immediate* effect of the innate schema is doubtless exerted not on the automatism itself but on the governing, inhibitory center of the central nervous system, where it brings about partial or complete abolition of the inhibition. Recognition of the fact that the inhibitory center intervenes *between* the innate releasing schema and the disinhibited motor pattern provides a causal physiological basis for clear conceptual separation of the receptor mechanism of the schema from the instinctive motor pattern that it releases.

Clarification of the physiological peculiarity of the instinctive motor pattern also contributed in another significant way to recognition of the functional, and doubtless physiological, independence of the innate schema. Sharp separation of endogenous automatisms from *other kinds* of behavior patterns that are similarly innate and had previously *also* sailed under the flag of "instincts" revealed an important fact: there are many innate schemata that do *not* disinhibit instinctive motor patterns in this new, exactly defined physiological sense. Instead, they do really release responses, particularly taxes, in a purely reflexive manner. Alternatively, they may give rise to specific *inhibitions.* Striking examples of the latter are provided by the quite common schemata that respond to *submissive postures* of conspecifics in social animals, arresting the most intensive attack responses with the typical predictability of an unconditioned reflex. Those remarkably selective mechanisms of stimulus filtering that we refer to as innate schemata reveal a high degree of functional independence through the manner in which they can be built into the action system at a *completely different place.* Evidently, *the same* physiological mechanism that functions as the "lock to the response," which we first encountered as the releasing process of instinctive motor patterns, can also serve as the "lock" to all other conceivable kinds of behavior patterns!

We have already explained in detail how much our understanding of the *summatory effect of stimuli* in the innate schema *depends on* our recognition of *cumulative processes of response-specific energy* and how the two developed hand in hand. It is therefore presumably unnecessary to discuss any further how the demonstration of endogenous stimulus production, by virtue of its physiological explanation of just those processes of accumulation, also provides an exact, causal-analytical foundation for our understanding of the innate schema, particularly with respect to quantification of its releasing, or disinhibiting, function. [Alfred] Seitz's meticulous demonstration of constant proportionality between stimulus strength and response intensity may be regarded as one of the first fruits of this new understanding.

The discovery of endogenous automatisms also obliges us to undertake certain analytical steps and to *modify* our previous conceptual approach with respect to *taxes* or orienting responses. In fact, on closer examination, all of the behavior patterns that have been labeled "instinctive motor patterns," which are actually largely based on automatisms, proved to be guided by the participation of spatial orientation responses that are surely genuinely reflexive in nature. Conversely, most of the behavior patterns that we have customarily labeled as taxes prove on closer inspection to be intermeshed with automatic processes. We shall first of all consider the first of these two possibilities involving a functional entity combining an instinctive motor pattern with an orienting response.

Because of its physiological peculiarity, the centrally coordinated sequence of impulses that is generated by an endogenous rhythm is, in itself, completely fixed and unaffected by external stimuli. To take a medical analogy, it would be rather like a human suffering from tabes dorsalis, in which reflexes are entirely lacking because of destruction of all of the dorsal, afferent nerve roots of the spinal cord. An organism would continually stumble over and collide with features in its spatial environment if there were no spatially orienting reflexes *superimposed* upon the impulses of the endogenous automatism. These intervene like an elastic buffer between the *fixed* coordination of the automatisms and the equally unyielding structure of the spatial environment. Erich von Holst therefore speaks of a *mantle of reflexes* that surrounds the fixed, automatic motor sequences from all sides. Through unequivocal observations and decisive experiments, he has demonstrated that in one and the same motor pattern there may be *superimposition* of a reflexive motor impulse on that from an endogenous automatism. We have already seen an example of this in the tracing shown on p. 292, although it is not clear from this whether the superimposition represents a real *addition* or, in the opposite direction, a *subtraction* of the two impulses coming from two such different sources. In many cases, a minor "addition" or "subtraction" in the movement, produced in a reflexive manner, provides for an adequate *adaptation* of the automatically produced sequence of impulses to the spatial conditions of the environment. For example, the leg movements of a running dog are demonstrably produced by endogenous and centrally coordinated automatisms. But every slight unevenness in the substrate releases, through the optical channel, small reflexive supplementary impulses. These bring about an adaptation of the individual leg movements by producing a small increase or decrease in the degree of extension of each leg. Even the simplest reflexes that we know, the "short" tendon reflexes of the muscles of the extremities, serve fundamentally similar functions. They are released whenever there is *sudden* stretching of the muscle concerned. In this way, they inhibit all "involuntary" deflecting movements, both with passive bending of the extremities following a sudden encounter with some resistance and with overshooting of a movement whose power impulse is calculated for contact with a resistance and "unexpectedly" encounters a void.

Accordingly, the mantle of reflexes that intervenes as a compensatory buffer between the automatism and the outside world, the reflexive movements that are superimposed on the fixed performance of the instinctive motor pattern—as in the above-mentioned example of slight raising or lowering of the legs of a trotting dog—by no means represent any *modification* or "plasticity" of the instinctive motor pattern itself. Instead, they simply involve an *addition* or *subtraction.* As von Holst has put it in a neat analogy, one could with equal justification speak of "plastic modification" of the form of a house roof when its external contours are changed and rounded off with a covering of snow in the wintertime.

The *extent* of superimposition of reflexes on instinctive motor patterns, one could say the "thickness" of the mantle of reflexes, shows marked *differ-*

ences between different instinctive motor patterns and, indeed, between *homologous* instinctive motor patterns of different animal species. With animals that live in the open sea, and also with many steppe-living forms, such superimposition is limited to relatively very small deflections of movement and their control through reflex-releasing stimuli is also very coarse and "approximate." With the walking movements of the great bustard (*Otis tarda*), which is a typical steppe-living animal, the range of influence of orienting reflexes is so limited that even the smallest obstacles bring about maladaption of the overall movement, with the result that the bird stumbles. This big, long-legged bird lacks the physiological and psychological capacity to raise one leg in a controlled fashion to walk over an obstacle such as an iron lawn fence just 10 inches high. Even after innumerable collisions, it never learns to do this. The walking movements of many animal species that inhabit flat expanses of steppe are similarly poorly supplied with guiding reflexes. By contrast, the extent and complexity of reflexively controlled adjusting movements can be enormously greater in quite closely related species that live in habitats with marked spatial structuring. Steppe antelopes stumble over small obstacles almost like a great bustard, whereas mountain-living antelopes, such as the chamois, are able to superimpose almost any required reflex-controlled movement of any magnitude on the impulse sequences of walking, trot, and gallop. As a result, every single step with any hoof is precisely *guided* to a suitable spot, as is illustrated by a chamois galloping across an expanse of boulders. In a special section of this book, it will be shown that there is a remarkably close relationship between the advanced development of such capacities for spatial orientation and the origin of the "higher" capacity for insightful behavior. Despite its naiveté, one's immediate impression that a great bustard repeatedly stumbling over a lawn fence is a particularly "stupid" bird proves to be completely justified following detailed comparison with a wide variety of animals. The reflexes controlling the walking movements of a great bustard, which are barely capable of producing a slightly greater elevation and forward swing above a stone lying in the way, represents one end of a spectrum. The range extends through all conceivable transitional stages up to the highly complex spatial representation of an anthropoid ape, which is still clearly based on taxes. In an anthropoid ape, literally every pace and every grasp is controlled by orienting responses and the automatism itself represents, so to speak, no more than an "inner skeleton" underlying locomotor movements.

The same features as those just described for automatisms involved in locomotion apply in a general fashion to most instinctive motor patterns. There are only a few, quite exceptional cases in which an actual "mantle of reflexes," in the sense of a layer of *superimposed* reflex movements is *lacking.* We are closely familiar with two forms of such "naked" automatisms. In the first place, there are some cases in which spatial orientation, a kind of *aiming,* is temporally separated and takes place *before* the performance of the instinctive motor pattern. Secondly, there are cases in which instinctive motor patterns and reflexive orienting movements are spatially separated in the animal's body, influencing the movements of different groups of muscles. For these two special kinds of interaction between the instinctive motor pattern and the taxis, we have introduced the term of *interlocking* of the two different kinds of motor pattern—in contrast to the phenomenon of superimposition discussed above. For easily understandable reasons, both represent particularly suitable objects for experimental analysis. Instinctive motor patterns that are free of superimposition because they are temporally preceded by an aiming taxis include very many of the *stimulus-transmitting* instinctive motor patterns to which innate schemata of conspecifics respond in an appropriate manner. The courtship patterns of many birds and bony fish provide us with the most conspicuous and longest-known examples of such motor patterns. These releasing "signals" operate purely *visually* and do not require any spatial adjustment to the location and form of an object with which *mechanical* interaction takes place, as is the case with the great majority of other kinds of instinctive motor patterns. For this reason, with such signaling motor patterns, the participation of an orienting response is limited to appropriate orientation of the animal's body *before* the motor pattern is released, such that the *operative* side of the body is turned toward the conspecific that is to receive the signal. In the jewelfish (*Hemichromis*), the motor pattern involving spreading of the gill covers is "calculated" for an anterior view, so the head is turned toward the partner. The same applies to a peacock displaying its tail feathers, to a *Cichlasoma meeckei* spreading its gill flaps, and to many similar displays. In the courtship of a fighting fish (*Betta splendens*), in a fin-spreading male *Astatotilapia,* in a courting golden pheasant (*Chrysylophus cristatus*) and in many other such cases, prior to performance of the releasing instinctive motor pattern, the body is oriented in such a way that the side of the body is held *perpendicular* to the partner's gaze. The motor pattern that then follows, which always generates a visual effect of conspicuous characters in coloration and body form, is *completely free of superimposed reflex movements,* apart from orientation to gravity, which of course continues to function during performance of the motor pattern. As has already been indicated elsewhere (p. 242), it is *historically* extremely interesting how Whitman and Heinroth, in their fundamental systematic studies, above all made use of precisely these simplest limiting cases of *pure* (i.e. taxis-free) instinctive motor patterns. This is utterly characteristic of the astonishing refinement of the physiological and systematic judgment of these two investigators. Other cases of instinctive motor patterns that follow on from a temporally separated orienting response are generally rare. One typical example that is familiar to everyone is provided by the *snapping* movements of many bony fish, amphibians, and reptiles, which similarly follow on from "targeting" taxes without any further participation of reflexive guidance.

A conveniently analyzed example of the second case of "interlocking" of the instinctive motor pattern with a taxis, in which the automatism and the reflex govern different muscle groups of the animal concerned, is provided by the *egg-rolling response* of geese and ducks. In this motor pattern, which serves for the retrieval of an egg that has rolled out of the nest hollow, the automatism and the taxis operate together in a quite peculiar fashion. The movement that takes place along the *median* or sagittal plane of the bird's body is purely automatic and centrally coordinated in a fixed pattern. By contrast, the lateral movements that take place *perpendicular* to this plane, which balance the egg on the tip of the beak by preventing any deviation to the right or left and thus steer it back into the nest hollow, are demonstrably brought about by orienting reflexes. The automatic instinctive motor pattern thus operates rather like a horse whose direct forward motion can be steered to the left or right by reins on either side. Taking this particularly favourable object, Tinbergen and I carried out a detailed experimental analysis of the interaction of the automatism and the taxis. This is discussed more fully in the general part of this book.

As has already been indicated, clear causal-physiological formulation of the concept of the instinctive motor pattern thus leads us inevitably to the conclusion that many, indeed the great majority of the behavior patterns that are still traditionally referred to as instinctive motor patterns, are, properly speaking, complex interactions between automatisms and reflexes. Some of them are of the type involving *superimposition,* while others are of the *interlocking* type. In most cases, the functional unit of behavior involves a finely balanced interaction between two complementary processes that are entirely different from one another in causal-physiological terms. Of course, recognition of this fact *provides no argument against a clear and precise separation of these processes.* Vitalists such as Bierens de Haan (1941) [1940] have repeatedly raised the objection that this conceptual separation is an infringement against the *holistic nature* of organismal behavior. This is just as unfounded as if somebody were to object to the conceptual separation between nerves and muscles on the grounds that the two always occur together as a functional unit and that each is simply unable to operate without the other! In what follows, we shall encounter a large number of concrete observational facts that forcefully demonstrate the necessity for this clear conceptual separation and the fruit that it has borne. Above all, we shall see how the gradual phylogenetic increase in orienting responses *at the cost* of fixed automatisms leads, through a progressive evolutionary process, to spatially insightful behavior and mental spatial representation in humans.

A sharp conceptual discrimination of endogenous-automatic components in animal and human behavior was made possible by von Holst's results and hence became obligatory for any investigator concerned with analysis. However, this is not only a necessity with respect to behavior patterns that are largely composed of instinctive motor patterns, as has just been explained. It is equally necessary for processes that consist largely of orienting responses. The great majority of taxes, most particularly those that are labeled as "positive" and "negative" because they induce the organism to move toward or away from a specific stimulus, almost always include automatisms. This is so because motor patterns of *locomotion* are, of course, always involved in such processes. As we have already seen from a variety of examples, the latter are almost always based on endogenous automatisms. Take, for instance, a tadpole that becomes "positively phototactic" when it lacks oxygen. In other words, it swims toward the light under the directional control of stimuli. The most important component of this behavior is doubtless represented by the reflexive movements which *steer* the organism toward the light and whose *extent* is controlled by stimuli. But in addition, a parallel part is also played by central disinhibition of the locomotor automatism that generates the animal's *forward motion.* A graphic illustration of the mechanism underlying the overall process is provided by the "positive Americotaxis" of an ocean-going steamship. Coordination of the movements of the forward-propelling machine, involving a fixed interaction between pistons, cams, connecting rods, valves, and so on, is in itself invariable. The central organ, the commander of the vessel, only has the power to throttle back the machine or unleash it. Even when running at full power, the machine cannot perform more rotations than is permitted by the steam produced in the boiler, which corresponds to endogenous stimulus production in our analogy. In contrast, the taxis—corresponding to the man at the wheel of the steamer—responds to *external* stimuli such as constellations, landmarks, deflections of the compass, and so on with small *turning movements* whose angular magnitude depends in a directly "reflexive" manner on the angle of incidence of the stimuli concerned. The nature of what we refer to as a *topical response* or taxis in the narrower sense, following [Alfred] Kühn, resides in such *quantitatively controlled* small turning movements [of an organism] upward or downward, to the left or the right. In this way, many processes that we generally refer to as taxes prove on closer examination to involve an interlocking, functional, interaction between reflexive and automatic components. However, the structure of this interaction is, in fact, fundamentally different from that found with the instinctive motor patterns interwoven with taxes that we have discussed above. The traditional distinction between instinctive motor patterns and taxes, still in use today, does actually reflect a fundamental distinction, but in both cases *both* components—the automatism and the reflex—play a decisive part.

The influence exerted by the discovery of endogenous stimulus production on our notion of the *conditioned reflex* was somewhat different from that exerted on conceptualization and analysis of the *unconditioned reflex* mechanisms of the innate releasing schema and the orienting response. On the one hand, it presented a quite different explanation for a vast number of processes that had been interpreted as reflexes by the great mechanistic schools of behavioral research. Compelling evidence banished forever the reflex, and particularly the conditioned reflex, from its previous dominance as the exclusive explanatory principle of animal and human behavior. On the other hand, however, it led us to an analysis of extraordinarily close functional *relationships* between the conditioned reflex and the endogenous-automatic instinctive motor pattern. Indeed, closer inspection of these relationships may well reveal that they provide a promising basis for a broader and deeper understanding of the conditioned reflex itself. As we have already seen in the chapter on the mind-body problem in the Biological Prolegomena [chapter 9] (p. 162 et seq.), very close relationships exist between the capacity for development of conditioned reflexes and the presence of *purposive behavior* in general, but most particularly *appetitive behavior* directed at the release of instinctive motor patterns. It was analysis of appetitive behavior promoted by recognition of the endogenous-automatic nature of the instinctive motor pattern that first led us to discovery of the following rule, the few exceptions to which are in all probability only *apparent: The only unconditioned reflexes that can serve as a basis for the development of conditioned reflexes are those that initiate purposive behavior of an organism.* Unconditioned stimuli which do not lead to purposive behavior, which are not the object of appetite or aversion in Craig's terminology, *can never be replaced by a conditioned stimulus!* Even if a very conspicuous signal is given thousands of times before, say, the release of the knee tendon reflex in a human being, this will never lead to development of a conditioned reflex that permits release of the tendon reflex by the acquired substitute stimulus. The few and—as we shall show—probably only apparent exceptions to this important rule are discussed in the general part.

The development of conditioned reflexes is thus associated with the requirement that purposive behavior must be present. Conversely, therefore, all purposive behavior, which originated from the simplest limiting cases of positive or negative taxes and blind seeking according to the principle of trial and error, *only acquires its essential, species-preserving function through the development of conditioned reflexes!* After all, this species-preserving function of all highly developed, purposive behavior resides in the individual capacity for adaptation to variable environmental conditions, which in turn depends on the organism's capacity for development of conditioned reflexes, in other words, on its *learning ability.* When seen from the experiential side, this close reciprocal functional dependence between the conditioned reflex and appetitive behavior is expressed in the following way: the existence of a purpose in a subjective, psychological sense—that is, the goal of *pleasurable feelings* associated with the stimulus situation that releases instinctive motor patterns and with their performance, and also feelings of aversion that are evoked by a disturbing stimulus that is to be avoided—doubtless serves its species-preserving function as a *conditioning factor.* It does so by providing motivation in the form of reward and punishment for subjective modification of behavior that is also purposive in the sense of promoting survival of the species. There is a crude, naive psychological interpretation that the bait of pleasure and the whip of aversion provide the indispensable precondition for the occurrence of any learning or conditioning. A highly significant confirmation of this interpretation is provided by the objectively recognizable close connection between the conditioned reflex and purposive behavior.

23

Conclusions

The foregoing, condensed account of the historical development and conceptual and practical methods of comparative behavioral research is presumably sufficient to serve its intended purpose. This is that of providing the reader with an outline sketch of our overall field of research and of ensuring sufficient familiarity with its theoretical approaches and concepts for an understanding of the immediately following special part of this book. But description of the path that led first to the discovery of instinctive motor patterns and eventually to an understanding of their specific physiological laws is also sufficient to demonstrate something else. Holistically appropriate analysis on a broad front, to which alone we owe our understanding of these laws, by no means leads us to *overestimate* the range of their applicability. We are by no means tempted to attribute an exaggerated significance to the explanatory principle that has been discovered and certainly are not inclined to declare that it is *exclusive* in the way that can so easily happen with the atomistic approach to research. Instead, precise extrication of the instinctive motor pattern as a quite peculiar physiological phenomenon has forced us to recognize that there are in addition several other processes that are physiologically entirely different. All of these processes together are involved, in an extremely complicated interaction, in the construction of the harmonic functional entities that we refer to as the *behavior patterns* of animals and humans! It is also far from our intention to claim that the mechanisms whose interaction is accessible to analysis in the present state of our knowledge are the *only* "elements" involved in the construction of the entity represented by the behavior of any higher organism. In no way would we maintain that *apart* from instinctive motor patterns, reflexes, taxes, innate releasing schemata, and conditioned reflexes, *there may not be other* psychophysiological processes with their own laws that are equally clearly identifiable but have so far completely escaped our attention. To the contrary, we believe that there must be a whole series of such processes, particularly at the highest and most complex levels of psychophysiological phenomena. Nevertheless, even at this stage we are fully justified in making a quite specific statement: In *the great majority* of animal and human behavior patterns that we observe everyday, *all* of the different kinds of individual processes that we have so far been able to analyze are involved *simultaneously* in harmonic interaction! The behavior patterns that we have cited as *examples* give a distorted picture of this situation. This is because the special instances that are particularly favorable for the analysis of any single mechanism among those discussed always represent limiting cases and exceptions. In such cases, the process concerned is relatively pure and lacks overlap or superimposition of other processes, and hence operates in a particularly *simple* fashion. Such extremely simple limiting cases consistently represent the "favorable objects" that originally indicated the *presence* of a previously unrecognized special law. The pure endogenous-automatic motor patterns of courtship discovered by Whitman and Heinroth play the same role in comparative behavioral research as that played in genetic research by the separation of characters in further breeding of hybrids that are heterozygous for just one inherited factor, which led Gregor Mendel to the discovery of his laws of inheritance.

Only certain exceptional cases in animal and human behavior, which are highly welcome for our attempts at analysis, are of simple construction, containing just one or very few components. The organic entity of the action system of any higher animal species, in contrast, is an immeasurably complex network of processes that are physiologically very different from one another. Recognition of this fact leads us back to the unconditional necessity for analysis on a broad front as the only method for investigation of organic entities. The entire history of development of comparative behavioral research, which is identical to the history of the discovery and analysis of the instinctive motor pattern, constitutes one long chain of evidence demonstrating this necessity. This entire history shows how an understanding of one part of the entity can be advanced only simultaneously and hand in hand with an understanding of all other parts. Conversely, every unraveling of a small segment of a *single* causal thread is always accompanied by a corresponding small advance in the unraveling of all the other threads that are interwoven with it. But the history of development of comparative behavioral research also shows something else: With this sole legitimate approach to analysis of the entity, we never end up with "pieces in the hand" that lack a "mental connection." Instead, we can see in wonderment, "how everything is woven tight, all acts and life as one unite!" On the basis of this knowledge, we can now confidently turn to a description of the organic entity itself, to a portrayal of the behavior of living organisms.

End of Part One

Bibliography

Entries marked with * are not cited in the text.

*Allen, F. H. The role of anger in evolution with particular reference to the colours and songs in birds. The Auk 51, 1934.

Altum, B. Lehrbuch der Zoologie. Freiburg 1883.

*Alverdes, F. Tiersoziologie. Leipzig 1925.

Antonius. O. Über Herdenbildung und Paarungseigentümlichkeiten der Einhufer. Z. Tierpsychol. 1, 1937.

———. Über Symbolhandlungen und Verwandtes bei Säugetieren. Ibid.

Baerends, G. P. Fortflanzungsverhalten und Orientierung der Grabwespe *Ammophila campestris.* Tijdschrift Ent. 84, 1941.

———. On the life-history of *Ammophila campestris iur.* Nederl. Akademie van Wetenschappen, Proceedings 44, 1941.

Baumgarten, E. Der Pragmatismus. Frankfurt/Main 1938.

Bell, Sir Charles. The Nervous System of the Human Body. 1830.

Bernard, Claude. Physiologie générale. 1872.

Bierens de Haan, J. A. Die tierischen Instinkte und ihr Umbau durch Erfahrung. 2 vols. 1940.

Brehm, A. E. Brehms Tierleben. 4th ed., Leipzig and Vienna 1921.

———. Gefangene Vögel. Leipzig and Vienna 1934.

*Brunswick, E. Wahrnehmung und Gegenstandswelt. Psychologie vom Gegenstand her. Leipzig and Vienna 1934.

Bühler, K. Handbuch der Psychologie, I. Teil: Die Struktur der Wahrnehmungen. Jena 1922.

*Buytendijk, F. J. J. Wege zum Verständnis der Tiere. Zürich and Leipzig 1940.

*Carmichael, L. The development in vertebrates experimentally removed from the influence of external stimulation. Psychological Review 33 (34, 35) 1926.

*Coburn, C. A. The behaviour of the crow. J. Animal Behaviour 4, 1914.

Craig, W. Appetites and aversions as constituents of instinct. Biol. Bull. 34/2, 1918.

———. A note on Darwin's work on the expressions of emotions etc. J. Abnormal and Social Psychology, 1921, 1922.

———. The voices of pigeons regarded as a means of social control. The American Journal of Sociology 14, 1908.

———. The expression of emotion in the pigeons: The blond ring-dove (*Turtur risorius*). J. Comp. Neur. and Psych. 19, 1901.

———. Observations of young doves learning to drink. J. Animal Behaviour 4, 1914.

———. Male doves reared in isolation. J. Animal Behaviour 4, 1914.

———. Why do animals fight? Intern. J. of Ethics 31, 1921.

Dacqué, E. Vergleichende biologische Formenkunde. 1921, 1922.

Darwin, C. Der Ausdruck der Gemütsbewegungen. Stuttgart 1874.

———. Das Variiren der Tiere und Pflanzen im Zustande der Domestication. 2 vols. Stuttgart 1873.

Dewey, J. Experience and Nature. Chicago and London 1925 (Open Court Publishing).

———. Reconstruction in Philosophy. New York 1936 (Henry Holt).

Driesch, H. Philosophie des Organischen. Leipzig 1928 (Quelle und Meyer).

Ehrenfels, C. v. Über Gestaltqualitäten. Vierteljschr. für wissenschaftliche Philosophie 14, 1904.

*Feuerborn, H. J. Der Instinktbegriff und die Archetypen C. G. Jungs. Biologia Generalis XIV, 1939.

Freud, S. Vorlesungen zur Einführung in die Psychoanalyse. Vienna 1930.

*Frisch, K. v. Über die Sprache der Bienen. Zool. Jb., Abt. Physiol., 40, 1923.

*Gehlen, A. Der Mensch, seine Natur und seine Stellung in der Welt. Berlin 1940 (Junker und Dünnhaupt).

Goethe, F. Beobachtungen und Untersuchungen zur Biologie der Silbermöwe (*Larus argentatus* Pontopp.) auf der Insel Memmerstand. Jb. f. Ornith. 85, 1935.

———. Beobachtungen und Versuche über angeborene Schreckreaktionen bei jungen Auerhühnern (*Tatrao u. urugallus* L.). Z. Tierpsychol. 4, 1940.

———. Beiträge zur Biologie des Iltis. Z. Säugetierkunde 15, 1940.

Hartmann, M. Allgemeine Biologie. Jena 1927 (G. Fischer).

Heinroth, O. Beiträge zur Biologie, insbesondere zur Ethnologie der Anatiden. Verh. d. 5 int. Ornith. Kongr., Berlin 1910.

———. Über bestimmte Bewegungsweisen bei Wirbeltieren. Sitzungsber. d. Ges. d. natforsch. Freunde, Berlin 1930.

——— and Heinroth, M. Die Vögel Mitteleuropas. 4 vols. Berlin 1928–1934 (Behrmüller).

Helmholtz, H. v. Handbuch der physiogischen Optik. 2 vols. 1856–1867.

Herder, J. G. Ideen zur Philosophie der Geschichte der Menschheit. Riga and Leipzig 1784 (Hartknoch).

Herrick, F. H. Instinct. Western Res. Univ. Bulletin 22.

———. Wild Birds at Home. New York and London 1935.

Holst, E. v. Alles oder Nichts—Block, Alternans, Bigemini und verwandte Phänomeme als Eigenschaften des Rückenmarks. Pflügers Archiv f. d. gesamte Physiologie 236, 1935.

———. Über den "Magnet-Effekt" als koordinierendes Prinzip im Rückenmark. Ibid 237, 1936.

———. Versuche zur Theorie der relativen Koordination. Ibid 237, 1936.

———. Vom Dualismus der motorischen und der automatisch-rhythmischen Funktion im Rückenmark und vom Wesen des automatischen Rhythmus. Ibid 237, 1936.

———. Neue Versuche zur Deutung der relativen Koordination bei Fischen. Ibid 240, 1938.

———. Entwurf eines Systems der lokomotorischen Periodenbildung bei Fischen; ein kritischer Beitrag zum Gestaltproblem. Z. vergl. Physiol. 26, 1939.

Hume, D. Eine Untersuchung über den menschlischen Verstand. Leipzig 1911 (Meyer).

———. Dialoge über die natürlische Religion. Leibzig 1905 (Dürr'sche Buchhandlung).

*Huxley, J. S. The courtship of the great crested grebe. Proc. Zool. Soc., London 1914.

Jennings, H. S. The Behavior of the Lower Organisms. New York 1906.

Kant, I. Prolegomena zur Kritik der reinen Vernunft. Complete version. E. Cassirer, 11 vols. Berlin 1918 (Bruno Cassirer).

*Kitzler, G. Die Paarungsbiologie einiger Eidechsen. Z. Tierpsychol. 4, 1942.

*Knoll, F. Insekten und Blumen. Abh. Zool. Bot. Ges. Wien, 12, 1926.

Koehler, O. Die Ganzheitsbetrachtung in der modernen Biologie. Verh. der Königsberger gelehrten Gesellschaft 1933.

Köhler, W. Intelligenzprüfungen an Anthropoiden. I. Abt. Preuß. Akad. Wiss., Berlin 1917.

Kortlandt, A. De uitdrukkingsbewegingen ein geluiden van *Phalacrocorax sinensis* Shaw and Nadder. Ardea 1938.

*Kramer, G. Beobachtungen über Paarungbiologie und soziales Verhalten der Mauereidechsen. Z. Morph. u. Ökol. d. Tiere 32, 1937.

Kretschmer, E. Körperbau und Charakter. 1921.

Kühn, A. Die Orientierung der Tiere im Raum. Jena 1919 (G. Fischer).

*Kuo, Z. Y. The genesis of the cat's response to the rat. J. comp. Psych. 11, 1930.

———. Further study of the behavior of the cat toward the rat. Ibid 25, 1938.

Leider, K. Das transzendentale Bewußtsein oder die Welt der Erscheinung. 1939 (K. Triltsch Verlag).

Lloyd Morgan, C. Instinkt und Erfahrung. Berlin 1913 (Julius Springer).

———. Instinkt und Ganzheit. Leipzig and Berlin 1909 (B. G. Teubner).

*Loeb, J. Die Tropismen. Handb. vergl. Physiol. 4, 1913.

Lorenz, K. Beobachtungen an Dohlen. J. Ornith. 75, 1927.

———. Beiträge zur Ethologie sozialer Corviden. Ibid 79, 1931.

———. Beobachtungen über das Erkennen der arteigenen Triebhandlungen bei Vögeln. Ibid 80, 1932.

———. Betrachtungen über das Fliegen der Vögel und über die Beziehungen der Flügel- und Steuerform zur Art des Fluges. Ibid 81, 1933.

———. A contribution to the comparative sociology of colony-nesting birds. Proc. 8th int. Ornith. Congr., Oxford 1934 (Oxford University Press).

———. Der Kumpan in der Umwelt des Vogels. J. Ornith. 83, 1935.

———. Über den Begriff der Instinkthandlung. Folia biotheoretica, Ser. B, 2, 1937.

———. Biologische Fragestellungen in der Tierpsychologie. Z. Tierpsychol. 1, 1937.

———. Über die Bildung des Instinktbegriffes. Naturwiss. 25, 1937.

———. Über Ausfallserscheinungen im Instinktverhalten und ihre sozialpsychologische Bedeutung. 16. Kongr. d. Dt. Ges. f. Psychol., Leipzig 1938 (Johann Ambrosius Barth).

———. Vergleichende Verhaltensforschung. Zool. Anz. Suppl. 12, 1939.

———. Die Paarbildung beim Kolkraben. Z. Tierpsychol. 3, 1940.

———. Durch Domestikation verursachte Störungen arteigenen Verhaltens. Z. angew. Psychol. u. Charakterkd. 59, 1940.

———. Kants Lehre vom Apriorischen im Lichte gegenwärtiger Biologie. Blätter f. Dt. Philosophie 15, 1941.

———. Vergleichende Bewegungsstudien an Anatiden. J. Ornith. 89, Ergänzungsbd. 3, Festschr. O. Heinroth, 1941.

———. Induktive und teleologische Psychologie. Naturwiss. 30, 1942.

———. Psychologie und Stammesgeschichte. In: Die Evolution der Organismen (Hrsg. G. Heberer), Jena 1943 (G. Fischer).

———. Die angeborenen Formen möglicher Erfahrung. Z. Tierpsychol. 5, 1943.

Lorenz, K., and N. Tinbergen. Taxis und Instinkthandlung in der Eirollbewegung der Graugans. Z, Tierpsychologie. 2, 1938.

McDougall, W. An Outline of Psychology. London 1923 (Methuen).

Magendie, F. Précis élémentaire de Physiologie. 2 vols. 1816/17.

Matthaei, R. Das Gestaltproblem. Munich 1943 (J. F. Bergmann).

Metzger, W. Gesetze des Sehens. Frankfurt 1936.

Meyer-Holzapfel, M. Triebbedingte Ruhezusände als Ziel von Appetenzhandlungen. Naturw. 28, 1940.

Molitor, A. Das Verhalten der Raubwespen, I and II. Tierpsychol. 3, 1939.

Müller, J. Handbuch der Physiologie des Menschen. 2 vols. Koblenz 1833–40.

Nice, M. Studies of the life-history of the song sparrow. Transact. Linnean Soc. 4, New York 1937.

Pavlov, I. P. Conditioned Reflexes. Oxford 1927.

———. Die höchste Nerventätigkeit bei Tieren. 3d ed. Munich 1926.

Planck, M. Sinn und Grenzen der exakten Wissenschaft. Nauturw. 30, 1942.

Portielje, A. F. J. Dieren zien en leeren kennen. Amsterdam 1928 (Nederlansdsche Keurboekerij).

Roux, W. Die Selbstregulation ein charakteristisches und nicht notwendig vitalistisches Vermögen aller Lebewesen. Nova Acta Abh. K. Leop. Carol. D. Akad. d. Natf. Halle 1914.

*Russell, E. R. The Behaviour of Animals. London 1934.

*Schjelderup-Ebbe, Th. Zur Sozialpsychologie des Haushuhns. Z. Psychol. 87, 1922/23.

Seitz, A. Die Paarbildung bei einigen Cychliden, I and II. Z. Tierpsychol. 4, 1940; 5, 1941.

Selous, E. Observations tending to throw light on the question of sexual selection in birds, including a day to day diary on the breeding habits of the rudd, *Machetes pugnax*. The Zoologist 4, 1905.

———. Observational diary on the nuptual habits of the blackcock, *Tetrao tetrix*. The Zoologist 13, 1909.

———. An observational diary of the domestic life of the little grebe or dabchick. Wild Life, 7.

———. Schaubalz und geschlechtliche Auslese beim Kampfläufer (*Philomachus pugnax* L.)J. Ornith. 77, 1929.

*Sievert, H. Bilder aus dem Leben eines Sperberpaars zur Brutzeit. J. Ornith. 77, 1930

———. Der Schreiadler. Ibid 80, 1932.

———. Beoachtungen am Horst des schwarzen Storches, Ciconia nigra L. Ibid 80, 1932.

———. Die Brutbiologie des Hühnerhabichts. Ibid 81, 1933.

Spemann, H. Experimentelle Beiträge zu einer Theorie der Entwicklung. Berlin 1936.

Spencer, H. Principles of Biology 1864–67; Principles of Psychology 1855–72; Principles of Sociology 1876–96; Principles of Ethics 1892–93.

Stresemann, E. Aves. In: Kükenthals Handbuch der Zoologie VII, Leipzig and Berlin 1927–34.

Tinbergen, N. On the analysis of social organization among vertebrates, with special reference to birds. Amer. Midl. Natural 21, 1939.

———. Die Ethologie als Hilfswissenschaft der Ökologie. J. Ornith. 88, 1940.

———. Die Übersprungbewegung. Z. Tierpsychol. 4, 1940.

———. An objectivist study of the innate behaviour of animals. Bibliotheca Biotheoretica Ser. D, 1, 1942.

Tinbergen, N., and D. J. Kuenen. Über die auslösenden und richtunggebenden Reizsituationen der Sperrbewegungen con jungen Drosseln (*Turdus m. merula* L. und *T. e. ericetorum* Turton). Z. Tierpsychol. 3, 1939.

Tolman, E. C. Purposive Behavior in Animals and Men. New York 1932 (Appleton Century).

Uexküll, J. v. Umwelt und Innenleben der Tiere. Berlin 1909.

———. Theoretische Biologie. 1920.

Vervey, J. Die Paarungsbiologie des Fischreihers. Zool. Jahrb., Abt. allg. Zool. 48, 1930.

Volkelt, H. Über die Vorstellungen der Tiere. Leipzig 1914.

———. Tierpsychologie als genetische Ganzheitpsychologie. Z. Tierpsychol. 1, 1937.

Wallace, A. R. On the law which has regulated the introduction of new species. Ann. and Mag. Nat. Hist. 1855.

Whitman, Ch. O. Animal behavior. Biol. Lect. Mar. Biol., Woods Hole/Mass., 1898.

———. The Behavior of Pigeons. Publ. Carnegie Inst. 257, 1919.

Windelband, W. Geschichte und Naturwissenschaft. Strassburg 1894 (Trübner).

Wundt, W. Vorlesungen über die Menschen- und Tierseele. Leipzig 1922 (Voss).

Index

Note: Page numbers appearing in italic type refer to pages that contain illustrations. This index does not include information from the translator's foreword.

Adaptation, 120
  law of irreversibility of, 128–130
  and mutation and selection, 153
Altum, Bernhard, 181
Amoebae, 124, 161
Amphibians, 102–103. *See also specific species*
Analysis on a broad front, xxxvii, 204, 207, 313
  vs. atomistic research, 144, 147
  and causation, 141–145
  and endocrine glands, 145–146
  and entities, 141–145, 179, 314
  and mechanism, 196–197, 198
Die angeborenen Formen möglicher Erfahrung (Lorenz), xxx
Animal keeping, 221–233
  and action systems, 223–226, 229–230
  analytic observation required for, 224–226
  and behavior as function of organs, 231
  and controllable conditions, 221, 223
  and environment reconstruction, 223, 225–227, 229–230
  vs. field observation, 221–222, 232
  Gestalt perception required for, 224
  and normal vs. pathological behavior, 221, 226–230
Anthropomorphism
  compensating for, 24
  and innate releasing schemata, 23–24, 160–161
Antonius, 120
Apes, 307
  evolution of humans from, 133, 134
Appetitive behavior, 259–270
  and conditioned reflexes, 267
  of humans, 265–266, 268
  and innate motor patterns, 259–270
  and pleasure, 265–268
A priori schematizations, xli. *See also* Innate releasing mechanisms
Archetypes, 274
Association psychology, 46, 147, 195, 197–199
Atomism. *See also* Materialism; Mechanism
  vs. analysis on a broad front, 144, 147, 313
  and Gestalt, 139
Automatic stimulus production, 287–301
  and additional stimulus, 290, 292, *293*
  and conditioned reflexes, 311
  as continuous, 293–294, 298
  as identical to innate motor patterns, 294–300

Automatic stimulus production (*cont.*)
and inhibition, 290, *292,* 294, 295, 297–298
and innate motor patterns, xxxviii, 180, 205–206, 241–243, 257, 270
physiological basis of, 294
and purposive behavior, 295–296
vs. reflexes, 288, 294–295, 304
and relative coordination, 289–290, *291*
and spinal contrast, 296–298
and spontaneity, 295
Awe and wonderment, 185–187, 190–191
Axolotl, 132, 133

Bacon, Francis, 14
Baerends, G. P., 222, 256
Baumgarten, Edvard, 74–75
Behaviorism
as atomistic, 147
as learning by trial and error, 181–182, 197, 199–201
as mechanistic, 195
as monistic, 181–182, 209–210
Behind the Mirror (*Die Rückseite des Spiegels;* Lorenz), ix
Bell, Sir Charles, 202
Berkeley, George, 14
Bernard, Claude, 74, 85, 188, 196
Bierens de Haan, J. A., 155, 211, 309
Biology. *See also* Life, defining characteristics of
adaptation, 93, 120, 128–130, 153
assimilation and dissimilation, 84–86
as blocked by idealism, 74
cancers, 97
complexity of organisms, 90–91, 95
differentiation and subordination of parts, 91, 97
ecological equilibrium, 87
individuality of organisms, 89
oxidation processes, 85
parasites, 97
and psychology, 43–44, 46, 95–96
viruses, 88–89, 98
Birds, 119–120, 235–237, 252
appetitive behavior of, 260–263, 267–268
in captivity, 218–219, 228–229
classification of, 103, 108, 120
ducks, 120, 235–237, 253, 309
eagles, 119–120
evolution from dinosaurs, 108, 112
fowl, 250
geese, 120, 236, 251, 309
innate motor patterns in, 272–273, 275–282, 299–300, 308, 309
jackdaws, 272–273, 275–280
ravens, 260–263
similarity to reptiles, 109
starlings, 281–282
taxes (orienting responses) in, 307, 308, 309
Brain, human. *See also* Nervous system
and limits of knowledge, 20, 22, 41, 192
organic functioning of, 14, 17–18
Brehm, Alfred Edmund, 181, 218
Bronn's Klassen und Ordnungen des Tierreiches (Stresemann), 120
Bunge, Gustav von, 8, 195
Busch, Wilhelm, 35
Bustard, great, 307

Causality, perception of, 18, 24, 40
Causation, 140–141. *See also* Atomism; Mechanism; Vitalism
and analysis on a broad front, 141–145
and psychophysical parallelism, 169–174
and species-preserving finality, 153, 154–156
Chaos, 30
Chemistry
endothermic processes, 85
exothermic processes, 85–86
and physics, 41–42
Chimpanzees, 166–167
Chordates
classification of, 103
evolution of, *104,* 113, 117
Christianity, 77, 157

Christoleit, 254
Civilization, decline of, 75–78
Club Against the Scandal of Double Logic (Baumgarten), 75
Coelenterates, 117, *118*
Communication, and intention movements, 252–253
Comparative morphology, 181
 and evolution, 111, 113–114
 and origins of vertebrates, 111–112
Conditioned reflexes, 201–203. *See also* Reflexology
 and appetitive behavior, 267
 and automatic stimulus production, 311
 discovery of, 147, 181–182, 201–202
 vs. innate releasing schemata, 202, 276–278
 and learning by trial and error, 202, 312 (*see also* Learning)
 and pleasure/aversion, 312
 and purposive behavior, 164, 311–312
 similarity of human/animal development of, 164–165
Confucius (K'ung Fu-tzu), 158
Constancy phenomena, 59–60
Convergence, 116–120, *118*
 genetic, 122–123
Craig, Wallace, 259
 on aversions, 269
 discovery of appetitive behavior, 265–270
Creation, 9
 vs. evolution, 100
Curiosity, human
 and collecting, 28–29, 216
 and Gestalt perception, 28–29, 216
 goal-directedness of, 6, 28–29
 and knowledge, 6, 216
 and learning capacity, 29

Dacqué, Edgar, 32
Darwin, Charles, 102, 253–254
Death drives, xxvi
Deduction
 and prediction, 32
 as secondary to induction, 32–37, 73
Descartes, René, 8, 168, 169
Deuterostomes, 117
Development, phylogenetic. *See* Evolution
Developmental mechanics, 192
Dewey, John, 189
Dialectical materialism, xxxi, 196
Differentiation, 91, 97
 and species-preserving finality, 153–154
Dinosaurs, evolution of birds from, 108, 112
Display patterns, 121–122
Dogs, 166–167
Dollo, Louis, 128–130
Domestication, xl, 97
 and instincts, 246–247
 and social decline, 75–76
Driesch, Hans, 192, 193, 196
Drive-governed behavior. *See* Appetitive behavior; Purposive behavior
Dualism, 15–16, 74–75, 168. *See also* Mind-body problem
Ducks, 120, 235–237, 253, 309

Eagles, 119–120
Earthworms, 287–289, 294
Ectropy, 92–94, 97–98
 and evolution, 92, 133–134
 and values, 94
Eels, 288–289, *289*, 294
Ego, 94, 157
Ehrenfels, Christian von, 60, 138
Embryonic development, 138, 192, 210
 and evolution, 112, 114, 116–117
Emotions, 248–254
Empiricists, English, 14, 18
Endocrine glands, 145–146, 296
Entity, 137–139
 and analysis on a broad front, 141–145, 179, 314
 endocrine glands as entities, 145–146
 finality of organs of, 144 (*see also* Finality, species-preserving)

Entity (*cont.*)
and Gestalt perception, 138, 139
hierarchy of entities, 172
as holistic regulatory system, 89–90, 97, 140–141
machine analogy, 141–143
objective nature of, 140
as primary over its parts, 138
relatively entity-independent components of, 146–147, 148, 179, 182
supersummation of, 138
transposability of, 138–139
Entropy, 93
Ethical laws, 75
and domestication, 76
organic basis of, 78–79
and sociology, 79
Ethics, philosophical, 187–188
and categorical imperatives, xxxvi
and comparative behavioral research, xxxiii, xxxv–xxxvii
and ideals, 10
inflexibility of, xxxvi
and social decline, 77
Ethics, religious
and categorical imperatives, xxxvi
inflexibility of, xxxvi
and social decline, 77
undermined by science, 78
Evolution, 49, 99–135, 158. *See also* Evolutionary tree; *individual species*
and adaptation, 120 (*see also* Adaptation)
anatomical evidence for, 111
and chance, 99
and characters, table of, 115
chemical evidence for, 112
comparative morphological evidence for, 111, 113–114
and comparison, 101
and convergence, 116–120, 122–123
vs. creation, 100
and display behavior, 121–122
and ectropy, 92, 133–134
embryological evidence for, 112, 114, 116–117
and genetic relatedness, 101–102
geographical evidence for, 112
Gestalt perception in study of, 115, 127–128
and homologous characters, 102, 103, 115, 235–238
human, 133, 134
of innate behavior patterns, 100, 121–122, 235–238
and language, 121–122
and mutation, 122
and neoteny, 130–132
of nervous system, *104*
as originating in zoological systematics, 100, 111
paleontological evidence for, 106, 111–112, 114
reliability of theory of, 113
and rudimentary organs, 129–130
and secondary reduction, 106, 114
and systematic finesse, 127–128
and vertebrates, classification of, 102–110
Evolutionary tree, 110
of chordates, *104,* 113
multidimensionality of, 108, 123–127, *125, 127*
of vertebrates, *107*
Experience, as organic functions, 12–14
Experience, subjective. *See* Subjective experience
Extroverts, vs. introverts, 54

Fatigue, 280–281, 282, 298
Finality, species-preserving, 151–156, 209
causes underlying, 153, 154–156
and constancy phenomena, 59–60
as defining characteristic of life, 91, 97
and differentiation, 153–154
and innate behavior patterns, 154, 231–232
of machines, 142
of organs, 144
and purposive behavior, 261, 263–264, 270

scope of, 151–152, 155
and vitalism, 151–152, 155, 209, 210
Fish, 224–225, 227, 253
automatic stimulus production in, 288–293, *289, 291–293,* 294, 296–298
classification of, 102, 103, 106
display behavior of, 121
eels, 288–289, *289,* 294
evolution of, 111
innate motor patterns of, 272, 281, 308, 309
jewelfish, 281, 308
sea horses, 296–298
taxes (orienting responses) in, 308, 309
wrasse, 288, 289–293, *291–293,* 296
Formenkreis (Kleinschmidt), 125–126
The Foundations of Ethology (*Vergleichende Berhaltensforschung—Grundlagen der Ethologie;* Lorenz), ix
Freud, Sigmund, 47–48

Galileo, 5
Geese, 120, 236, 251, 309
Genetic convergence, 122–123. *See also* Convergence
Genetics, 101–102, 180, 193, 314. *See also* Evolution
Genome, 122
Gestalt perception, 168. *See also* Intuition
and abstraction, 30, 60–61
and atomistic explanation, 139
capacity for, 60
constancy phenomena as, 59
and curiosity, 28–29, 216
Ehrenfels criteria of, 60, 138
and entities, 138, 139
in evolution research, 115, 127–128
as ignoring inessential features, 58–60
and images, 273, 277
incorrigibility of, 68–69
and innate releasing schemata, 272–277
intuition as, 56–57, 62
and learning, 165
and nervous system, 165–166
and objectivity, 31
vs. perception of details, 54
reliability of, 63
and values, 165
and vitalism, 198
Gestalts, physical, 89–90, 139
Goal-directed behavior. *See* Appetitive behavior; Purposive behavior
Goethe, Johann Wolfgang von, 19, 187, 222
on development as differentiation, 91
on inductive science, 54–55
on life processes, 84
on propagation, 86

Hartmann, Max, xxxviii n., 212
on evolution vs. development, 92
on physical vs. experiential processes, 95, 168
Heck, 120
Heinroth, Oskar, 120, 217, 229, 254–258
bird keeping of, 218–219, 225, 233, 235
on drive-governed patterns, xxxviii, 242–243, 255
on Gestalt perception, 274
on induction, 247–248
on innate motor patterns, 308–309
on intention movements, 252, 271–272
mood theory of, 248, 250, 254, 255–257
on philosophy, 247
on phylogenetically homologous behavior, 115–116, 215, 237–238, 239, 241, 242
on species-preserving finality, 152
Helmholtz, Hermann von, 57, 63, 159
Heraclitus, 10
Herder, Johann Gottfried von, 101–102
Herrick, F. H., 272
Holst, Erich von, ix, x
discovery of automatic stimulus production, 287–301 (*see also* Automatic

Holst, Erich von (*cont.*)
stimulus production)
on mantle of reflexes, 306
Holzapfel, Monika, 270
Homologous mutation. *See* Genetic convergence
Hormones, 296
Human beings, xxvi. *See also* Perception, human
appetitive behavior of, 265–266, 268
evolution of, 133, 134
language of, 121–122
learning capacity of, 29
as object of scientific research, 12–14, 22–23
Humanities, synthesis with natural sciences, 53–79
consistency between theory and practice needed for, 73–75
impartiality evaluation of facts needed for, 68–69, 70, 71–73
philosophy needed as guiding authority for, 68, 75
Hume, David, 14

Ichthyosaurs, 117, 118
Idealism, philosophical, 5–25
consistency lacking in, 65–67, 73
and correspondence of intersubjective experience, 66
and creation, 9
Kantian, 13, 14–16, 24–25, 66
vs. materialism, 9–10, 65–67
and mind/soul, 157
natural science blocked by, 74
Platonic, 9, 10, 11, 74
and wonderment, 186–187
Ideals, 10
Ideism. *See* Idealism, philosophical
Idiographic stage of scientific development, 28, 29, 100, 213–215
Induction, 27–37. *See also* Gestalt perception
accountability of, 64
and advances in knowledge, 36–37
bias in, 31–32
comparative ethology as based on, 213–219
comprehensive knowledge required for, 215–217
and conflicting facts, 68
deduction as secondary to, 32–37, 73
definition of, 30–31
and experiments, 32–33
as idiographic stage of scientific development, 28, 29
and intuition, 55–65
and probability, 31, 32, 35, 64
shortsightedness of, 65
and statistical evaluation of facts, 71
and veridicality of human perception, xxx
The Innate Forms of Possible Experience (Lorenz), xxx
Innate motor patterns, 166, 174, 271–285. *See also* Appetitive behavior; Innate releasing schemata
and automatic stimulus production, xxxviii, 205–206, 241–243, 257, 270 (*see also* Automatic stimulus production)
and emotions, 248–254
evolutionary homology of, 235–239
impairment of, 228
increasing intensity of, 250–251, 256, 297–298
and intention movements (motivation), 248, 250, 251–254, 255–257, 266, 271–272, 300
vs. learning by trial and error, 200–201
as organlike, 238
and pleasure/aversion, 164–165
and proprioceptors, 288
and purposive behavior, 262–270
reduction of, 249
refractoriness of, 298–300
terminology for, 238
Innate releasing schemata, 121–122, 146, 182, 206–207
and anthropomorphism, 23–24, 160–161
and a priori schematizations, xli

vs. conditioned reflexes, 202, 276–278
disruption of, 228
and fatigue, 280–281, 282, 298
and Gestalt perception, 272–277
incorrigibility of, 160
and inhibition, 304–305
and internal responsiveness, 271, 279–285
physiological basis of, 273, 304
and strength of stimulus, 271, 277–279, 280, 281–285, *284*, 305
and subjective experience, 159–161
and substitute objects, 281
taxes (orienting responses), 269–270, 304, 305–311
and vacuum responses, 281–283, 295
and value assessments, xxxv, xlii
Innate social behavior patterns, human. *See also* Ethical laws; Innate releasing schemata
vs. cultural norms, xxxiv
disrupted by domestication, 76
and evolution, 100
inflexibility of, xxxiv
malfunction of, xxxiv–xxxv
normativity of, xxxiii
and species-preserving finality, 154
Innate working hypotheses, 33–35. *See also* Intuition
Insects, 131–132
Instincts, xxvi, xxxiv, 182–183, 230–231. *See also* Innate releasing schemata; Innate motor patterns
and domestication, 246–247
and learning, 245–246
and vitalism, 155, 191, 193, 195, 210–211
Intentions. *See* Purposive behavior
Introverts, vs. extroverts, 54
Intuition. *See also* Gestalt perception
a priori schematizations of, xli, 14–15, 167–168
bias in, 71
as Gestalt perception, 56–57, 62
inaccessibility to consciousness, 62–63, 64
incorrigibility of, 64
and induction, 55–65
and innate working hypotheses, 33–35
and knowledge of external world, 18–19
organic functioning of, 16–18, 33–34
reliability of, 62–63
role in scientific research, 56
scope of, 63–64

Jackdaws, 272–273, 275–280
Jennings, Herbert S., 213–215, 217
Jewelfish (red African cichlid fish), 281, 308
Journal of Comparative Psychology, 101, 181
Jung, Carl, 47–48, 274

Kant, Immanuel
on a priori schematizations, xli
on ethics, 187–188
idealism of, 13, 14–16, 24–25, 66
on soul, 159–160
Kant-Laplace theory, 187
Kants Lehre vom Apriorischen im Lichte gegenwärtiger Biologie (Lorenz), xxx, 25
"Kant's Theory of the *a priori* in the Light of Modern Biology" (Lorenz), xxx, 25
Karroo Formation (Cape Province, South Africa), 111
King Solomon's Ring (Lorenz), x
Kleinschmidt, Otto, 125–126
Knowledge
and curiosity, 6, 216
and intuition, 18–19
limits of, 20, 22, 41, 69, 192
as organic functions, 12–14
refusal of, 48, 71–72, 210
and self, 6
and thought, 18–19
Koehler, Otto, 143
Köhler, Wolfgang
on Gestalt perception, 28–29, 165, 273–274
on physical Gestalts, 89–90, 139
Kortlandt, A., 222

Kramer, Gustav, ix, x
Kretschmer school, 47
Kühn, Alfred, 196, 269
K'ung Fu-tzu (Confucius), 158

Lampreys, 106, 108
Lancelets, 105, 106, 124
Language, human, 121–122
Laws, of natural science, 192, 313
as consistent, hierarchical system, 39–51, 70
and experiments, 32–33
formulation of, 29–30 (*see also* Nomothetic stage of scientific development)
Learning. *See also* Knowledge
and curiosity, 29
and Gestalt perception, 165
human capacity for, 29
and instinct, 245–246
and pleasure/aversion, 164, 312
and purposive behavior, 162–164
by trial and error, 181–182, 199–201 (*see also* Conditioned reflexes)
Leider, Kurt, 11, 187
Life, defining characteristics of, 83–98
assimilative expansion, 84–86, 87–88, 96–97
ectropy, 92–94, 97–98
growth and reproduction, 86–89
historically unique origins, 86, 91, 97 (*see also* Evolution)
holistic regulation, 89–90, 97
inexplicability, 188
metabolism, 84–86, 96
possession of mind, 158
subjective experience, 94–95
system-preserving finality of organic structures, 91, 97
will to power, 86
Linnaeus, Carolus (Carl von Linné), 61–62, 102–103
Lloyd Morgan, C., 218, 239, 245
Locke, John, 14
Lorenz, Konrad
Die angeborenen Formen möglicher Erfahrung, xxx
Kants Lehre vom Apriorischen im Lichte gegenwärtiger Biologie, xxx, 25
King Solomon's Ring, x
lectures by, xiii
Man Meets Dog, x
as prisoner of war in Russia, xi–xii
reception to work of, xxx
as recycler, xii
Die Rückseite des Spiegels (*Behind the Mirror*), ix
Russian manuscript, x–xi, xiv
Vergleichende Berhaltensforschung—Grundlagen der Ethologie (*The Foundations of Ethology*), ix

McDougall, William, 155, 218, 239
on emotions and instincts, 248–249
on innate perceptory patterns, 271
rejection of mechanism, 259
on species-preserving finality, 261
on spontaneity and purposive behavior, 259, 261, 263
Machines
as analogous to entities, 141–143
finality of, 142
relative complexity of, 141–142
Magendie, François, 202
Mammals. *See also specific species*
classification of, 103
evolution of, 108–109, 111
Man Meets Dog (Lorenz), x
Marx, Karl, 138
Materialism. *See also* Atomism; Mechanism
and correspondence of intersubjective experience, 66–67
vs. idealism, 9–10, 65–67
life processes, applied to, 42–43, 49–50
vs. mechanism, 196
Matthaei, Ruprecht, 145
Max Planck Foundation, x
Max Planck Institute for Behavioral Physiology (Seewiesen, Bavaria), x

Max Planck Institute for Marine Biology (Wilhelmshaven, Germany), x
Mechanism, 195–208. *See also* Atomism; Materialism; Reflexology
and analysis on a broad front, 196–197, 198, 207
and association psychology, 195, 197–199
as atomistic, 209
and behaviorism, 195, 197, 199–201, 209–210
as dogmatic, 196, 209–210
and instincts, 211
vs. materialism, 196, 209
scientists as mechanists, 195–196
vs. vitalism, 139, 209–212, 217–218
Medicine (field), 152, 156
Mendel, Gregor, 314
Mental processes. *See* Psychology
Metamorphosis, 131
Methodology. *See also* Analysis on a broad front; Animal keeping; Induction
action systems, study of, 214, 223–226, 314
comparison, 101
didactic representation of organic systems, xxxviii
direction of research, origin of, 179–180
experiments, 32–33
extrapolating from animals to humans, xxvii
hierarchy of research, xxxii–xxxiii, 199
and human dignity, xxvii
and knowledge quest, 215–217
nomenclature, 27
from particular to general, 30
working hypothesis, 33
Metzger, Wolfgang, 11
Mice, 269
Mind-body problem, 157–175. *See also* Subjective experience
and anthropomorphism, 160–161
and conditioned reflexes, 164–165
and Gestalt perception, 165–166, 168
and innate releasing schemata, 160–161
and learning, 162–164
and memory, 163
mind, dependence on body, 157–158
nervous system, 161–162, 165–166
and pleasure/aversion, 164–165, 166, 167
psychophysical parallelism, 169–174, 203–204
purposive behavior, 162–163, 174
receptor organs, 159, 167–168, 175
Molitor, A., 222
Morgenstern, Christian, 8, 33
Motivation. *See* Innate motor patterns, and intention movements
Müller, Johannes, 74, 196, 202
vitalism of, 188–189, 190, 192, 210–211
Mutation, 122, 130. *See also* Genetic convergence
and adaptation, 153

Necessary truths, 33–35
Neoteny, 130–133
Nervous system, 229. *See also* Innate motor patterns
and analysis on a broad front, 207
automatic stimulus production in, 180, 241–243 (*see also* Automatic stimulus production)
complexity of, 215
evolution of, *104*
and Gestalt perception, 165–166
and innate motor patterns, 205
integrating function, 172–174
mediating agencies of, 172
and mind-body problem, 161–162, 165–166
and perception, 13
receptors, 19–20, 159, 167–168, 175 (*see also* Innate releasing schemata)
and subjective experience, 161–162
world-imaging structures of, 19
Newts, 131

Nice, Mrs. M. Morse, 222
Nietzsche, Friedrich, 86
Nomothetic sciences, 41
Nomothetic stage of scientific development, 29–30, 100, 241–243, 241

Objectivity, 23–24
Organs
as adapted to inorganic outside world, 19
behavior as function of, 231–232
finality of, 144 (*see also* Finality, species-preserving)
receptor, 159, 167–168, 175
rudimentary, 129–130
Origin, of organisms. *See* Evolution

Paleontology, 106, 111–112
Pavlov, Ivan Petrovich
discovery of conditioned reflex, 147, 201–202, 204–205 (*see also* Reflexology)
on goal-oriented reflexes, 206
on psychophysical parallelism, 203–204
Perception, human. *See also* Gestalt perception; Intuition; Thought
auditory, 60–61
of causality, 18
of constancy of form, 59–60
empiricist view of, 14, 18
and images, 273
inference as analogous to, 57–59
Kant on, 14–15
and model of external world, 19–22
and nervous system, 13
as object of scientific research, 12–14
as organic function, 15, 16, 18–19, 59–60
spatial, 16, 18
of substantiality, 18
temporal, 18
veridicality of , xxx
and world image, 7–8
Phenomenal world vs. real world, 13, 19–22
Philology, 121–122
Philosophy. See also Ethics, philosophical; Idealism, philosophical; Materialism
empiricism, 14, 18
factual basis needed by, 67–68
as guiding authority for natural science, 68, 75
impartiality evaluation of facts needed by, 68–69, 70, 71–73
as intuitive, 53
as inward-looking, 53
natural science blocked by, 7, 10–12, 74
and reflection, 5–7
solipsism, 67
and tension with science, 54–55
theory dependence of, 69
Phylogenetic approach, origins of, 235–239
Phylogenetic development. *See* Evolution
Physiology (field), 74
Physics
and chemistry, 41–42
entropy, 93
as most basic science, 41, 44
Planck, Max, 37, 174
neo-Kantian rejection of, 72–73
on quantity, 24
on quest for knowledge, 27
on real vs. phenomenal world, 25
Platonic idealism, 9, 10, 11, 74
Pleasure/aversion, 269–270
and appetitive behavior, 265–268
and learning, 164, 312
and mind-body problem, 164–165, 166, 167
Population explosion, 87
Preservation. *See* Finality, species-preserving
Protostomes, 117
Psychiatry, 74
Psychoanalysis, xxvi
Psychology. *See also* Behaviorism; Gestalt perception; Reflexology
association psychology, 46, 147, 195, 197–199

and biology, 43–44, 46, 95–96
and comparative behavioral research, 49–50
depth psychology, 47–48
history of, 47–48
and idealism, 74
perceptual psychology, 46
scientific method lacking in, 46
Psychophysical parallelism, 169–174, 203–204
Purposive behavior, 209, 261. *See also* Appetitive behavior
and automatic stimulus production, 295–296
and conditioned reflexes, 164, 311–312
and experiential processes, 162–163
and innate motor patterns, 262–270
and learning, 162–164
and memory, 163
and mind-body problem, 162–163, 174
and species-preserving finality, 261, 263–264, 270
and spontaneity, 259–270
and taxes (orienting responses), 269–270
Purposivity, species-preserving. *See* Finality, species-preserving

Quantity, 24, 35

Ravens, 260–263
Realism, naive, 21–22, 23
Real world vs. phenomenal world, 13, 19–22
Reflection, philosophical, 5–7
Reflexology, 197, 201–202. *See also* Conditioned reflexes; Innate releasing schemata
as atomistic, 147
and images, 273
as mechanistic, 195
as monistic, 182, 204–207, 211, 218
Religion
Christianity, 77, 157
creation, 9, 100
and ethics, xxxvi, 77, 78
soul/mind, 157
Reptiles. *See also specific species*
classification of, 103, 108
evolution of, 111–112
mammals evolved from, 108–109
similarity to birds, 109
Researchers. *See* Scientists
Research Station for Behavioral Physiology, Max Planck Institute for Marine Biology (Wilhelmshaven, Germany), x
Response patterns, human, xxvi
Roux, Wilhelm, 192
Die Rückseite des Spiegels (*Behind the Mirror;* Lorenz), ix

Salamanders, 131
Schiller, Friedrich von, 188
Schopenhauer, Artur
on perception, 57
skepticism of, 66, 73
on soul, 159–160
Scientific development
idiographic stage of, 28, 29, 100, 213–215
nomothetic stage of, 29–30, 100, 241–243, 241
systematic stage of, 28–29, 100
Scientists
as detail-oriented, 54
and interaction between fields, 44–45
intuition used by, 56
as mechanists, 195–196
as naive realists, 21–22
as outward-looking, 53–54
overview, difficulty achieving, 67
vs. philosophers, 54–55
subjective experience of, 161, 185–186, 190
and working hypotheses, 69
Sea horses, 296–298
Seitz, Alfred, 284, 305
Self, 6–8
Selous, E., 222
Sherrington, Sir Charles Scott, 296
Skeletal system, 146–147, 148

Skepticism, about external world, 8
Slowworms, 114
Sociology, xxxii–xxxv, 77, 79
Solipsism, 67
Soul, 157–158, 159. *See also* Mind-body problem; Subjective experience
Space, perception of, 16, 18, 39–40
Spemann, Hans, 192
Spencer, Herbert, 218, 239, 245
Spengler, Oswald, 76
Spontaneity
  and automatic stimulus production, 295
  physiological basis of, xxxviii
  and purposive behavior, 259–270
Starlings, 281–282
Station for Comparative Ethology (Altenberg, Austria), ix–x
Steinmann, 117
Stentors, 163
Stimulus production, automatic. *See* Automatic stimulus production
Stresemann, Erwin, 120
Subjective experience, 173–174. *See also* Mind-body problem
  as anthropomorphic, 160–161
  of emotions, 248–254
  in higher animals, 166–167
  as innate releasing patterns, 159–161
  and nervous system, 161–162
  pain, 166
  as physiological, 167
  and pleasure/aversion, 164–165, 166, 167
  and purposive behavior, 162–163
Substantiality, perception of, 18, 24, 39–40
Survival of the species. *See* Evolution; Finality, species-preserving
Systematic sciences, 41
Systematic stage of scientific development, 28–29, 100

Taxes (orienting responses)
  and innate releasing schemata, 269–270, 304, 305–311
  and purposive behavior, 269–270
Technology, danger from, 77
Teleology, 155
Tetrapods, 103, 113–114, 128
Thorndike, Edward Lee, 199
Thought
  a priori schematizations of, xli, 14–15
  and knowledge of external world, 18–19
  organic functioning of, 16–18, 33–34
Tiergarten (Berlin), 236
Time, perception of, 18
Tinbergen, Nikolaas, 222, 256, 309
Tolman, Edward Chace, 162, 261
Truths, necessary, 33–35

Uexküll, Jakob von, 19, 196
  on adaptation, 93
  on rudimentary organs, 129

Values
  and ectropy, 94
  and Gestalt perception, 165
  and innate releasing schemata, xxxv, xlii
Vergleichende Berhaltensforschung–Grundlagen der Ethologie (*The Foundations of Ethology;* Lorenz), ix
Vertebrates
  acraniates, 105
  amniotes, 108, 109
  amphibians, 102–103
  birds, 103, 108 (*see also* Birds)
  chordates, 103, 104
  classification of, 102–110
  cotylosaurs, 109
  craniates, 105–106
  cyclostomes, 108, 110
  evolution of, 111–112, *107*
  fish, 102, 103, 106 (*see also* Fish)
  Gnathostomata, 106, 108
  vs. invertebrates, 105
  lampreys, 106, 108
  lancelets, 105, 106
  mammals, 103, 108–109, 111
  marsupials, 109–110
  monotremes, 109
  placentals, 109–110

Placodontia, 109
platypus, 109
reptiles, 103, 108–109, 111–112
Sauropsida, 108, 109
tetrapods, 103, 113–114, 128
theromorphs, 108
tunicates, 103, 105
Vervey, J., 222
Viennese Institute for Science and Art, xiii
Vitalism, xxxviii, 42, 129, 185–193
and directional influence, 192
and entelechy, 193
of Gestalt psychology, 198
and inexplicability, 188, 189–190, 191–192, 210–211
and instincts, 155, 191, 193, 195, 210–211
vs. mechanism, 139, 209–212, 217–218
and Platonic idealism, 74
and restoration, 193
and soul, 159
and species-preserving finality, 151–152, 155, 209, 210
and wonderment, 185, 190–191
Volkelt, Hans, 165, 172

Wallace, Alfred, 102
Watson, John B., 200–201
Weber, Hermann, 223
Westenhöfer, Max, 32
Whales, 117, 118
Whitman, Charles Otis, 215, 217
bird keeping of, 218–219, 233, 235
on innate motor patterns, 308–309
on instincts, xxxviii, 115–116, 238, 242
on instinct vs. intelligence, 246–247
philosophical background of, 245
on phylogenetically homologous behavior, 237–238, 239, 241, 242
Windelband, Wilhelm, 28–30
Wonderment and awe, 185–187, 190–191
World image, xxx
and central nervous system, 19
and dialectical materialism, xxxi
in natural science, 27
Wundt, Wilhelm, 47, 180, 195
on association, 197
Wrasse, 288, 289–293, *291–293*, 296

Zoological systematics, 61–62, 100, 111